THE BEHAVIOURAL ECOLOGY OF PARASITES

THE BEHAVIOURAL ECOLOGY OF PARASITES

Edited by

E.E. Lewis

Department of Entomology
Virginia Polytechnic Institute and State University
USA

J.F. Campbell

USDA, ARS, Grain Marketing and Production Research Center
Manhattan
Kansas
USA

and

M.V.K. Sukhdeo

Department of Ecology, Evolution and Natural Resources
Rutgers University
New Jersey
USA

CABI *Publishing*

CABI *Publishing* is a division of CAB *International*

CABI Publishing
CAB International
Wallingford
Oxon OX10 8DE
UK

Tel: +44 (0)1491 832111
Fax: +44 (0)1491 833508
E-mail: cabi@cabi.org
Web site: www.cabi-publishing.org

CABI Publishing
10 E 40th Street
Suite 3203
New York, NY 10016
USA

Tel: +1 212 481 7018
Fax: +1 212 686 7993
E-mail: cabi-nao@cabi.org

A catalogue record for this book is available from the British Library, London, UK.

Library of Congress Cataloging-in-Publication Data
The behavioural ecology of parasites / edited by E.E. Lewis, J.F. Campbell, and M.V.K.
Sukhdeo.
     p.   cm.
Includes bibliographical references.
   ISBN 0-85199-615-9 (alk. paper)
   1. Parasites--Behavior.   2. Host-parasite relationships.   I. Lewis, E.E. (Edwin
E.)   II. Campbell, J.F. (James F.)   III.   Sukhdeo, M.V.K.
   QL757.B42 2002
   577.8'57--dc21   2002001335

ISBN 0 85199 615 9

Typeset by AMA DataSet Ltd, UK
Printed and bound in the UK by Biddles Ltd, Guildford and King's Lynn

Contents

Contributors

A.D. Bansemir (née Keating), *Division of Life Sciences, Rutgers University, 604 Allison Rd, Piscataway, NJ 08904, USA*

P. Bartoli, *Centre d'Océanologie de Marseille, UMR CNRS 6540, Campus Universitaire de Luminy, 13288 Marseille cedex 9, France*

J.F. Campbell, *Grain Marketing and Production Research Center, USDA-ARS, 1515 College Avenue, Manhattan, KS 66502, USA*

C. Combes, *Laboratoire de Biologie Animale, UMR CNRS-UP 5555, Centre de Biologie et d'Ecologie Tropicale et Méditerranéenne, Université Perpignan, 52, Av. de Villeneuve, 66860 Perpignan cedex, France*

J.G.C. Hamilton, *Centre for Applied Entomology and Parasitology, School of Life Sciences, Keele University, Keele, Staffordshire ST5 5BG, UK*

L. Hemerik, *Biometris, Department of Mathematical and Statistical Methods, Wageningen University, Wageningen, The Netherlands*

H. Hurd, *Centre of Applied Entomology and Parasitology, School of Life Sciences, Keele University, Keele, Staffordshire ST5 5BG, UK*

K.D. Lafferty, *USGS, Marine Science Institute, University of California, Santa Barbara, CA 93106-6150, USA*

E.E. Lewis, *Department of Entomology, Virginia Polytechnic Institute and State University, Blacksburg, VA 24061, USA*

F.J. Messina, *Department of Biology, Utah State University, 5305 Old Main Hill, Logan, UT 84322-5305, USA*

R.E.L. Paul, *Biochimie et Biologie Moléculaire des Insectes, Institut Pasteur, 25 Rue du Dr Roux, 75724 Paris Cedex 15, France*

R. Poulin, *Department of Zoology, University of Otago, PO Box 56, Dunedin, New Zealand*

A.F. Robinson, *USDA-ARS, Southern Crops Research Laboratory, 2765 F&B Road, College Station, TX 77845, USA*

K. Rohde, *School of Biological Sciences, University of New England, Armidale, NSW 2351, Australia*

J.J. Schall, *Department of Biology, University of Vermont, Burlington, VT 05405, USA*

M.R. Strand, *Department of Entomology, 420 Biological Sciences Building, University of Georgia, Athens, GA 30602-2603, USA*

R.J. Stuart, *Citrus Research and Education Center, Institute of Food and Agricultural Sciences, University of Florida, 700 Experiment Station Road, Lake Alfred, FL 33850, USA*

M.V.K. Sukhdeo, *Department of Ecology, Evolution and Natural Resources, Rutgers University, 14 College Farm Road, New Brunswick, NJ 08901, USA*

S.C. Sukhdeo, *Department of Ecology, Evolution and Natural Resources, Rutgers University, 84 Lipman Drive, New Brunswick, NJ 08901, USA*

A. Théron, *Laboratoire de Biologie Animale, UMR CNRS-UP 5555, Centre de Biologie et d'Ecologie Tropicale et Méditerranéenne, Université Perpignan, 52, Av. de Villeneuve, 66860 Perpignan cedex, France*

L.E.M. Vet, *Netherlands Institute of Ecology, Centre of Terrestrial Ecology, Heteren, The Netherlands, and Laboratory of Entomology, Wageningen University, PO Box 8031, 6700 EH Wageningen, Wageningen, The Netherlands*

M.E. Viney, *School of Biological Sciences, University of Bristol, Woodland Road, Bristol BS8 1UG, UK*

M.E. Visser, *Netherlands Institute of Ecology, Centre of Terrestrial Ecology, Heteren, The Netherlands*

F.L. Wäckers, *Netherlands Institute of Ecology, Centre of Terrestrial Ecology, Heteren, The Netherlands, and Laboratory of Entomology, Wageningen University, PO Box 8031, 6700 EH Wageningen, The Netherlands*

Preface

For the purpose of this work, we define parasitism as a two-trophic-level interaction where one participant causes harm to another through a symbiotic relationship. Although great strides have been made in understanding parasitic behaviour, studies have been uneven across taxa and hampered by lack of communication among research disciplines. The last comprehensive treatment of parasite behaviour was published as a special edition in the journal *Parasitology* in 1994, entitled 'Parasites and behavior', edited by M.V.K. Sukhdeo, but this treatise focused only on parasites of vertebrates. There has never been a book devoted to the behavioural adaptations of parasites in general. Thus, *The Behavioural Ecology of Parasites* addresses the behavioural adaptations of parasites across a broad range of taxa, and expands beyond the traditional realm of parasitology to include invertebrate parasites, such as parasitoid wasps, entomopathogenic nematodes, seed-feeding insects and plant-parasitic nematodes. The chapters in this book emphasize the fundamental principles of parasitism, which apply across taxa, and address the diversity and sophistication of parasite behavioural adaptations. The chapters are arranged in four sections: Foraging for Hosts, Host Acceptance and Infection, Interactions among Parasites within Hosts, and Parasite–Host Interactions.

This project could not have been completed without a lot of assistance. We thank the authors for the time and effort they put into their contributions. We are particularly grateful for the assistance of Anne Keating, Stacey Hicks, Simone Sukhdeo and Suzanne Sukhdeo for copy-editing and compiling the book and Alex Hernandez for all the computer help.

Edwin Lewis, James F. Campbell and Michael Sukhdeo
30 November 2001

Trematode Transmission Strategies 1

Claude Combes,[1] Pierre Bartoli[2] and André Théron[1]
[1]Laboratoire de Biologie Animale, UMR CNRS-UP 5555, Centre de Biologie et d'Ecologie Tropicale et Méditerranéenne, Université Perpignan, 52, Av. de Villeneuve, 66860 Perpignan cedex, France; [2]Centre d'Océanologie de Marseille, UMR CNRS 6540, Campus Universitaire de Luminy, 13288 Marseille cedex 9, France

Introduction

Trematodes all have a heteroxenous life cycle. Typically, they exploit three hosts in succession; the first host is always a mollusc and the third one is always a vertebrate, whereas the second (herein called the vector) can belong to nearly any group of metazoans. In some particular cases, the life cycle comprises only two hosts (the mollusc and the vertebrate) and exceptionally only one (the mollusc). One of us (Combes, 1991, 2001) proposed using the term upstream host (USH) for the host where the parasite originates and downstream host (DSH) for the one that the parasite goes to, at any step of the life cycle.

Although life cycles of trematodes seem extremely diverse at first glimpse, with few exceptions they exhibit three fundamental strategies:

1. When they leave the vertebrate host that harbours the adults, trematodes follow a complicated route through the ecosystem(s), with at least one intermediate host and usually two, with two different free-living stages transferring between hosts.
2. Along with classical sexual reproduction, which takes place in the vertebrate host, there is an additional phase of multiplication that occurs in the mollusc host.
3. In all three-host life cycles, the transmission between the second intermediate host (in fact, a vector) and the definitive host involves a link in a food-chain: the third host eats the second.

©CAB International 2002. The Behavioural Ecology of Parasites
(eds E.E. Lewis, J.F. Campbell and M.V.K. Sukhdeo)

The Optimal Transmission Strategy

Since there are at least two host resources (and generally three) that are exploited during trematode development, there is an equivalent number of optimal foraging strategies that can be positively selected.

In the classic three-host life cycle, the miracidium forages for a mollusc, the cercaria forages for a vector and the metacercaria, although not mobile, forages for a vertebrate. A narrow host range is the rule, especially for the miracidium, which forages for molluscs. Host specificity is often less marked for the cercaria, which forages for the vector, and the metacercaria, which forages for the vertebrate host.

Even though natural selection acts separately at each step of the life cycle, the end result is an 'optimal transmission strategy' (OTS). Taken as a whole, the OTS is selected to maximize the number of individuals that reach the appropriate definitive host, where reproduction and genetic recombination occur. It is likely that trade-offs exist in trematode life cycles. For instance, one may forecast that a limited production of eggs might be compensated for by a large production of cercariae and vice versa. Heavy losses at a given stage should be compensated for by better success elsewhere in the cycle. Regulatory mechanisms may, if necessary, intervene to reduce the success of transmission.

As Combes *et al.* (1994) pointed out, behavioural adaptations of trematode free stages are largely independent of the actual presence of a potential adequate host. Although there are various reports in the literature that miracidia and cercariae can to some extent respond to physical and chemical stimuli originating in their target host (Haas, 1994; Haas *et al.*, 1995), these processes invariably act at a very short distance. Responses to host cues do constitute a part of the OTS, but may at most give the finishing touch to host finding. There is considerable theoretical literature on the concept of sequential steps in host finding. This is especially true for parasitoid insects, where host finding is performed by the adult stages, which possess sophisticated sensory structures. One of us (Combes, 2001) considers that there are three main origins of stimuli that may be perceived by parasite infective stages: habitat of the host, effective presence of the host and the host itself. Once the encounter between an infective stage and a particular organism occurs, the latter can be susceptible (the parasite develops), resistant (the host opposes active processes to infection) or unsuitable (the host simply cannot be exploited as a resource).

OTS involves selective adaptations that increase the probability of the free-living stages to localize themselves in what Combes *et al.* (1994) called 'host space' and 'host time'. As stated above, selection tends to increase the fitness of parasites by retaining genes at each step of the life cycle. Available data mostly concern cercariae. Spectacular examples are the distribution of cercariae at various depths in water (Bartoli and Combes, 1986) and cercarial shedding at various times of the circadian period (Théron, 1984). These cercarial emergence rhythms tend to limit the temporal dispersion of the short-lived infective cercariae by concentrating them in the period of time when the chances of meeting the host are the greatest. Interspecific variability of cercarial emergence rhythms correlated with different periods of host activity is well documented (Combes *et al.*, 1994). Genetic differences in the timing of cercarial shedding have also been demonstrated between sympatric populations of parasites belonging to the same species (Théron and Combes, 1988). Such chronobiological polymorphism might occur when two species of DSH with contrasted periods of activity are involved in the parasite life cycle. This is the case for *Schistosoma mansoni* on the island of Guadeloupe, where humans, with diurnal water contacts, and rats, with nocturnal behaviour, are both exploited by the parasite (Théron and Pointier, 1995). A strong disruptive selection maintains distinct populations of schistosomes with early cercarial shedding adapted to human behaviour and with crepuscular cercarial shedding adapted to rat behaviour (Théron and Combes, 1995). Success in reaching host space and host time depends on the selection of adaptive responses to stimuli coming from the environment. This selection is most important when the USH and the DSH are at different points in space and time (Fig. 1.1).

One may wonder why foraging strategies are more variable among trematode species compared with predator–prey systems (and probably most parasites, parasitoids excepted), while targets can be similar. The reason is not in the signals (visual, chemical or acoustic cues) emitted by the targets, but in the limited physical ability of the infective stages to detect and respond to the corresponding information. Infective stages are minute compared with their hosts. They therefore lack sufficient energy and morphological adaptations to learn or to search for and pursue their hosts in the way that predators search for and pursue prey. This explains why natural selection may act to retain more general responses to environmental cues: the infective stages do not search for potential hosts directly, but rather for the space and time where the probability of meeting them is the highest. Whereas data and models are available in parasitoid–host systems, the individual probability of a trematode infective stage to transmit its genes still represents a black box.

For both transmission phases, this 'generalist' strategy of dispersal towards the potential host space is reinforced by the high reproductive capacity of trematodes, which leads to a quasi-permanent flow of numerous infective stages into the environment (with the exception of seasonal fluctuations). High productive rates nearly always characterize the miracidial and cercarial phases of the life cycle. This can be seen as a typical adaptation of parasites. However, in the case of trematodes, this strategy has two additional advantages.

First, two different processes of multiplication occur in parasitic stages that exploit independent resource patches, which may act as a mechanism to avoid competition between larval and adult stages. This can be compared to the case of many insects and amphibians, where larvae and adults have a totally different diet in different habitats.

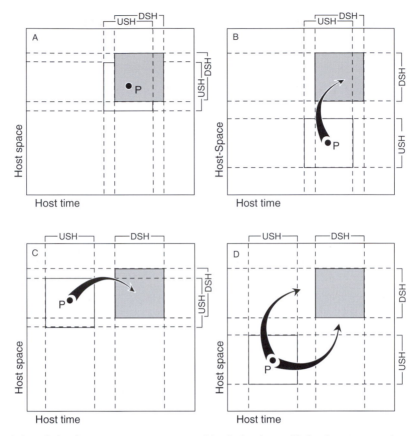

Fig. 1.1. Selective pressures accounted for behaviour of infective stages of trematodes (USH, upstream host; DSH, downstream host). USH host time and USH host space do not obviously coincide with DSH host time and DSH host space. The figure shows the four possible selective pressures that may act on parasite behaviour. In (A), there is no need for a particular behaviour. In (B), the infective stages (P) are selected for to move into DSH host space. In (C), the infective stages are selected for to move into DSH host time. In (D), the infective stages are selected for to move into both DSH host space and DSH host time.

Secondly, the multiplication in the mollusc is, as far as we know, clonal. This means that, while an individual larval stage has a unique chance for meeting the right host, the genome it contains has thousands of possibilities to ensure its transmission either at the same time or several days or weeks later. For instance, the unique genome of a schistosome miracidium can be spread in tropical water bodies for weeks (an infected *Biomphalaria* mollusc may shed some 100,000 cercariae over a period of 2–3 months). A potential cost of this strategy is that, if several larval stages having the same genome succeed in infecting the same host, this may significantly reduce the genetic diversity within parasite infrapopulations, as demonstrated by Mulvey *et al.* (1991) for *Fascioloides magna* infecting the white-tailed deer.

Foraging for the First and Second Intermediate Hosts

Transmission to the first and second intermediate hosts is, in both cases, accomplished by free-living stages, which are usually aquatic. However, marked differences characterize these two life-cycle phases depending on the USH and DSH involved in this process. Regarding transmission to the first intermediate host, USHs, where trematode eggs are produced and dispersed, are very diverse and include all classes of vertebrates. They are also highly mobile and located in a variety of different environments (aquatic, terrestrial, aerial). The target DSHs are all members of a single taxonomic group (molluscs) and are sedentary benthic organisms (Fig. 1.2). In contrast, during transmission to the second intermediate host, the USHs (where cercariae are produced, but not or little dispersed) are homogeneous (molluscs), while the target DSHs are highly diverse and include many classes of invertebrates and vertebrates, which are mobile and spatially dispersed (Fig. 1.3).

This may explain why morphological adaptations and behaviours differ remarkably between the two infective stages, miracidia and cercariae: (i) miracidia exhibit narrowly diversified morphology (Fig. 1.2), whereas cercariae exhibit a fantastic diversity of sizes and shapes (Fig. 1.3); and (ii) it seems to us that, in most species, miracidia explore the

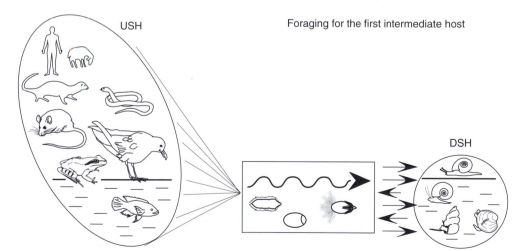

Fig. 1.2. Transmission to the first intermediate host is characterized by: (i) a highly diversified pool of mobile USHs, producing and dispersing the infective stages (eggs and miracidia) within the environment; (ii) poorly morphologically diversified free-living miracidia exploring the host space transversally; and (iii) target DSHs that are taxonomically homogeneous (molluscs) with little mobility. USH, upstream host; DSH, downstream host.

Foraging for the second intermediate host   DSH

USH

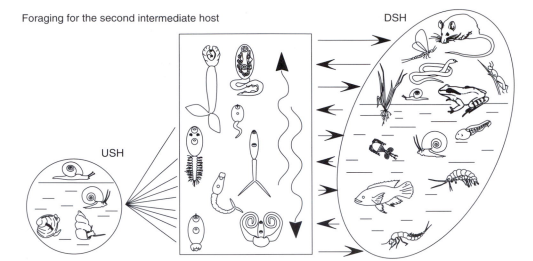

Fig. 1.3. Transmission to the second intermediate host is characterized by: (i) a taxonomically homogeneous group of USHs (molluscs) emitting the infective stage (cercaria) within the environment; (ii) highly morphologically diversified free-living cercariae exploring the host space vertically; and (iii) target DSHs that are taxonomically diversified, mobile and spatially dispersed. USH, upstream host; DSH, downstream host.

host space 'transversally', moving more or less straight on at a relatively constant speed (Fig. 1.2), whereas cercariae tend to explore it 'vertically' moving up and down with frequent resting phases (Fig. 1.3). Transverse movement appears to be adapted to the exploration of an almost two-dimensional environment (the habitat of molluscs represented by the bottom of water bodies). Vertical movement appears to be adapted to the exploration of a three-dimensional environment (the water column). There are some exceptions to this scheme, especially when the miracidium is not free and when the cercaria remains attached to the substrate. Moreover, these two strategies can be explained by the behaviour of the targets: miracidia have to encounter almost immobile targets, while cercariae infect mobile hosts, either by active searching or by being consumed by them. In the case of cercariae, it happens that some of these morphological or behavioural adaptations are signals that make the larval stages attractive to the DSH (prey mimetism). This is never the case with the miracidia.

Foraging for the Definitive Host

In two-host life cycles, the definitive host takes the place of the vector and the transmission strategies are of the same type as those described above.

In three-host life cycles, trematodes make use of food-chains or food webs (Fig. 1.4). In doing so, trematodes invade a prey–predator system where the relationship has undergone selection, i.e. the predator is very likely to eat the species prey that is infected by the parasite. Qualitatively, this gives the metacercariae the maximum probability of reaching the correct host. However, quantitatively, everything will depend on the number of prey items that are actually consumed by the target host, the proportion of the prey population infected, the diversity of the predator's diet and the number of different species that feed upon the prey. For instance, a metacercaria of an amphibian trematode that is carried by an insect will certainly end in voles (microtine rodents), lizards and spiders as well. Thus it is in the interest of the parasite that natural selection maximizes the probability of metacercariae being ingested by an acceptable host. This is the case when favourization processes (see definition in Combes, 1991, 2001) modify the morphology, colour or behaviour of the vector (see Moore and Gotelli, 1990; Poulin, Chapter 12, this volume). Little is known as to how 'risky' such a strategy is, because, when a selection process makes a vector more conspicuous in its environment, this may increase its consumption by many predators and not only by suitable hosts. For instance, the metacercariae of *Clinostomum golvani* cause the formation of yellow spots on the backside of small fishes in tropical ponds, which render infected individuals more obvious. This may increase ingestion by fish-eating birds. However, nothing indicates that all these birds are suitable hosts. One may only suppose that, if natural

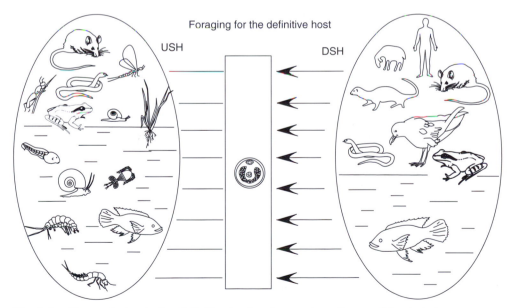

Fig. 1.4. Transmission to the definitive host makes systematic use of food-chains. Modification of morphology, colour or behaviour of infected USHs may occur, significantly increasing their predation by the DSHs. USH, upstream host; DSH, downstream host.

selection has retained such a favourization process, it means that the advantages are greater than the inconveniences. Such imprecision can also provide opportunities for lateral transfers (host switching) during the course of evolution (Combes and Théron, 2000).

In many cases, selection has been advocated as a result more of intuition than of experiments. However, limited predation tests have demonstrated that infected vectors are more prone to consumption than uninfected individuals. For instance, Helluy (1982) estimated that the probability of gammarids being ingested by gulls was multiplied by a factor of four when they harboured metacercariae of *Microphallus papillorobustus*. Additional convincing results were obtained by Lafferty and Morris (1996) with *Euhaplorchis californiensis*. The metacercariae of this trematode accumulate (up to 1500) in the brain of the killifish *Fundulus parvipinnis*. The authors have shown in semi-open conditions that heavily infected fishes exhibit several conspicuous abnormal behaviours and were 40 times more susceptible to predation by final host birds than uninfected controls.

It is difficult to be certain that favourization always results from natural selection processes. However, it is also difficult to attribute only to chance the fact that metacercariae often occupy 'strategic locations' in the host. In the examples cited above, one concerns an invertebrate and the other a fish, but, in both cases, the metacercariae are located in the brain.

An Example: Trematode Strategies in a Marine Ecosystem

In this section, we focus our attention on cercarial transmission because, as stated above, this is the trematode life stage at which transmission strategies are most diversified. As suggested by various authors (see, for instance, Combes, 2001), the signals involved in cercarial transmission can be classified as follows: (i) the information is received by the cercaria and originates from either the environment of the DSH or from the DSH itself; and (ii) the information is received by the DSH and comes either from the cercaria itself or from an intermediate host manipulated by the encysted cercaria (the cercaria may remain unencysted in some cases). We shall illustrate these exchanges of information by a series of little-known examples taken from a marine ecosystem.

Information originating from the environment received by the cercaria

Bartoli and Combes (1986) have shown that cercariae of various species of trematodes, all emitted by benthic molluscs at the bottom of a lagoon, were sensitive to different signals of the environment (light, gravity, current, temperature). They respond with different behaviours (positive or negative phototactism, photokinetism, geotactism, rheotactism) as a function of the spatial localization of the second intermediate host. The cercariae of *Cardiocephalus longicollis* (Strigeidae) produced by a benthic prosobranch (*Amyclina corniculum*) show positive phototactism associated with negative geotactism. The resulting swimming behaviour locates the cercariae approximately halfway between the bottom and water surface, where the second intermediate hosts (fishes of the family Sparidae) usually reside. In contrast, cercariae of *Deropristis inflata* (Acanthocolpidae), also produced by a benthic prosobranch (*Hydrobia acuta*), show negative phototactism associated with positive geotactism and remain on the substrate, where their second intermediate hosts, which are annelids (*Nereis diversicolor*), are present.

Information originating in the DSH received by the cercaria

Cercariae can be sensitive to physical (shadow, turbulence, contact) or chemical signals (excreted, secreted substances) produced by the host (Rea and Irwin, 1995; Haberl *et al.*, 2000). Such stimuli are more or less specific and generally act at a short distance between the parasite infective stage and the host. Cotylomicrocercous cercariae (cercariae with a short tail transformed into a sticky sucker) of *Opecoeloides columbella* emerge just after sunrise from a prosobranch snail (*Columbella rustica*) that lives among photophilic algae. The cercariae remain immobile and attached to the rocky substrate by their ventral sucker. When turbulence is produced by a potential host passing by, the cercaria immediately becomes erect, attached by the tip of its sticky tail, and awaits host contact (Jousson and Bartoli, 2000).

Information originating in the cercaria received by the DSH

Cercariae are able to produce signals that can be perceived by the DSH. This is the case for two species of trematodes (*Knipowitschiatrema nicolai* and *Condylocotyla pilodorus*) whose macrocercous cercariae (cercariae with an unusually long and large tail), emitted by a benthic mollusc (*Cerithium vulgatum*), mimic a polychaete annelid and are usually consumed by the fish intermediate host (*Belone belone*). The cercariae mimic both morphological and behavioural characteristics of intermediate hosts' prey items: thanks to their large tail, they exhibit the swimming movement and the violet colour that characterize planktonic organisms (Prévot, 1974; Pearson and Prévot, 1985). Thus, shape, movement and colour all seem to be involved in the information diffused by the cercaria. Bartoli and Gibson (1998) described an unusual visual signal emitted by the cercariae of *Cantharus dorbignyi*. The body of the cercaria resembles a flattened disc, which moves and turns in such a way that it makes the cercaria alternately very visible and practically invisible. During this gyration, the cercaria emits a succession of flashes, which attract the attention of microphagous fishes.

Information originating in the manipulated USH received by the DSH

This is an example where an extension of the parasite's phenotype into the host's phenotype occurs (see Dawkins, 1982). The exchange of signals between the parasitic stage and the target organism is thus indirect. By their localization around the nerves of the locomotory legs of the sand crab (*Carcinus mediterraneus*), the metacercariae of *Microphallus bittii* handicap the movements of the USH, facilitating its predation by gulls, the definitive host (Prévot, 1974). Interestingly, two other microphallid species parasitize the same crab species, but the localization of their metacercariae is different. Those of *Megalophallus carcini* accumulate in the digestive gland and those of *Gynaecotyla longiintestinata* in the urinary lobules. There is no evidence that *M. carcini* or *G. longiintestinata* manipulate their hosts, but they obviously benefit from the extended phenotype of *M. bittii*, since they seek the same DSH. Associations of manipulative and non-manipulative species in the same host have been described by Helluy (1984) and more recently by Thomas *et al.* (1998). The latter discuss to what extent the non-manipulative species are 'hitch-hikers' or simply 'lucky passengers'.

Concluding Remarks and Future Directions

The above examples show that, in a single ecosystem, the circulation of information that characterizes trematode cercarial transmission is highly diversified. If we try to compare this diversity with the taxonomy and phylogenetic relationships either of trematodes or of hosts, it is almost impossible to find a correlation. For instance, the response of cercariae to vibratory cues, the location of metacercariae in a particular organ, etc., have been selected in many different trematode families. Conversely, closely related species of trematodes can use different strategies and even different types of hosts. A striking example is that of *Meiogymnophallus nereicola*, whose cercariae settle in annelids, and *Meiogymnophallus fossarum*, whose cercariae settle in molluscs (Bartoli, 1972). Both are parasites of oyster-catcher definitive hosts. This suggests that natural selection has frequently modified trematode transmission strategies in the course of evolution. Obviously, using an annelid or a mollusc constitutes equivalent strategies in terms of the fitness of these parasites. It is difficult to account for the fact that each of two related trematode species (classified into a single genus) uses only one of these strategies, but avoiding competition at this stage of the life cycle (metacercariae) could be a working hypothesis. Also, an OTS is vital for the species and, because the environmental conditions, composition of faunas, densities and behaviour of potential hosts have been continuously changing on an evolutionary scale, transmission strategies could be one of the first and more sensitive targets of selection. Changes imply both host switching and 'signal switching'. While it is possible to design phylogenetic trees of species (trematodes and hosts), it seems impossible to think of a phylogeny of strategies. That no phylogenetic pattern of strategies emerges may result either from a lack of research or from the fact that environment has more influence than developmental or phyletic constraints. This is a fruitful area for future work.

In most cases, the mechanisms of transmission are better understood than the determinants of the strategies, and the latter (e.g. why such a host, why such a signal?) remain to be clarified. Another important avenue of investigation (Poulin, 2000) will be to examine to what extent the selection of a given trematode strategy may modify host phenotype in such a way that host evolution can itself be modified.

References

Bartoli, P. (1972) Les cycles biologiques de *Gymnophallus nereicola* Rebecq et Prévot, 1962 et de *G. fossarum* Bartoli, 1965, espèces jumelles parasites d'oiseaux de rivages marins (Trematoda, Digenea, Gymnophallidae). *Annales de Parasitologie Humaine et Comparée* 47, 193–223.

Bartoli, P. and Combes, C. (1986) Stratégies de dissémination des cercaires de trématodes dans un écosystème marin littoral. *Acta Oecologica Oecologia Generalis* 7, 101–114.

Bartoli, P. and Gibson, D.I. (1998) A new acanthocolpid cercaria (Digenea) from *Cantharus dorbignyi* (Prosobranchia) in the Western Mediterranean. *Systematic Parasitology* 40, 175–184.

Combes, C. (1991) Ethological aspects of parasite transmission. *American Naturalist* 138, 866–880.

Combes, C. (2001) *Parasitism. The Ecology and Evolution of Intimate Interactions.* University of Chicago Press, Chicago, 728 pp.

Combes, C. and Théron, A. (2000) Metazoan parasites and resource heterogeneity: constraints and benefits. *International Journal for Parasitology* 30, 299–304.

Combes, C., Fournier, A., Moné, H. and Théron, A. (1994) Behaviours in trematode cercariae that enhance parasite transmission: patterns and processes. *Parasitology* 109 (suppl.), S3-S13.

Dawkins, R. (1982) *The Extended Phenotype.* Oxford University Press, Oxford, 307 pp.

Haas, W. (1994) Physiological analyses of host-finding behaviour in trematode cercariae: adaptations for transmission success. *Parasitology* 109, S15-S29.

Haas, W., Haberl, B., Kalbe, M. and Körner, M. (1995) Snail-host finding by miracidia and cercariae: chemical host cues. *Parasitology Today* 11, 468–472.

Haberl, B., Körner, M., Spengler, Y., Hertel, J., Kalbe, M. and Haas, W. (2000) Host-finding in *Echinostoma caproni*: miracidia and cercariae use different signals to identify the same snail species. *Parasitology* 120, 479–486.

Helluy, S. (1982) Relations hôte–parasite du trématode *Microphallus papillorobustus* (Rankin, 1940). II. Modifications du comportement des *Gammarus* hôtes intermédiaires et localisation des métacercaires. *Annales de Parasitologie Humaine et Comparée* 58, 1–17.

Helluy, S. (1984) Relations hôte–parasite du trématode *Microphallus papillorobustus* (Rankin, 1940). III. Facteurs impliqués dans les modifications du comportement des *Gammarus* hôtes intermédiaires et tests de prédation. *Annales de Parasitologie Humaine et Comparée* 59, 41–56.

Jousson, O. and Bartoli, P. (2000) The life cycle of *Opecoeloides columbellae* (Pagenstecher, 1863) n. comb. (Digenea, Opecoelidae): evidence from molecules and morphology. *International Journal for Parasitology* 30, 747–760.

Lafferty, K.D. and Morris, A.K. (1996) Altered behavior of parasitized killifish increases susceptibility to predation by bird final hosts. *Ecology* 77, 1390–1397.

Moore, J. and Gotelli, N.J. (1990) A phylogenetic perspective on the evolution of altered host behaviours: a critical look at the manipulation hypothesis. In: Barnard, C.J. and Behnke, J.M. (eds) *Parasitism and Host Behaviour*. Taylor and Francis, London, pp. 193–233.

Mulvey, M., Aho, J.M., Lydeard, C., Leberg, P.L. and Smith, M.H. (1991) Comparative population genetic structure of a parasite (*Fascioloides magna*) and its definitive host. *Evolution* 45, 1628–1640.

Pearson, J.C. and Prévot, G. (1985) A revision of the subfamily Haplorchinae Looss, 1899 (Trematoda: Heterophyidae). III – Genera *Cercarioides* and *Condylocotyla* n.g. *Systematic Parasitology* 7, 169–197.

Poulin, R. (2000) The evolutionary ecology of parasite-induced changes in host phenotype. *Bulletin of the Scandinavian Society for Parasitology* 10, 44–48.

Prévot, G. (1974) Recherches sur le cycle biologique et l'écologie de quelques trématodes nouveaux parasites de *Larus argentatus michaellis* Naumann dans le Midi de la France. Thèse, Doctorat d'Etat, Université d'Aix-Marseille, 319 pp.

Rea, J.G. and Irwin, S.W.B. (1995) The effects of age, temperature and shadow stimuli on activity patterns of the cercariae of *Cryptocotyle lingua* (Digenea: Heterophyidae). *Parasitology* 111, 95–101.

Théron, A. (1984) Early and late shedding patterns of *Schistosoma mansoni* cercariae: ecological significance in transmission to human and murine hosts. *Journal of Parasitology* 70, 652–655.

Théron, A. and Combes, C. (1988) Genetic analysis of cercarial emergence rhythms of *Schistosoma mansoni*. *Behavior Genetics* 18, 201–209.

Théron, A. and Combes, C. (1995) Asynchrony of infection timing, habitat preference, and sympatric speciation of schistosome parasites. *Evolution* 49, 372–375.

Théron, A. and Pointier, J.-P. (1995) Ecology, dynamics, genetics and divergence of trematode populations in heterogenous environments: the model of *Schistosoma mansoni* in the insular focus of Guadeloupe. *Research and Reviews in Parasitology* 55, 49–64.

Thomas, F., Poulin, R. and Renaud, F. (1998) Nonmanipulative parasites in manipulated hosts: 'hitch-hikers' or simply 'lucky passengers'? *Journal of Parasitology* 84, 1059–1061.

Entomopathogenic Nematode 2 Host-search Strategies

J.F. Campbell[1] and E.E. Lewis[2]

[1]Grain Marketing and Production Research Center, USDA-ARS, 1515 College Avenue, Manhattan, KS 66502, USA; [2]Department of Entomology, Virginia Polytechnic Institute and State University, Blacksburg, VA 24061, USA

Introduction

An organism's foraging behaviour is a critical component of its life history because acquisition of resources is closely linked with fitness, and foraging mode can be correlated with a suite of ecological, behavioural, physiological and morphological traits. The study of foraging behaviour from both theoretical and empirical perspectives has been central to the development of modern behavioural ecology (Stephens and Krebs, 1986; Perry and Pianka, 1997). For most parasites, foraging for resources occurs during two distinct phases: (i) the search for sites within a host by parasitic stages; and (ii) the search for potential hosts by infective stages. Parasitic stages of many species need to find suitable locations within a host to feed, avoid host defences, locate mates or facilitate transmission (Sukhdeo et al., Chapter 11, this volume). Parasites also need a mechanism to bridge the gap between hosts. For many species, this mechanism involves production of a free-living infective stage. Host finding by infective stages may be passive (e.g. the infective stage is an egg) or active. For those species with active free-living infective stages, proximate and ultimate questions about their foraging behaviour can be addressed. In this chapter, we focus on the diversity and sophistication of infective juvenile host-search strategies of insect-parasitic nematodes.

Foraging Strategies

A number of conceptual models for the processes by which searchers locate resources have been developed. Two broad categories of conceptual models are: (i) those that are based on the behavioural responses to sequentially presented stimuli that vary in the quality of the information they convey; and (ii) those that are based on how searchers move through their environment. Both types of models can be useful in facilitating our understanding of the host-search process. The division of host search into a hierarchical process of host habitat location, host location, host acceptance and host suitability (Salt, 1935; Laing, 1937; Doutt, 1964) has been a widely adopted model of the first type. This conceptual model has proved useful for organizing information about parasitoid host-search behaviour (Godfray, 1994), and its application to other parasite species is apparent in this and other chapters in this book. This model does not imply that foragers use a static hierarchical set of behaviours (Vinson, 1981), and more recent models have emphasized the more dynamic nature of the process (Lewis *et al.*, 1990; Vet *et al.*, 1990; Godfray, 1994; Vet *et al.*, Chapter 3, this volume). These recent models have emphasized the ranking of stimuli based on how closely they are associated with the host; the influence of forager internal state, experience and genetics on the parasites' response; and the amount of directional information provided by the stimuli.

Another conceptual model of host search is based on the way that foragers move through the environment and when they scan for cues. In this model, foraging strategies are classified into two broad categories, cruise (widely foraging) and ambush (sit and wait), which represent end-points on a continuum of strategies (Pianka, 1966; Schoener, 1971; Eckhardt, 1979; Huey and Pianka, 1981; McLaughlin, 1989). Into which category an organism fits results mechanistically from differences in how foraging time is allocated to motionless scanning and moving through the environment (Huey and Pianka, 1981; O'Brien *et al.*, 1989). Cruise foragers allocate more of their foraging time to scanning for resource-associated cues when moving through the environment or during short pauses. Ambush foragers scan during long pauses, which are interrupted by repositioning bouts of shorter duration. These differences are significant because the length of scanning pauses influences the types of resources that the organism is likely to encounter. Cruise foragers have a higher probability of finding sedentary and cryptic resources than ambushers, and ambush foragers have a higher probability of finding resources with high mobility than cruise foragers.

This dichotomous view of the foraging mode has been criticized as arbitrary and an oversimplification of what is really a continuum of strategies (Regal, 1978; Taigen and Pough, 1983). However, the dichotomous view of foraging has been studied extensively over the last few decades from field, laboratory and theoretical perspectives and has been applied to a taxonomically diverse group of organisms, including birds (Eckhardt, 1979), lizards (Pianka, 1966; Pietruszka, 1986), fish (O'Brien *et al.*, 1989), arthropods (Inoue and Matsura, 1983; Caraco and Gillespie, 1986) and, as discussed in more detail in this chapter, parasites. These studies have indicated that foraging strategies often have a bimodal distribution that justifies the utilization of the ambusher/cruiser dichotomy (McLaughlin, 1989). The ambush-foraging strategy may be particularly relevant for parasite infective stages, which often search for hosts that are larger and more mobile, e.g. parasitic nematodes (Rogers and Sommerville, 1963; Hernandez and Sukhdeo, 1995) and trematodes of vertebrates (Combes *et al.*, 1994), arthropod ectoparasites (Lees and Milne, 1950) and insect-parasitic nematodes (Campbell and Gaugler, 1993).

The adoption of a particular foraging mode influences a range of related characters, forming what has been termed an adaptive syndrome (Root and Chaplin, 1976). Under this perspective, species that are distantly related taxonomically may share similar adaptive syndromes due to similarity in foraging mode (Eckhardt, 1979). For example, some postulated correlates of foraging mode have been developed for lizards (Huey and Pianka, 1981). These include that ambush foragers should eat mobile prey, capture few prey per day and have a low daily metabolic expenditure, a low probability of encountering prey, limited endurance, limited learning ability and stocky morphology and use primarily visual cues. Cruise foragers are predicted to eat sedentary and unpredictable prey that are clumped or large, capture high numbers of prey per day and have a high daily metabolic expenditure, a high probability of encountering prey, a high endurance capacity, enhanced learning and memory and streamlined morphology and use visual and chemical cues. Clearly, many of these characters may not apply to parasite infective-stage foragers, but the idea that there may be adaptive syndromes associated with the way that parasites search is a fruitful area for research. The presence of adaptive syndromes means that assigning a forager to a particular category provides insight into aspects of an organism's biology other than just host search.

Foragers typically respond behaviourally to stimuli from the environment in ways that improve the probability or rate of encounter with a resource. The method of scanning, the relative importance of different stimuli and the nature of the response will be influenced by foraging strategy (O'Brien *et al.*, 1989, 1990). Many foragers use multiple sensory modalities and cues (e.g. mechanical, chemical or auditory) during search and their relative importance may depend on foraging strategy. Research into behavioural mechanisms has been heavily biased toward the more active cruise foragers – for example, the use of chemotaxis and localized search patterns in patches to facilitate finding resources (e.g. Bell, 1985; Huettel, 1986; Ramaswamy, 1988; Vet and Dicke, 1992). However, stimuli from the environment have also been demonstrated to be important for ambush foragers. Cues are used for selecting ambush sites (Greco and Kevan, 1994, 1995), assessing patch quality (O'Brien *et al.*, 1990; Sonerud, 1992) and triggering resource-capture behaviours (Bye *et al.*, 1992). Foragers may be able to improve their foraging efficiency by adjusting their allocation of time to pausing and moving in response to external stimuli (i.e. shift along the continuum between ambush and cruise foraging) (O'Brien *et al.*, 1990). This ability has been demonstrated for a number of species of insects, fish, birds and lizards (Akre and Johnson,

1979; Formanowicz and Bradley, 1987; Inoue and Matsura, 1983; O'Brien
et al., 1990).

Nematode–Insect Associations

Many nematode species have intimate heterospecific associations with
vertebrates, invertebrates and plants (i.e. symbiotic associations) (Cheng,
1991). Behavioural adaptations have probably played a critical role in
enabling nematodes to exploit this diversity of habitats (Croll, 1970;
Dusenberry, 1980). The symbiotic associations between nematodes and
insects are diverse and can be classified into three categories: phoretic
relationships, facultative parasitism and obligate parasitism (Poinar,
1975). Parasitic associations occur in many nematode families and are
likely to have evolved multiple times (Poinar, 1975; Blaxter *et al.*, 1998).
Behavioural adaptations to increase the probability of encountering
insects were probably critical for facilitating this diversity of associations,
but our understanding of the behavioural interactions between nematodes
and insects is limited for most associations. The most thoroughly studied
species tend to be those with potential as biological control agents.

Phoresy is a phenomenon in which a life stage of one species is
transported by an animal of a different species (Farish and Axtell, 1971;
Houck and O'Connor, 1991). The degree of symbiosis in these phoretic
associations can vary considerably and transport can be external or internal. Phoresy may be the most common nematode–insect association, with
the phoretic life stage typically being a dauer juvenile (Massey, 1974;
Sudhaus, 1976; Kiontke, 1996). Phoresy is an important strategy for
species with limited mobility that exploit patchy and ephemeral
resources, such as rotting logs and animal waste. For example, the nematode *Diplogaster coprophila* is a free-living species associated with animal
manure piles that produces a dauer juvenile phoretically associated with
some species of sepsid flies (Kiontke, 1996). Dauer juveniles aggregate on
fly pupae and initiate waving behaviour, which facilitates attachment to
emerging flies. Phoresy is likely to have been an important first step in the
evolution of many parasitic associations.

Nematodes that are facultative insect parasites can have free-living
generations, which are typically saprophagous. The degree of symbiosis
and harm caused to the host varies considerably among species (Poinar,
1972). A typical example of this type of association is *Rhabditis
insectivora*, which can complete its life cycle in the external environment
feeding on bacteria, but dauer juveniles that encounter larvae of
the cerambycid beetle, *Dorcus parallelopipedus*, enter the insect via the
digestive system and develop and mate in the haemocoel (Poinar, 1975).
Females leave the host to lay eggs, and only high rates of infection
are lethal to the insect. A more symbiotic facultative association occurs
between the nematode *Deladenus siricidicola* and the wood wasp *Sirex
noctilio* (Bedding, 1972, 1993). During free-living generations, which can

go on indefinitely, the nematodes feed on the wasp's symbiotic fungi in the traceid system of trees. The wood wasp also feeds on the fungi. When juvenile nematodes encounter the microenvironment surrounding a sirex larva, they mature into a different form of adult. After mating, males die and females penetrate into the sirex larva's haemocoel. Female nematode reproductive systems do not develop until the host larva begins to pupate; then juvenile nematodes are produced that migrate to the host's reproductive system and enter the host eggs. The now sterile female wasp then deposits the nematodes back into trees to resume the free-living life cycle. Juvenile nematodes will invade the testes of male wasps, but transmission does not occur and males are considered to be dead-end hosts.

Obligate parasites typically occupy the insect's body cavity, but some species occur in the digestive tract, reproductive organs or cuticle. Some of the best-studied examples of obligate parasites of the insect body cavity occur in the family Mermithidae (Poinar, 1975). Nematodes in this family infect a wide range of invertebrate species in both terrestrial and aquatic ecosystems. Typically, eggs are deposited into the external environment, where second-stage juvenile nematodes hatch and seek out and penetrate into the host. The parasitic stages generally remain in the insect haemocoel and absorb nutrients through their cuticle. One individual per host is typical, and the nematode can become quite large and is usually coiled up within the haemocoel in species-specific locations. The process of exiting the host is typically lethal to the host, due to the large exit hole produced. The postparasitic juveniles enter the soil or the aquatic sediment and mature to adults, mate and lay eggs. Another important group of obligate parasites is the entomopathogenic nematodes, which will be discussed in detail below.

Entomopathogenic Nematode Biology

Entomopathogenic nematodes (Heterorhabditidae and Steinernematidae) have mutualistic associations with bacteria (*Photorhabdus* and *Xenorhabdus* spp.). These nematodes are lethal endoparasites, capable of infecting a broad range of insect species (Poinar, 1975; Kaya and Gaugler, 1993). These two families are discussed together because of their similar life cycles and bacterial associations, but these similarities are probably the result of convergent evolution (Poinar, 1993; Forst and Nealson, 1996; Blaxter *et al.*, 1998). The nematode–bacteria complex represents a mutualistic relationship, because the bacterium creates a favourable environment for nematode growth and development within the insect host and the bacteria are vectored between insects by the nematode. Each nematode species is associated with only one bacteria species, although some bacteria species are associated with more than one *Steinernema* species (Akhurst, 1993).

Although these nematodes occur in soil and epigeal habitats throughout most regions of the world, their population ecology and host associations in the field are poorly understood. Nevertheless, natural populations of these nematodes can play an important role in ecosystems (Sexton and Williams, 1981; Akhurst *et al.*, 1992; Campbell *et al.*, 1995; Strong *et al.*, 1996). Entomopathogenic nematodes have been the focus of considerable research because of their potential as biological control agents against a wide range of insect species in a wide range of crops (Gaugler and Kaya, 1990; Kaya and Gaugler, 1993).

There is only one free-living stage, the infective juvenile, which is non-feeding, non-developing and non-reproductive; its sole function is to bridge the gap between a depleted and a new host. Infective juveniles carry their symbiotic bacteria in their digestive tracts. When a potential host is located, infective juveniles enter the host through natural body openings, such as spiracles, mouth or anus, but in some cases can penetrate areas of thin cuticle. After penetration into the insect haemocoel, the nematodes begin feeding and development. The symbiotic bacteria are released into the haemocoel, where they multiply, and the host is usually killed by septicaemia within 24–48 h. Nematodes feed, develop, mate and lay eggs within the host. The size and species of the host and the number of founding nematodes influence the number of generations that can occur within a host and the number of new infective-stage juveniles produced. Adult *Steinernema* spp. are amphimictic and therefore require at least one individual of each sex to infect a host for reproduction. In contrast, *Heterorhabditis* spp. infective juveniles develop into hermaphroditic adults, so only a single individual is needed to initiate an infection. In subsequent parasitic generations of *Heterorhabditis*, both amphimictic and hermaphroditic individuals are produced (Strauch *et al.*, 1994; Koltai *et al.*, 1995). As time passes in the infection, nematode density increases and nutrient levels decrease, which lead to the formation of infective-stage juveniles instead of parasitic third-stage juveniles. When the host becomes depleted, these infective juveniles emerge and seek new insects to infect.

Entomopathogenic Nematode Foraging Behavior

Behavioural repertoire

Nematodes' small size, lack of appendages and limited sensory organs constrain their host-seeking behaviour. Many nematode infective stages are often looking for hosts that are considerably larger and more mobile than they are. In addition, the surface tension of the water film in which they move holds them firmly to the substrate. The surface-tension forces of water films on nematodes have been estimated to be 10^4–10^5 times the force of gravity (Crofton, 1954).

Three behaviours – crawling, standing and jumping – play important roles in determining where along the continuum between ambush and cruise foraging a nematode infective stage lies. Most nematode species crawl by sinusoidal movement on the substrate, using the surface-tension forces associated with the water film to propel them forwards or backwards (Croll, 1970). Nematodes perceive their environment primarily by chemosensation, thermosensation and mechanosensation. How environmental cues such as chemical and temperature gradients influence crawling nematodes has been the most extensively studied area of nematode behaviour (Croll, 1970; Dusenberry, 1980; Zuckerman and Jansson, 1984; Huettel, 1986; Bargmann and Mori, 1997). Crawling nematodes may scan for environmental cues while crawling on the substrate or during short pauses. Crawling locomotion and the use of various kineses and taxes to locate hosts is consistent with a cruise type of foraging strategy.

The infective stages of some entomopathogenic nematode species exhibit two additional behaviours that facilitate ambush foraging: standing and jumping. Most nematodes can raise the anterior portion of their body off the substrate and wave it back and forth. However, some species can elevate more than 95% of their body off the substrate and balance on a bend in their tail (Reed and Wallace, 1965; Ishibashi and Kondo, 1990; Campbell and Gaugler, 1993). This behaviour has been termed '*Winken*' (Völk, 1950), 'nictation' (Ishibashi and Kondo, 1990; Campbell and Gaugler, 1993) and most recently 'standing' (Campbell and Kaya, 1999a,b, 2000). Standing behaviour is restricted to the free-living infective or dauer stages of certain species (Campbell and Gaugler, 1993; 1997; Campbell and Kaya, 2002). Among species that exhibit standing behaviour, variation occurs in the duration of standing bouts and nematode activity while standing. Some species have a stable standing behaviour, in which the nematode becomes straight and immobile and can maintain this posture, with interspersed periods of waving, for extended periods of time (can exceed 2 h). Standing behaviour facilitates attachment to mobile hosts by reducing the surface tension holding the nematode to the substrate (Campbell and Gaugler, 1993). Standing bouts are ended by the nematode falling, by touching the anterior portion of its body to a surface during waving or by jumping. Standing behaviour may function as both an immobile scanning bout and a mechanism to attack passing hosts.

Jumps occur when an infective juvenile is standing; the nematode forms a loop with its body, which, when released, propels the nematode many times its body length through the air (Reed and Wallace, 1965; Campbell and Kaya, 1999a,b, 2000). The forces generated by *Steinernema carpocapsae*'s jumping mechanism propel individuals an average distance of 4.8 ± 0.8 mm (nine times the nematode's body length) and an average height of 3.9 ± 0.1 mm (seven times body length) (Campbell and Kaya, 1999a,b). The frequency of jumping, like standing behaviour, varies among species of *Steinernema*, but jumping has not been observed in *Heterorhabditis* (Campbell and Kaya, 2002). Jumping can function as a means of dispersal and also as an ambush attack mechanism.

Ambush vs. cruise foraging

Differences in entomopathogenic nematode foraging strategies can be most easily identified by comparing the ability of infective juveniles to locate mobile versus sedentary insects (Campbell and Gaugler, 1997). More infective juveniles of cruise-foraging species are able to find sedentary hosts compared with mobile hosts and the opposite is true for ambush foragers (Fig. 2.1). Intermediate foragers are similar in their ability to find both types of hosts. By comparing attachment to mobile and constrained individuals of the same insect species, confounding factors associated with host-specific differences in nematode behaviour are reduced. Variation in the time allocated to crawling and standing among species of *Steinernema* is consistent with variation along a continuum between ambushing and cruising (Fig. 2.1), but all tested *Heterorhabditis* spp. appear to be cruise foragers. Cruise foragers do not exhibit standing or jumping behaviour, ambush foragers have stable standing periods and exhibit high rates of jumping and intermediate foragers have lower rates of standing and jumping (Campbell and Kaya, 2002; J.F. Campbell *et al.*, unpublished data). It is also apparent that some species along the continuum of the host-finding mode vary in their response to host cues, which will be discussed in more detail later.

Although some foragers change foraging strategy in response to experience or changes in internal state, this does not appear to be true for entomopathogenic nematode infective stages. Time after emergence does influence foraging behaviour, but there is no evidence that infective juveniles change foraging strategies as they age. The general pattern is for infective juveniles to become less effective at their original foraging mode (i.e. become less mobile and less able to stand) and less able to infect a host over time and this decline in infectivity has been correlated with a decline in lipid levels (Lewis *et al.*, 1995b, 1997).

Cruising is a more energetically expensive foraging strategy than ambushing. Because infective juveniles are non-feeding and have a fixed amount of stored nutrients, primarily in the form of lipids, we predicted that cruise foragers will be larger (i.e. store more lipids) than ambush foragers to compensate for their more energetically expensive search strategy. Campbell and Kaya (2002) have noted that the *Steinernema* spp. cruise foragers tend to have longer infective juveniles than ambush foragers. Selvan *et al.* (1993a) reported that larger *Steinernema* infective juveniles had more stored lipids: the estimated energy content of a single infective juvenile was 0.123 J for *S. glaseri* (a cruise forager), 0.065 J for *S. feltiae* (an intermediate forager) and 0.030 and 0.029 J for *S. carpocapsae* and *S. scapterisci*, respectively (ambush foragers). Lipid levels have been shown to influence parasite infectivity and movement (Croll, 1972; Lee and Atkinson, 1977). Lewis *et al.* (1995b) demonstrated how lipid levels declined with infective juvenile age and how this was correlated with changes in nematode behaviour and infectivity. In some cases, lipid level was a better predictor of infectivity than infective

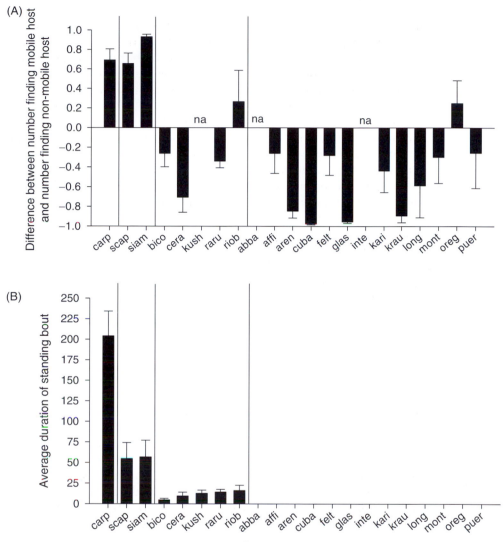

Fig. 2.1. Variation among species of *Steinernema* infective juveniles in (A) the ratio between number of infective juveniles finding a mobile host (unrestrained *Galleria mellonella* larva on the surface of a sand-filled 90 mm Petri dish) minus the number finding a non-mobile host (*G. mellonella* larva buried beneath the surface of a sand-filled 90 mm Petri dish) divided by the total number of nematodes finding both types of host; and (B) the average duration of standing bouts. In (A), a value of 1.0 indicates that nematodes only found the mobile host, a value of −1.0 indicates that nematodes only found the sedentary host and a value of 0.0 indicates equal ability to find both types of host. The vertical lines indicate how species were grouped along the continuum between ambush and cruise foraging based on duration of standing bouts, from left to right the groups are long, medium, short and none. The labels on the *x* axes are the first four letters of the species name tested.

juvenile age alone (Patel *et al.*, 1997). *S. carpocapsae*, an ambush forager, had a slower metabolic rate than *S. glaseri* and *Heterorhabditis bacteriophora*, the two cruise-foraging species tested (Lewis *et al.*, 1995b). *H. bacteriophora* had the highest metabolic rate and the shortest lifespan. *S. glaseri* had an intermediate metabolic rate, but lived the longest due to its larger size and corresponding lipid reserves. *Heterorhabditis* spp. tend to be cruise foragers (Campbell and Kaya, 2002), but have low energy contents (Selvan *et al.*, 1993b), are similar in size to ambush-foraging *Steinernema* and have a short lifespan in the laboratory (Lewis *et al.*, 1995b) and persistence in the field (Bauer and Kaya, 2001). Clearly, it is difficult to draw any firm conclusions from these comparisons, but the results within *Steinernema* are consistent with the predictions. The comparison between *Heterorhabditis* spp. and *Steinernema* spp. is difficult, since these genera are distantly related.

Host-habitat Selection

The point when an infective juvenile first emerges from a depleted host is that at which we regard the process of host infection as starting. Infective juveniles emerge from a depleted host over a period of time that ranges from several days to several weeks, depending upon host quality, nematode species, number of founders that initiated the infection and environmental conditions. While the interaction between infective juveniles and a depleted host has not received as much attention as the interaction with a potential host, recent studies suggest that the depleted host influences nematode ecology and behaviour. There are two phases where there are potentially significant interactions between infective juveniles and depleted hosts. Firstly, during the process of emergence and dispersal, there are endogenous and exogenous factors that determine the timing of emergence. Secondly, the depleted host and/or its exudates can continue to influence infective juvenile behaviour during the process of host finding.

Variation in the sex ratio of nematodes that emerge at different times from a host cadaver has been reported. Infective juveniles of *S. glaseri* that disperse early from a depleted host cadaver tend to be larger and to have a more male-biased sex ratio (Lewis and Gaugler, 1994; Nguyen and Smart, 1995; Stuart *et al.*, 1996). Stuart *et al.* (1996) found that there was a genetic basis to *S. glaseri* emergence time and that there was selection for a highly skewed early emergence in a population recently isolated from the field. Increased size is likely to be correlated with improved fitness due to increased lipid storage and longevity. However, it is not clear if larger size is correlated with being male or if infective juveniles that are formed first, and presumably developed under the best environmental conditions, disperse first. Lewis and Gaugler (1994) proposed that this phenomenon is analogous to adult males emerging before females (i.e. protandry). By dispersing first, males are more likely to infect uninfected hosts before females from the same cadaver and therefore be in a better position to mate with females when they infect. Lewis and Gaugler (1994) found that male *S. glaseri* infective juveniles were more sensitive to volatile cues from uninfected hosts than female infective juveniles and that females were more attracted to infected hosts than to non-infected hosts. However, when a population of male and female *S. glaseri* infective juveniles was exposed to *Galleria mellonella*, there was no trend for males to infect first (Stuart *et al.*, 1998). *S. carpocapsae*, an ambush forager, showed no sex bias in early emergers, and in *S. feltiae*, an intermediate forager, females tended to emerge first (E.E. Lewis, unpublished data). *Heterorhabditis megidis* infective juveniles that emerge early from a depleted host were better host finders, more tolerant of warm temperatures and less tolerant of desiccation than later-emerging infective juveniles (O'Leary *et al.*, 1998). Further research is needed to assess how differences in emergence patterns are related to foraging strategy and infection dynamics.

Environmental conditions influence the emergence pattern of entomopathogenic nematodes. Koppenhöfer *et al.* (1997) found that emergence was delayed under low soil-moisture conditions and that they could survive longer in dehydrated cadavers than they could exposed to dry conditions in the soil. It is not clear if infective juveniles are trapped in the cadavers or if this represents an adaptation to persist in low-moisture conditions. The pattern for *S. carpocapsae* not to be recovered at or near the soil surface during daylight hours also suggests that infective juvenile emergence from cadavers may be timed to favourable environmental conditions (Campbell *et al.*, 1996).

Exudates from a depleted host influence nematode behaviour (Shapiro and Glazer, 1996; Shapiro and Lewis, 1999). *H. bacteriophora* and *S. carpocapsae* infective juveniles allowed to emerge into sand from their host cadaver dispersed farther than infective juveniles first collected in water and then applied to the sand (Shapiro and Glazer, 1996). *H. bacteriophora* infective juvenile infectivity was also reduced if they were collected in water and applied artificially, but infectivity could be restored by applying the nematodes with an infected host (Shapiro and Lewis, 1999). Many questions remain about the adaptive value of this response. As currently understood, it suggests that *H. bacteriophora* infective juveniles are stimulated to disperse and infect by materials associated with the depleted host. It may be that cues are picked up from the depleted host that may be useful in identifying encountered insects as acceptable hosts. This phenomenon has been noted for some parasitoid wasps that learn cues from their natal host (Turlings *et al.*, 1992). However, given the many changes in the host during the infection process of entomopathogenic nematodes, it is not clear what relevance these cues might have.

Entomopathogenic nematodes live in soil and epigeal environments, where potential hosts are likely to be clumped or patchy in distribution. Location in the soil profile may be one of the most significant host-habitat selection behaviours. Some species of entomopathogenic nematode tend to be located at or near the soil surface and others prefer deeper locations (Moyle and Kaya, 1981; Campbell *et al.*, 1996). Ambush foraging requires a substrate favourable for standing and is most effective near the soil surface. Mobile insects are also most prevalent at the surface. In turf grass, endemic populations of *S. carpocapsae*, an ambush forager, were greater at the soil surface and nematodes were most prevalent during crepuscular and dark periods, when environmental conditions were favourable for nematode persistence (Campbell *et al.*, 1996). In contrast, *H. bacteriophora* were recovered throughout the upper soil profile. In sand columns containing growing grass, most ambush and intermediate foragers preferred the region above the soil surface and most cruise-foraging species preferred the subsurface regions (J.F. Campbell, unpublished data).

Host Finding

Cues that indicate the proximity of a potential host are most likely to occur in the host habitat. These cues may be contact or volatile stimuli from a host or its immediate environment. How foragers respond behaviourally to these stimuli varies depending on the type of search strategy they use, the information contained in the stimuli and their internal state. Here we shall evaluate how infective juveniles interact with stimuli associated with hosts within the context of their foraging strategy. We focus on *Steinernema* species, which show the greatest variation and have been the most extensively studied.

Cruise-forager infective juveniles move using relatively linear movement patterns that are typical of ranging search (Lewis *et al.*, 1992). Ranging search maximizes the search area. During ranging search, species characterized as cruise foragers respond to volatile cues from hosts. For example, Lewis *et al.* (1993) found that *S. glaseri* responded positively to volatile cues from an insect host in a y-tube olfactometer and that this response was eliminated if CO_2 was absorbed. Grewal *et al.* (1994) found a similar level of response to volatile cues for other cruise-foraging species in *Steinernema* (*S. arenarium*) and in *Heterorhabditis* (*H. bacteriophora* and *H. megidis*). Recently, this strong response to volatile cues was extended to many *Steinernema* spp. that are effective at finding sedentary hosts (E.E. Lewis, unpublished data). Cruise foragers also respond to host contact cues. Lewis *et al.* (1992) found that *S. glaseri* switched to localized search (e.g. speed decreased, distance travelled decreased, proportion of time spent moving decreased) after contact with host cues (e.g. cuticle, faeces). Localized search can maximize the chance that a searcher will remain in a patch or re-establish contact with a host. Thus, the scenario for cruise foragers is that they range through the environment, move toward the source of volatile cues and switch to localized search after they contact host-associated cues. These mechanisms are consistent with the behavioural responses of a diverse range of cruise-type foragers (Bell, 1990).

Ambush foragers, in contrast, have a very different response to host cues. First, although on smooth substrates ambush foragers like *S. carpocapsae* exhibit ranging search, they are not attracted to host volatile cues like CO_2 (Lewis *et al.*, 1993). This response does not appear to result from a constraint in their sensory systems, but to the absence of selection pressure that would lead to the character, because artificial selection can produce lines that are attracted to CO_2 (Gaugler *et al.*, 1989). Also, ambush foragers do not switch to area-concentrated search in response to contact with host cues such as cuticle and faeces, as *S. glaseri* does (Lewis *et al.*, 1992). Although initially it was assumed that ambush foragers were just not as responsive to chemical cues as cruise foragers (Lewis *et al.*, 1992), it has since become apparent that they are responsive to chemical cues but their response is fundamentally different.

Lewis *et al.* (1995a) first demonstrated that ambush foragers respond to volatile cues, but that their response was context-specific. They found that, although *S. carpocapsae* infective juveniles when crawling on the substrate were not attracted to host volatiles, they did respond to volatile cues after contact with host cuticle. Lewis *et al.* (1995a) hypothesized that this change in response occurred because, after a nematode contacted a host while standing, it would be triggered to seek routes of entry into the host (e.g. triggered to orientate towards the spiracles). They proposed that ambush foragers are presented with cues in a more sequential fashion than cruise foragers and therefore only responded to cues if they were presented in the appropriate sequence.

It was subsequently determined that ambush foragers do respond to volatile cues, even prior to host contact. Their behavioural response to host cues needed to be investigated from within the context of their foraging strategy. Cruise foragers scan while moving through the environment or during short pauses, but ambush foragers scan the environment during long standing pauses. If host attachment is not a passive process, then we expect to find that standing infective juveniles change behaviour in response to host-associated cues. This has been demonstrated in two ways. First, ambush-foraging infective juveniles presented with host-associated cues (air movement and the presence of volatile cues) when standing wave back and forth and jump towards the source of the cue (Campbell and Kaya, 1999a,b, 2000). This appears to result from two types of cues: volatile cues and air movement. This behavioural response increases the host-attack area surrounding the standing infective juvenile and can increase the probability of attaching to a host. This result indicates that standing infective juveniles are scanning the environment and respond in a way that can increase the probability of host encounter. However, as discussed later, not all nematode species that stand respond to host cues.

Host volatile cues can also cause ambush foragers to increase their giving-up time during standing. It is likely that entomopathogenic nematode hosts have patchy distributions at the soil surface. Therefore, some locations where nematodes stand will have a higher probability of host encounter than other areas. When a forager arrives at a patch, it has to decide how long to remain in that patch. If no host arrives within a certain period of time, there are two possible explanations: it is a poor patch or it is a good patch but by chance no hosts have yet arrived. Many ambush foragers appear to use relatively simple decision rules about when to leave a patch in the absence of host encounter (e.g. Janetos and Cole, 1981; Janetos, 1982; Kareiva *et al.*, 1989). A number of rules of thumb for leaving a patch have been proposed, including: a fixed number rule, a fixed time rule, a giving-up time rule (leave after a certain amount of time without success) and an encounter rate rule (leave when intake drops below a certain rate) (Stephens and Krebs, 1986). The relative performance of a particular rule is strongly influenced by the distribution of prey within and between patches (Iwasa *et al.*, 1981; McNair, 1982; Green, 1984). The only rules that are appropriate for parasites are the giving-up time rule or the fixed time rule, which become functionally the same because parasites only use a single prey item. Infective juveniles, unlike predators or female parasitoids, do not have information on patch quality obtained directly by experience, but may use chemical cues to evaluate patch quality. Chemical cues are used by some parasitoids to influence patch leaving (e.g. Waage, 1979; Hemerik *et al.*, 1993). Information gained within a patch can influence both the tendency to leave the current patch and the tendency to leave subsequently encountered patches (Shettleworth, 1984).

For *S. carpocapsae,* the leaving tendency when standing tended to decrease as the time standing increased (J.F. Campbell *et al.*, unpublished data). *S. carpocapsae* infective juveniles that recently initiated standing bouts were more likely to terminate standing (0.081 terminations min^{-1} for bouts less than 10 min) than individuals that have been standing for longer periods of time (0.041 terminations min^{-1} for bouts longer than 10 min). The survival function that described the patch leaving was consistent with a Weibull function – rate changes over time – rather than the exponential distribution associated with leaving at a constant rate. What generates this pattern and its adaptive value is unclear, but the termination of standing bouts was not consistent with a fixed time rule for patch exploitation. When host-associated cues were present, *S. carpocapsae* was less likely to terminate a standing bout; the nematode's average patch-leaving rate was 0.0064 terminations min^{-1}. The tendency to use a fixed rate of leaving rule and to stay in patches longer than optimal has been reported for several other ambusher species (Janetos, 1982; Kareiva *et al.,* 1989). Kareiva *et al.* (1989) predicted that for an ambush-foraging spider the prey arrival rates may be so variable that no one patch exploitation strategy is markedly superior. Although we do not know the distribution of arrival rates of insects at nematode ambush locations, it is reasonable to speculate that arthropod arrival rates at the soil surface are also highly variable, both temporally and spatially.

    The suite of behavioural traits associated with intermediate foraging is more variable. Many intermediate foragers are attracted to host volatiles and also switch to localized search in response to contact with host cuticle (J.F. Campbell *et al.*, unpublished data). However, most species also exhibit standing and jumping behaviour. One intermediate forager, *Steinernema riobrave*, was found to have short-duration standing bouts, and standing giving-up time was not influenced by the presence of host cues (J.F. Campbell *et al.*, unpublished data). In this case, termination of standing bouts appears to be a random process, which generates an exponential distribution of leaving times. Intermediate foragers also tend not to be triggered to jump by the sudden introduction of host cues and not to jump towards the source of cues (Campbell and Kaya, 2000). All of these characters suggest that intermediate foragers use host cues in a manner consistent with cruise foragers, although perhaps not as effectively, but because they stand and jump may encounter moving hosts as well. Standing and jumping behaviours are unlikely to be adaptations specifically for ambush foraging and use of these behaviours to attach to passing insects may be unintentional but, given the opportunistic nature of entomopathogenic nematodes, still an effective infection strategy. Stable standing, chemical cues triggering changes in standing-bout duration and controlling of the timing and direction of jumps are likely to be adaptations to ambush foraging and these tend only to be found in extreme ambush foragers.

Host Acceptance

Host acceptance for a parasite infective stage culminates in entering the host haemocoel and the transition from the free-living infective stage to the parasitic stage. Host acceptance is typically not reversible, so selection of an appropriate host is a critical decision for an infective stage. Encountered hosts will vary in quality, due to differences in suitability for growth and development. Clearly nematodes should infect hosts that maximize nematode fitness, but how nematode infective stages assess host quality is not well understood. There have been a number of host-acceptance models developed for parasitoids (Godfray, 1994), which, while not directly applicable to parasite infective stages, do highlight some of the important factors in the host-acceptance process. These factors include host profitability (i.e. potential fitness benefits), risk of mortality and infective juvenile age. Infection can be a risky undertaking, but a large part of the risk is associated with the host immune response, so this risk could be reduced by preferentially infecting hosts that are already infected.

    Multiple cues are involved in the process of host acceptance, perhaps beginning with detection of host volatile cues, but, since most species will initiate development if injected into the haemocoel, these cues are not obligatory. For example, differences in giving-up time (i.e. return to ranging search) after contact with different species of host were observed and may be due to differences in host preference or suitability (Glazer and Lewis, 2000). Cues that trigger penetration and acceptance behaviour are not well understood, but infective stages are known to respond to a number of host-associated materials, such as faeces (Grewal *et al.*, 1993a), gut contents (Grewal *et al.*, 1993b) and haemolymph (Khlibsuwan *et al.*, 1992). For example, contact with host faeces increased head-thrusting behaviour, which may be involved in penetrating through the insect's digestive system (Grewal *et al.*, 1993a). The functional benefit of these behavioural responses, however, has not been tested.

There is evidence that entomopathogenic nematodes can evaluate host quality and make infection decisions about which hosts to infect. This is most apparent in the host-acceptance decision associated with infecting an infected versus an uninfected insect. For steinernematid nematodes, there appears to be a selective advantage to infecting a host that is already infected. The risk of immediate mortality is reduced due to degradation of the host immune response and the probability of encountering a parasite of the opposite sex is increased. Over time postinfection, the fitness advantages of infecting a host that is already infected will probably diminish. In addition, there is an Allee effect on the number of infective juveniles produced and a decline in infective juvenile size as a function of increasing number of founders (Selvan *et al.*, 1993b).

Several species of entomopathogenic nematode infective juveniles are more strongly attracted to infected hosts 24 h after infection than to uninfected hosts (Lewis and Gaugler, 1994; Grewal *et al.*, 1997). Infective juveniles also distinguished between conspecific infections and hetero-specific infections, but there was considerable variation in this ability among species (Grewal *et al.*, 1997). For example, *S. feltiae* was not attracted to hosts infected by any of the tested heterospecifics, but *S. glaseri* was attracted to all infected hosts except those infected by *S. riobrave*. In some cases, infective juveniles were actually repelled by the heterospecific-infected hosts. J.F. Campbell *et al.* (unpublished data) found that, when given a choice between a host infected by conspecifics 24 h earlier and an uninfected host, *S. feltiae* preferentially infected the previously infected hosts, but showed no preference for hosts infected by conspecifics 2, 4 or 8 h prior (Fig. 2.2). In contrast, Glazer (1997) found that 9 h after injecting nematodes into an insect's haemocoel, the number of infective juveniles that would infect that insect was reduced. Exposure of *S. carpocapsae*, *S. riobrave* and *S. feltiae* infective juveniles to a surface previously exposed to an infected host suppressed their infectivity when exposed to an uninfected host, and this suppression was removed by washing the infective juveniles (Glazer, 1997).

It has been widely reported that only a small proportion of infective juveniles, typically less than 40%, infect a host when exposed under laboratory conditions. The explanation most commonly invoked was that this was due to differences in internal state, i.e. not all individuals

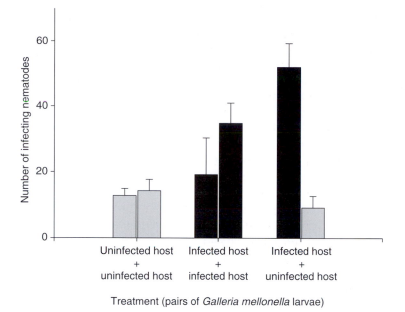

Treatment (pairs of *Galleria mellonella* larvae)

Fig. 2.2. Differences in *Steinernema feltiae* infection of two *Galleria mellonella* larvae paired in a 1.5 cm diameter arena. Infected hosts (black bars) had been exposed to 100 *S. feltiae* infective juveniles 24 h prior to the start of the experiment and uninfected hosts (grey bars) had not been previously exposed to nematodes. Control insects exposed to 100 nematodes 24 h prior to the start of the experiment were dissected and the number of nematodes present was determined. Treatments of two uninfected hosts, two infected hosts, and one infected host and one uninfected host were tested. After exposure, insects were dissected and the number of nematodes that infected was determined. The number of nematodes in the previously infected host treatments was corrected using the average number of nematodes present in the control hosts.

are infectious at the same time (Hominick and Reid, 1990). Bohan and Hominick (1995) developed an infection model based on this hypothesis and Hay and Fenlan (1995) proposed that there are primary and secondary invaders in a population. Campbell *et al.* (1999) found that, for three *Steinernema* species, the cause of the low infection was not that some infective juveniles were not infectious, but that the presence of acceptable hosts was the limiting factor. As the number of insects available to infect increased, so did the number of infective juveniles that infected. The distribution of nematodes among hosts was clumped, suggesting that some hosts are more likely to be infected than others. The reasons for this are not clear, but may be due in part to a preference to infect hosts already infected. In contrast, results of similar experiments with *H. bacteriophora* suggest that a large portion of the population may not be infectious at a given time (Campbell *et al.*, 1999).

Evolution of Foraging Strategy

Comparison of traits among species is a critical part of investigating adaptations, but it is important to take into account the evolutionary relationships among the species being compared (Harvey and Pagel, 1991). Two species may share a trait due to common descent or because both species evolved the same response to similar selection pressure. The considerable variation in foraging strategy within the genus *Steinernema* leads us to questions about how these strategies may have evolved and if the suites of traits associated with foraging strategy represent adaptations. The mapping of behavioural traits on to a recent molecular and morphological phylogeny of *Steinernema* (Stock *et al.*, 2001) provides a means of developing hypotheses about the evolution of foraging strategies within this group. Mapping the ability to find mobile versus sedentary hosts on to the phylogeny of *Steinernema* suggests that the ancestral species was an intermediate forager and that ambush and cruise foraging both evolved at least once in the genus (Fig. 2.3). We can also hypothesize that the

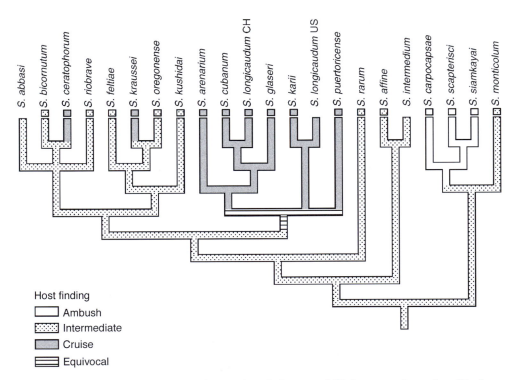

Fig. 2.3. Foraging strategy mapped on to the phylogeny of *Steinernema* based on Stock *et al.* (2001). Foraging-strategy categories were based on ratio between ability to find mobile versus non-mobile hosts as shown in Fig. 2.1. Behavioural data are lacking for two species; in one case this results in an equivocal mapping.

ancestral *Steinernema* had low or no standing behaviour and a low frequency of jumping, occupied epigeal habitats, was of medium size and had a low level of attraction to host volatiles and no change in behaviour after host contact (J.F. Campbell *et al.*, unpublished data). The lack of multiple clades with ambush- or cruise-foraging specialization makes it difficult to determine whether or not the suite of behavioural traits associated with each strategy represents adaptations. However, the diversity in Steinernematidae has probably been undersampled and, as new species are described, hypotheses about the evolution of foraging strategy can be further tested and refined.

Conclusions

The importance of parasites in ecosystem function and the influence of parasites on host behavioural ecology have become more widely acknowledged in recent years and we have seen a surge of research in these areas. However, research on the behavioural ecology of the parasites themselves is still limited. This is unfortunate. Parasite infective stages make very useful models for addressing behaviour questions and there are many important fundamental and applied questions that need to be addressed. There is a strong foundation of behavioural research that has been conducted on entomopathogenic nematode infective stages, but many questions still remain to be addressed. In this chapter we have tried to cover in a general way the different steps involved in foraging for hosts within the framework of the ambusher/cruiser continuum. This conceptual framework has proved useful for addressing behavioural questions for this group and has enabled us to develop adaptive syndromes for the different foraging strategies. These syndromes still need to be tested and refined as new species are discovered and new aspects of the parasite–host interaction are elucidated.

References

Akhurst, R.J. (1993) Bacterial symbionts of entomopathogenic nematodes – the power behind the throne. In: Bedding, R., Akhurst, R.J. and Kaya, H.K. (eds) *Nematodes and the Biological Control of Insect Pests.* CSIRO Publications, East Melbourne, Australia, pp. 127–135.

Akhurst, R.J., Bedding, R.A., Bull, R.M. and Smith, D.R.J. (1992) An epizootic of *Heterorhabditis* spp. (Heterorhabditidae: Nematoda) in sugar cane scarabaeids (Coleoptera). *Fundamental and Applied Nematology* 15, 71–73.

Akre, B.G. and Johnson, D.M. (1979) Switching and sigmoid functional response curves by damselfly naiads with alternate prey available. *Journal of Animal Ecology* 48, 703–720.

Bargmann, C.I. and Mori, I. (1997) Chemotaxis and thermotaxis. In: Riddle, D.L. Blumenthal, T., Meyer, B.J. and Priess, J.R. (eds) *C. elegans II.* Cold Spring Harbor Laboratory Press, Cold Spring Harbor, New York, pp. 717–739.

Bauer, M.E. and Kaya, H.K. (2001) Persistence of entomopathogenic nematodes. In: Bauer, M.E. and Fuxa, J. (eds) *Environmental Persistence of Entomopathogens and Nematodes*. Southern Cooperative Series Bulletin 398, Louisiana Agricultural Experiment Station, Baton Rouge, Louisiana.

Bedding, R.A. (1972) Biology of *Deladenus siricidicola* (Neotylenchidae) an entomophagous nematode parasitic in siricid woodwasps. *Nematologica* 18, 482–493.

Bedding, R.A. (1993) Biological control of *Sirex noctilio* using the nematode *Deladenus siricidicola*. In: Bedding, R.A., Akhurst, R.J. and Kaya, H.K. (eds) *Nematodes and the Biological Control of Insect Pests*. CSIRO Publications, East Melbourne, Australia, pp. 11–20.

Bell, W.J. (1985) Sources of information controlling motor patterns in arthropod local search orientation. *Journal of Insect Physiology* 31, 837–847.

Bell, W.J. (1990) *Searching Behavior: the Behavioral Ecology of Finding Resources*. Chapman & Hall, London, 359 pp.

Blaxter, M.L., De Ley, P., Garey, J.R., Liu, L.X., Schheldeman, P., Vierstraete, A., Vanfleteren, J.R., Mackey, L.Y., Dorris, M., Frisse, L.M., Vida, J.T. and Thomas, W.K. (1998) A molecular evolutionary framework for the phylum Nematoda. *Nature* 392, 71–75.

Bohan, D.A. and Hominick, W.M. (1995) Examination of *Steinernema feltiae* (Site 76 strain) infection interaction with *Galleria mellonella* host, using an infection model. *Parasitology* 111, 617–625.

Bye, F.N., Jacobson, F.V. and Sonerud, G.A. (1992) Auditory prey location in a pause–travel predator: search time and attack range of Tengmalm's owls (*Aegolius funereus*). *Behavioral Ecology* 3, 266–276.

Campbell, J.F. and Gaugler, R. (1993) Nictation behavior and its ecological implications in the host search strategies of entomopathogenic nematodes (Heterorhabditidae and Steinernematidae). *Behaviour* 126, 155–170.

Campbell, J.F. and Gaugler, R. (1997) Inter-specific variation in entomopathogenic nematode foraging strategy: dichotomy or variation along a continuum? *Fundamental and Applied Nematology* 20, 393–398.

Campbell, J.F. and Kaya, H.K. (1999a) How and why a parasitic nematode jumps. *Nature* 397, 485–486.

Campbell, J.F. and Kaya, H.K. (1999b) Mechanism, kinematic performance, and fitness consequences of entomopathogenic nematode (*Steinernema* spp.) jumping behavior. *Canadian Journal of Zoology* 77, 1947–1955.

Campbell, J.F. and Kaya, H.K. (2000) Influence of insect associated cues on the jumping behavior of entomopathogenic nematodes (*Steinernema* spp.). *Behaviour* 137, 591–609.

Campbell, J.F. and Kaya, H.K. (2002) Variation in entomopathogenic nematode (Steinernematidae and Heterorhabditidae) infective stage jumping behavior. *Nematology* (in press).

Campbell, J.F., Lewis, E.E., Yoder, F. and Gaugler, R. (1995) Entomopathogenic nematode (Heterorhabditidae and Steinernematidae) seasonal population dynamics and impact on insect populations in turfgrass. *Biological Control* 5, 598–606.

Campbell, J.F., Lewis, E., Yoder, F. and Gaugler, R. (1996) Entomopathogenic nematode (Heterorhabditidae and Steinernematidae) spatial distribution in turfgrass. *Parasitology* 113, 473–482.

Campbell, J.F., Koppenhöfer, A.M., Kaya, H.K. and Chinnasri, B. (1999) Are there temporarily non-infectious dauer stages in entomopathogenic nematode populations: a test of the phased infectivity hypothesis. *Parasitology* 118, 499–508.

Caraco, T. and Gillespie, R.G. (1986) Risk-sensitivity: foraging mode in an ambush predator. *Ecology* 67, 1180–1185.

Cheng, T.C. (1991) Is parasitism symbiosis? A definition of terms and the evolution of concepts. In: Toft, C.A., Aeschlimann, A. and Bolis, L. (eds) *Parasite–Host Associations: Coexistence or Conflict.* Oxford University Press, Oxford, pp. 15–36.

Combes, C., Fournier, A., Moné, H. and Théron, A. (1994) Behaviours in trematode cercariae that enhance parasite transmission: patterns and processes. *Parasitology* 109, S3-S13.

Crofton, H.D. (1954) The vertical migration of infective larvae of strongyloid nematodes. *Journal of Helminthology* 28, 35–52.

Croll, N.A. (1970) *The Behaviour of Nematodes: Their Activity, Senses, and Responses.* Edward Arnold, London, 117 pp.

Croll, N.A. (1972) Energy utilization of infective *Ancylostoma tubaeforme* larvae. *Parasitology* 64, 355–368.

Doutt, R.L. (1964) Biological characteristics of entomophagous adults. In: DeBach, P. (ed.) *Biological Control of Insect Pests and Weeds.* Reinhold, New York, pp. 145–167.

Dusenberry, D.B. (1980) Behavior of free-living nematodes. In: Zuckerman, B.M. (ed.) *Nematodes as Biological Models*, Vol. 1. Academic Press, New York, pp. 127–196.

Eckhardt, R.C. (1979) The adaptive syndromes of two guilds of insectivorous birds in the Colorado Rocky Mountains. *Ecological Monographs* 49, 129–149.

Farish, D.J. and Axtell, E.E. (1971) Phoresy redefined and examined in *Macrocheles muscad mesticae* (Acarina: Macrochelidae). *Acarologia* 13, 16–29.

Formanowicz, D.R., Jr and Bradley, P.J. (1987) Fluctuations in prey density: effects on the foraging tactics of scolopendrid centipedes. *Animal Behavior* 35, 453–461.

Forst, S. and Nealson, K. (1996) Molecular biology of the symbiotic-pathogenic bacteria *Xenorhabdus* spp. and *Photorhabdus* spp. *Microbiological Reviews* 60, 21–43.

Gaugler, R. and Kaya, H.K. (1990) *Entomopathogenic Nematodes in Biological Control.* CRC Press, Boca Raton, Florida, 365 pp.

Gaugler, R., Campbell, J.F. and McGuire, T. (1989) Selection for host-finding in *Steinernema feltiae. Journal of Invertebrate Pathology* 54, 363–372.

Glazer, I. (1997) Effects of infected insects on secondary invasion of steinernematid entomopathogenic nematodes. *Parasitology* 114, 597–604.

Glazer, I. and Lewis, E.E. (2000) Bioassays for entomopathogenic nematodes. In: Navon, A. and Ascher, K.R.S. (eds) *Bioassays of Entomopathogenic Microbes and Nematodes.* CAB International, Wallingford, UK, pp. 229–238.

Godfray, H.C.J. (1994) *Parasitoids: Behavioral and Evolutionary Ecology.* Princeton University Press, Princeton, New Jersey, 473 pp.

Greco, C.F. and Kevan, P.G. (1994) Contrasting patch choosing by anthophilous ambush predators: vegetation and floral cues for decisions by a crab spider (*Misumena vatia*) and males and females of an ambush bug (*Phymata americana*). *Canadian Journal of Zoology* 72, 1583–1588.

Greco, C.F. and Kevan, P.G. (1995) Patch choice in the anthophilous ambush predator *Phymata americana*: improvement by switching hunting sites as part of the initial choice. *Canadian Journal of Zoology* 73, 1912–1917.

Green, R.F. (1984) Stopping rules for optimal foragers. *American Naturalist* 123, 30–40.

Grewal, P.S., Gaugler, R. and Selvan, S. (1993a) Host recognition by entomo-pathogenic nematodes: behavioral response to contact with host feces. *Journal of Chemical Ecology* 19, 1219–1231.

Grewal, P.S., Gaugler, R. and Lewis, E.E. (1993b) Host recognition behavior by entomopathogenic nematodes during contact with insect gut contents. *Journal of Parasitology* 79, 495–503.

Grewal, P.S., Lewis, E.E., Gaugler, R. and Campbell, J.F. (1994) Host finding behaviour as a predictor of foraging strategy in entomopathogenic nematodes. *Parasitology* 108, 207–215.

Grewal, P.S., Lewis, E.E. and Gaugler, R. (1997) Response of infective stage parasites (Nematoda: Steinernematidae) to volatile cues from infected hosts. *Journal of Chemical Ecology* 23, 503–515.

Harvey, P.H. and Pagel, M.D. (1991) *The Comparative Method in Evolutionary Biology*. Oxford University Press, Oxford, UK, 239 pp.

Hay, D.B. and Fenlon, J.S. (1995) A modified binomial model that describes the infection dynamics of the entomopathogenic nematode *Steinernema feltiae* (Steinernematidae: Nematoda). *Parasitology* 111, 627–633.

Hemerik, L. Driessen, G. and Haccou, P. (1993) Effects of intra-patch experiences on patch time, search time and searching efficiency of the parasitoid *Leptopilina clavipes*. *Journal of Animal Ecology* 62, 33–44.

Hernandez, A.D. and Sukhdeo, M.V.K. (1995) Host grooming and the transmission strategy of *Heligmosomoides polygyrus*. *Journal of Parasitology* 81, 865–869.

Hominick, W.M. and Reid, A.P. (1990) Perspectives on entomopathogenic nematology. In: Gaugler, R. and Kaya, H.K. (eds) *Entomopathogenic Nematodes in Biological Control*. CRC Press, Boca Raton, Florida, pp. 327–345.

Houck, M.A. and O'Connor, B.M. (1991) Ecological and evolutionary significance of phorecy in the Astigmata. *Annual Review of Entomology* 36, 611–636.

Huettel, R.N. (1986) Chemical communicators in nematodes. *Journal of Nematology* 18, 3–8.

Huey, R.B. and Pianka, E.R. (1981) Ecological consequences of foraging mode. *Ecology* 62, 991–999.

Inoue, T. and Matsura, T. (1983) Foraging strategy of a mantid, *Paratenodera angustipennis* S.: mechanisms of switching tactics between ambush and active search. *Oecologia* 56, 264–271.

Ishibashi, N. and Kondo, E. (1990) Behavior of infective juveniles. In: Gaugler, R. and Kaya, H.K. (eds) *Entomopathogenic Nematodes in Biological Control*. CRC Press, Boca Raton, Florida, pp. 139–152.

Iwasa, Y., Higashi, M. and Yamamura, N. (1981) Prey distribution as a factor determining the choice of optimal foraging strategy. *American Naturalist* 117, 710–723.

Janetos, A.C. (1982) Foraging tactics of two guilds of web-spinning spiders. *Behavioral Ecology and Sociobiology* 10, 19–27.

Janetos, A.C. and Cole, B.J. (1981) Imperfectly optimal animals. *Behavioral Ecology and Sociobiology* 9, 203–209.

Kareiva, P., Morse, D.H. and Eccleston, J. (1989) Stochastic prey arrivals and crab spider giving-up times: simulations of spider performance using two simple 'rules of thumb.' *Oecologia* 78, 542–549.

Kaya, H.K. and Gaugler, R. (1993) Entomopathogenic nematodes. *Annual Review of Entomology* 38, 181–206.

Khlibsuwan, W., Ishibashi, N. and Kondo, E. (1992) Response of *Steinernema carpocapsae* infective juveniles to the plasma of three insect species. *Journal of Nematology* 24, 156–159.

Kiontke, K. (1996) The phoretic association of *Diplogaster coprophila* Sudhaus & Rehfeld, 1990 (Diplogastridae) from cow dung with its carriers, in particular flies of the family Sepsidae. *Nematologica* 42, 354–366.

Koltai, H., Glazer, I. and Segal, D. (1995) Reproduction of the entomopathogenic nematode *Heterorhabditis bacteriophora* Poinar, 1976: hermaphroditism vs amphimixis. *Fundamental and Applied Nematology* 18, 55–61.

Koppenhöfer, A.M., Baur, M.E., Stock, S.P., Choo, H.Y., Chinnasri, B. and Kaya, H.K. (1997) Survival of entomopathogenic nematodes within host cadavers in dry soil. *Applied Soil Ecology* 6, 231–240.

Laing, J. (1937) Host-finding by insect parasites. I. Observations on the finding of hosts by *Alysia manducator, Mormoniella vitripennis* and *Trichogramma evanescens. Journal of Animal Ecology* 6, 298–317.

Lee, D.L. and Atkinson, H.J. (1977) *Physiology of Nematode*, 2nd edn. Macmillan Press, London, 215 pp.

Lees, A.D. and Milne, A. (1950) The seasonal and diurnal activities of individual sheep ticks (*Ixodes ricinus* L.). *Parasitology* 41, 189–208.

Lewis, E.E. and Gaugler, R. (1994) Entomopathogenic nematode sex ratio relates to foraging strategy. *Journal of Invertebrate Pathology* 64, 238–242.

Lewis, E.E., Gaugler, R. and Harrison, R. (1992) Entomopathogenic nematode host finding: response to host contact cues by cruise and ambush foragers. *Parasitology* 105, 309–315.

Lewis, E.E., Gaugler, R. and Harrison, R. (1993) Response of cruiser and ambusher entomopathogenic nematodes (Steinernematidae) to host volatile cues. *Canadian Journal of Zoology* 71, 765–769.

Lewis, E.E., Grewal, P.S. and Gaugler, R. (1995a) Hierarchical order of host cues in parasite foraging: a question of context. *Parasitology* 110, 207–213.

Lewis, E.E., Selvan, S., Campbell, J.F. and Gaugler, R. (1995b) Changes in foraging behaviour during the infective juvenile stage of entomopathogenic nematodes. *Parasitology* 110, 583–590.

Lewis, E.E., Campbell, J.F. and Gaugler, R. (1997) The effects of aging on the foraging behaviour of *Steinernema carpocapsae* (Rhabdita: Steinernematidae). *Nematologica* 43, 355–362.

Lewis, W.J., Vet, L.E.M., Tumlinson, J.H., van Lenteren, J.C. and Papaj, D.R. (1990) Variations in parasitoid foraging behavior: essential element of a sound biological control theory. *Environmental Entomology* 19, 1183–1193.

McLaughlin, R.L. (1989) Search modes of birds and lizards: evidence for alternative movement patterns. *American Naturalist* 133, 654–670.

McNair, J.N. (1982) Optimal giving-up times and the marginal value theorem. *American Naturalist* 119, 511–529.

Massey, C.L. (1974) *Biology and Taxonomy of Nematode Parasites and Associates of Bark Beetles in the United States*. Agriculture Handbook 446, United States Department of Agriculture, 233 pp.

Moyle, P.L. and Kaya, H.K. (1981) Dispersal and infectivity of the entomogenous nematode, *Neoaplectana carpocapsae* (Nematoda: Steinernematidae). *Journal of Nematology* 13, 295–300.

Nguyen, K.B. and Smart, G.C., Jr (1995) Morphometrics of infective juveniles of *Steinernema* spp. and *Heterorhabditis bacteriophora* (Nemata: Rhabditida). *Journal of Nematology* 27, 206–212.

O'Brien, W.J., Evans, B.I. and Browman, H.I. (1989) Flexible search tactics and efficient foraging in saltatory searching animals. *Oecologia* 80, 100–110.

O'Brien, W.J., Browman, H.I. and Evans, B.I. (1990) Search strategies of foraging animals. *American Scientist* 78, 152–160.

O'Leary, S.A., Stack, C.M., Chubb, M.A. and Burnell, A.M. (1998) The effect of day of emergence from the insect cadaver on the behavior and environmental tolerances of infective juveniles of the entomopathogenic nematode *Heterorhabditis megidis* (strain UK211). *Journal of Parasitology* 84, 665–672.

Patel, M.N., Stolinski, M. and Wright, D.J. (1997) Neutral lipids and the assessment of infectivity in entomopathogenic nematodes: observations on four *Steinernema* species. *Parasitology* 114, 489–496.

Perry, G. and Pianka, E.R. (1997) Animal foraging: past, present and future. *Trends in Ecology and Evolution* 12, 360–364.

Pianka, E.R. (1966) Convexity, desert lizards, and spatial heterogeneity. *Ecology* 47, 1055–1059.

Pietruszka, R.D. (1986) Search tactics of desert lizards: how polarized are they? *Animal Behavior* 34, 1742–1758.

Poinar, G.O., Jr (1972) Nematodes are facultative parasites of insects. *Annual Review of Entomology* 17, 103–122.

Poinar, G.O., Jr (1975) *Entomogenous Nematodes*. E.J. Brill, Leiden, the Netherlands, 317 pp.

Poinar, G.O., Jr (1993) Origins and phylogenetic relationships of the entomophilic rhabditids, *Heterorhabditis* and *Steinernema*. *Fundamental and Applied Nematology* 16, 333–338.

Ramaswamy, S.B. (1988) Host finding by moths: sensory modalities and behaviors. *Journal of Insect Physiology* 34, 235–249.

Reed, E.E. and Wallace, H.R. (1965) Leaping locomotion by an insect-parasitic nematode. *Nature* 206, 210–211.

Regal, P.J. (1978) Behavioral differences between reptiles and mammals: an analysis of activity and mental capabilities. In: Greenberg, N. and MacLean, P.D. (eds) *Behavior and Neurology of Lizards: an Interdisciplinary Colloquium*. Publication 77–491, Department of Health, Education, and Welfare, Rockville, Maryland, pp. 183–202.

Rogers, W.P. and Sommerville, R.I. (1963) The infective stage of nematode parasites and its significance in parasitism. *Advances in Parasitology* 1, 109–177.

Root, R.B. and Chaplin, S.J. (1976) The life-styles of tropical milkweed bugs, *Oncopeltus cingulifer* and *Oncopeltus unifasciatellus* (Hemiptera: Lygaeidae) utilizing the same hosts. *Ecology* 57, 132–140.

Salt, G. (1935) Experimental studies in insect parasitism. III. Host selection. *Proceedings of the Royal Society of London B* 114, 413–435.

Schoener, T.W. (1971) Theory of feeding strategies. *Annual Review of Ecology and Systematics* 2, 369–404.

Selvan, S., Gaugler, R. and Lewis, E.E. (1993a) Biochemical energy reserves of entomopathogenic nematodes. *Journal of Parasitology* 79, 167–172.

Selvan, S., Campbell, J.F. and Gaugler, R. (1993b) Density dependent effects on entomopathogenic nematodes (Heterorhabditidae and Steinernematidae) within an insect host. *Journal of Invertebrate Pathology* 62, 278–284.

Sexton, S.B. and Williams, P. (1981) A natural occurrence of parasitism of *Graphognathus leucoloma* (Boheman) by the nematode *Heterorhabditis* sp. *Journal of the Australian Entomological Society* 20, 253–255.

Shapiro, D.I. and Glazer, I. (1996) Comparison of entomopathogenic nematode dispersal from infected hosts versus aqueous suspension. *Environmental Entomology* 25, 1455–1461.

Shapiro, D.I. and Lewis, E.E. (1999) Infectivity of entomopathogenic nematodes from cadavers vs. aqueous applications. *Environmental Entomology* 28, 907–911.

Shettleworth, S.J. (1984) Learning and behavioral ecology. In: Krebs, J.R. and Davies, N.B. (eds) *Behavioral Ecology: an Evolutionary Approach.* Blackwell Scientific, Oxford, UK, pp. 170–194.

Sonerud, G.A. (1992) Search tactics of a pause–travel predator: adaptive adjustments of perching times and move distances by hawk owls (*Surnia ulula*). *Behavioral Ecology and Sociobiology* 30, 207–217.

Stephens, D.W. and Krebs, J.R. (1986) *Foraging Theory.* Princeton University Press, Princeton, New Jersey, 247 pp.

Stock, S.P., Campbell, J.F. and Nadler, S.A. (2001) Phylogeny of *Steinernema* Travassos, 1927 (Cephalobina: Steinernematidae). *Journal of Parasitology* 87, 877–889.

Strauch, O., Stoessel, S. and Ehlers, R.-U. (1994) Culture conditions define automictic or amphimictic rhabditid nematodes of the genus *Heterorhabditis*. *Fundamental and Applied Nematology* 17, 575–582.

Strong, D.R., Kaya, H.K., Whipple, A.V., Child, A.L., Kraig, S., Bondonno, M., Dyer, K. and Maron, J.L. (1996) Entomopathogenic nematodes: natural enemies of root-feeding caterpillars on bush lupine. *Oecologia* 108, 167–173.

Stuart, R.J., Lewis, E.E. and Gaugler, R. (1996) Selection alters the pattern of emergence from the host cadaver in the entomopathogenic nematode, *Steinernema glaseri. Parasitology* 113, 183–189.

Stuart, R.J., Hatab, M.A. and Gaugler, R. (1998) Sex ratio and the infection process in entomopathogenic nematodes: are males the colonizing sex? *Journal of Invertebrate Pathology* 72, 288–295.

Sudhaus, W. (1976) Vergleichende Untersuchungen zur Phylogenie, Systematik, Ökologie, Biologie und Ethologie der Rhabditidae (Nematoda). *Zoologica* 43, 1–229.

Taigen, T.L. and Pough, F.H. (1983) Prey preference, foraging behavior, and metabolic characteristics of frogs. *American Naturalist* 122, 509–520.

Turlings, T.C.J., Wäckers, F.L., Vekt, L.E.M., Lewis, W.J. and Tumlinson, J.H. (1992) Learning of host-location cues by hymenopterous parasitoids. In: Lewis, A.C. and Papaj, D.R. (eds) *Insect Learning: Ecological and Evolutionary Perspectives.* Chapman & Hall, New York, pp. 51–78.

Vet, L.E.M. and Dicke, M. (1992) Ecology of infochemical use by natural enemies in a tritrophic context. *Annual Review of Entomology* 37, 141–172.

Vet, L.E.M., Lewis, W.J., Papaj, D.R. and van Lenteren, J.C. (1990) A variable response model for parasitoid foraging behavior. *Journal of Insect Behaviour* 3, 471–491.

Vinson, S.B. (1981) Habitat location. In: Nordlund, D.A., Jones, R.L. and Lewis, W.J. (eds) *Semiochemicals, Their Role in Pest Control.* John Wiley & Sons, New York, pp. 51–78.

Völk, J. (1950) Die Nematoden der Regenwürmer und aasbesuchenden Käfer. *Zoologische Jahrbücher Abteilung Systematik* 79, 1–70.

Waage, J.K. (1979) Foraging for patchily-distributed hosts by the parasitoid, *Nemeritis canescens. Journal of Animal Ecology* 48, 353–371.

Zuckerman, B.M. and Jansson, H.-B. (1984) Nematode chemotaxis and possible mechanisms of host/prey recognition. *Annual Review of Phytopathology* 22, 95–113.

Flexibility in Host-search and Patch-use Strategies of Insect Parasitoids

3

Louise E.M. Vet,[1,2] Lia Hemerik,[3] Marcel E. Visser[1] and Felix L. Wäckers[1,2]

[1]Netherlands Institute of Ecology, Centre of Terrestrial Ecology, Heteren, The Netherlands; [2]Laboratory of Entomology, Wageningen University, PO Box 8031, 6700 EH Wageningen, The Netherlands; [3]Biometris, Department of Mathematical and Statistical Methods, Wageningen University, Wageningen, The Netherlands

Introduction

Parasitoids, typically parasitic wasps, deserve to be the subject of a(nother) book, not a chapter. Anyone who has ever watched a searching, egg-laying or feeding parasitoid has fallen in love with the beauty of these elegant and often colourful insects. But, then again, some people may be disturbed by their gruesome life history. After all, these lovely parasitoids are insects that lay their eggs in or on other insects. The parasitoid larva feeds on the body of the victim, euphemistically called the host insect, which is eventually killed. A dying caterpillar with parasitic larvae crawling out of its body is perhaps too much hard-core biology for some people.

So we restrict ourselves to a chapter on insect parasitoids, with a focus on flexibility in their host-search and patch-use strategies. For extensive reviews on the ecology and life history of insect parasitoids, we refer to the excellent books by Askew (1971) and especially Godfray (1994) – a 'must' if you catch some of our enthusiasm for these fascinating creatures.

The majority of parasitoids belong to the Hymenoptera and Diptera. The parasitic habit has most probably evolved from the predatory lifestyle of ancestral species that ate dead or living insects. Through their intimate relationship with their hosts, evolution has led to explosive speciation and adaptive radiation in the way parasitoids locate, parasitize and develop in other insects. Their diversity in exquisite, often weird lifestyles and their adaptive behaviour are overwhelming and probably unequalled in the animal world. In addition to being interesting objects for fundamental biological study, parasitoids are economically and ecologically very important insects. In agricultural systems they play a leading role in the biological control of insect pests. In natural ecosystems they also play a crucial role, both in number of species and in number of individuals. Taxonomists estimate that parasitoids make up 20–25% of all 10 million to 100 million insect species that probably exist. After plants and herbivores, parasitoids are the third trophic level (or higher) and this high position in the food web, combined with their often specialized host association, makes them vulnerable: disturbance of natural habitats can easily reduce population size or cause species extinction. Parasitoids are considered to function as keystone species in almost every terrestrial ecosystem (LaSalle and Gauld, 1993). Together with other natural enemies, such as predators, they keep our planet green by limiting herbivore populations (Hairston *et al.*, 1960).

Like all parasites, insect parasitoids can only develop in or on another organism. However, insect parasitism differs from classic parasitism in several ways. First, the adult parasitoid is a free-living insect and the parasitic lifestyle of parasitoids is thus limited to the immature stages only. In parasitoids, it is generally the adult female that makes foraging and 'infection' decisions that influence the fitness of her progeny. This contrasts with classic parasites, where each individual infective stage makes its own foraging and infection decision and host infection is initiated by the stage that will become established. Secondly, because the host has been killed by the time of the adult parasitoid's emergence, the emerging female wasp is challenged to search for new hosts to produce her offspring. This task demands an active process of host searching by the female wasp in an environment that is highly variable in space and time.

These characteristics have played a major role in shaping the host-searching strategies of parasitoids. Searching for new hosts is not an easy task, since hosts are small organisms in an often very complex natural environment and they are under strong selection to remain inconspicuous to their enemies. In some cases, parasitoids can solve this problem by spying on the communication system of their host. Some parasitoids use the host's sex pheromone or aggregation pheromone to locate potential sites where host females may be laying eggs. But to most parasitoids this solution is not available and they are thus forced to use indirect cues to find their victims. Such indirect cues, such as odours from the food of their host, create the problem of unreliability. A potential host-food plant may be easy to find, but the location of the plant does not guarantee the presence of a suitable host insect. Reliable host cues are undetectable and detectable indirect cues are unreliable. This reliability–detectability problem has constrained the evolution of parasitoid foraging behaviour (Vet *et al.*, 1991; Vos *et al.*, 2001).

Another important difference between insect parasitoids and other parasites is the parasitoid's high degree of behavioural flexibility. We define behavioural flexibility as a change over time in the behaviour of an individual, due to changes in the animal's physiological state (hunger, egg load) or to different learning processes (e.g. classical associative learning).

This flexibility in behaviour is strikingly common in insect parasitoids and allows them to adaptively respond to crucial variation in their environment. We make it the focus of our chapter.

The success in encountering hosts directly determines the female wasp's Darwinian fitness: her searching strategy can make or break her. During her lifetime, each individual female has to make repeated decisions on whether to forage for hosts or for food, which cues to use and how to search, where to go and how long to stay. Having found a host, she needs to decide whether to use it for host feeding or for oviposition. If she accepts it for oviposition, she has to decide how many eggs to lay and which sex ratio. Figure 3.1 depicts these aspects of decision-making as discussed in the chapter. Through this direct link between behaviour and fitness, insect parasitoids are highly suitable model systems for asking evolutionary questions about behaviour and they have been very important in the development of behavioural ecological methods (Godfray, 1994). Behavioural ecology would not be where it is now without insect parasitoids. Throughout the chapter, we shall address both the mechanism and the function of flexibility in behaviour, since we strongly believe a combination of both approaches is needed in behavioural ecology. We shall point to the importance of learning and the leading role of olfaction and plants in patch and host finding. We shall also address the food ecology of parasitoids and how the need to forage for food interacts with foraging for hosts. Furthermore, we specify which factors play a role

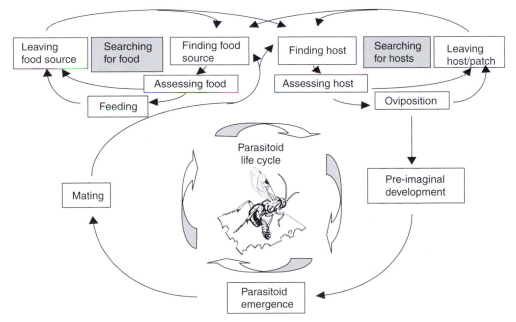

Fig. 3.1. The life cycle of a parasitoid with food and host search as main behaviours during adult life.

in host acceptance and what determines patch-leaving decisions. Since the audience is a general one, we have chosen not to extensively review the abundance of literature that is available on (some of) these subjects but instead illustrate key points by specific examples with key references. At the end of our chapter we shall discuss the future of a behavioural ecology approach and its value for understanding the ecology and evolution of insect parasitoids.

Patch Finding

Parasitoids generally do not emerge in the vicinity of a population of insects that can serve as hosts for their offspring. Hence, emerging wasps have to search for new sites. This first step in the searching process is patch finding – finding the location of a site where suitable hosts are likely to occur (Fig. 3.1). Parasitoids cannot simply detect new hosts as bees detect new flowers. Since hosts are under severe selection to avoid being found and devoured, they are difficult-to-detect objects. This low detectability of hosts constrains the use of direct host cues, and parasitoids have to deal with 'surrogate' indirect environmental cues. Parasitoids primarily focus on olfactory cues to locate potential host sites, although there is growing evidence for the use of visual cues (Wäckers and Lewis, 1999). The food of the host is a primary source of information during patch finding. Host-food chemicals can attract parasitoids even in the absence of the host itself. Parasitoids of flies, for example, are attracted to odours from the decaying ephemeral materials their hosts are feeding from: carrion, rotting plants and fermenting fruits, fungi and so on. To find their first host patch, parasitoids rely on innate responses to a library of cues that have been shown to be of importance to the parasitoid over evolutionary time (Vet *et al.*, 1990, 1995). We use the term innate in the sense of 'unlearned', without suggesting an instinct–learning dichotomy. The 'innate part' of a response is shown by naïve individuals, without the animal having had apparent experience with the stimuli concerned. However, as soon as a first host is found, the parasitoid's task becomes easier. They learn through experience which environmental cues reliably predict host presence and which do not, where we define learning by the following criteria: (i) behaviour can change in a repeatable way through experience; and (ii) learned responses can be forgotten (wane) or disappear as a consequence of another experience (see Vet *et al.* (1995) for a discussion on the definitions of learning, on the adaptive value of learning and on variation in how learning is expressed). During the last 20 years, many examples of learning have been reported for more than 25 species (for reviews, see Turlings *et al.*, 1993; Vet *et al.*, 1995). In often classic 'Pavlov-like' experiments, parasitoids are typically given experience with hosts (the unconditioned stimulus) in the presence or absence of a conditioned stimulus and their response to this conditioned stimulus is subsequently tested in single- or dual-choice experiments in olfactometers, wind-tunnels or flight set-ups or under field conditions. Naïve animals are often used as a control, or animals that have experienced different conditioned stimuli are compared in the same bioassay. The conditioned stimulus is often the odour of a food substrate of the host to which the parasitoid initially shows little or no behavioural response (α- and β-conditioning, respectively (see Vet *et al.*, 1995)). Studies have also shown sensitization, whereby responses to biologically relevant cues are enhanced without reinforcement, through mere exposure of the animal to the stimulus. For several species, it was shown that learned responses wane without repeated positive reinforcement, whereby responses return to those of the naïve state. The hymenopteran parasitoid *Leptopilina heterotoma* can be used as an example, since it has been the subject of many studies on both the mechanism and the function of parasitoid learning (e.g. Papaj and Vet, 1990; Vet and Papaj, 1992; Vet *et al.*, 1998). This species attacks larvae of several drosophilid fly species, living in a variety of microhabitats, such as decaying mushrooms, fermenting fruits and decaying plant materials. Laboratory and field experiments have shown that *L. heterotoma* females readily learn the odour of the micro-habitat by associating substrate odours with the presence of hosts, and strongly prefer these odours in choice situations. Female wasps were marked and released in a forest where apple–yeast and mushroom baits had been set out. An oviposition experience with *Drosophila* larvae in an apple–yeast or mushroom microhabitat had three effects. First, experienced females were more likely to find a bait than naïve ones. Secondly, for those females that found a bait, experienced females found it significantly faster than naïve females (i.e. experience reduced travel times). Thirdly, females who had experienced a particular microhabitat were more likely to find that microhabitat than an alternative one, i.e. experience strongly influenced microhabitat choice (Papaj and Vet, 1990). Experience with hosts in substrates affected movement in odour plumes of each substrate. Using a locomotor compensator, it was shown that females walked faster and straighter, made narrower turns and walked more upwind towards the source in a plume of odour they were experienced with, when compared with movement in odour plumes from the alternative substrate (Vet and Papaj, 1992). Associative learning-induced preferences disappear without repeated reinforcement after 72 h. Both positive (finding hosts) and negative reinforcement (not finding hosts in a substrate) play a role in habitat preference and substrate odour discrimination in this species (Vet *et al.*, 1998). Learning is not restricted to but is more pronounced in this generalist species, as compared with related microhabitat-specialist *Leptopilina* species. Microhabitat specialists have a strong innate response to the odours of their natural substrates. It is fascinating how well tuned parasitoid innate responses can be to these complex odours. *Leptopilina clavipes*, for example, a parasitoid of *Drosophila* larvae living in decaying fungi, is only attracted to mushrooms in a specific stage of decay, the stage likely to be infested with host larvae (Vet, 1983). Simons *et al.* (1992) showed how microhabitat specialists can also learn to prefer the odour of another substrate, but, unlike with the generalist *L. heterotoma*, visits to these substrates always remained shorter than visits to the natural substrates. Similar learning abilities may be used for different purposes and therefore the effect of learning may differ. The generalist *L. heterotoma* can achieve a great flexibility in substrate selection, while the specialists may employ learning to temporarily divert to less preferred substrates.

A high degree of flexibility through learning thus provides parasitoids with an adaptive mechanism, and this mechanism expands beyond patch finding. Both positive and negative reinforcement during foraging shapes the subsequent decisions of the parasitoids on where to go and how long to stay there. It helps parasitoids to assess spatial and temporal variation in patch profitability, i.e. to estimate host quantity, quality and distribution and to track changes in these important parameters. In this phase of searching, learning is essential for optimizing foraging decisions.

A multitrophic approach to parasitoid behaviour

Parasitoids of plant-feeding insects function in a multitrophic world in which plants play a central role in determining parasitoid development and foraging behaviour (Vet and Dicke, 1992). Parasitoids of herbivores are attracted to potential food plants of their hosts and, since the early 1990s, it has become clear that plants can play an active role in guiding parasitoids to their potential victims (Turlings *et al.*, 1990). Although an uninfested plant may already be attractive to parasitoids, this attraction is a magnitude larger when the plant is being attacked by herbivores, suggesting evolutionary mutualism between desperately defending plants and eagerly searching parasitoids. Vet (1999) discusses this important role of the plant for the searching parasitoid, stating that it has important evolutionary and ecological consequences for all trophic levels. For the plant, selection for good signalling ability can potentially affect all components of a plant's defence strategy: the composition of the odour blend, signal transduction and biosynthetic pathways, systemic responses to herbivore damage or the relative investments in direct and indirect defence. For the herbivore, the conspiracy between plant and parasitoid is based on the feeding activity of the herbivore itself and selection will act on the herbivore's diet choice and feeding behaviour to reduce the conveyance of information that reveals its presence. In addition, herbivores can influence their chance of being parasitized through selection of feeding sites, when plant species, cultivars, genotypes or even plant parts vary in their attractiveness to parasitoids. A less attractive plant is searched less, which creates a partial refuge for the host, affecting the spatial pattern of parasitism and thus the dynamics of the herbivore and parasitoid populations. For the parasitoid, in the evolutionary setting of real plant–parasitoid mutualism, parasitoids are selected to optimize their response to those plant signals that indicate the presence of suitable herbivores. They can do so by increasing their sensory and behavioural ability to perceive the signal, by efficient information-processing and decision-making, by enhanced discrimination between signal and noise and by other behavioural adaptations. The tritrophic approach to parasitoid and predator behaviour is now a strong research interest (Dicke and Vet, 1999; Sabelis *et al.*, 1999).

Plant cues

Through learning, parasitoids can specialize temporarily on available and profitable host plants. *Cotesia glomerata*, a parasitoid of *Pieris* caterpillars, learns to fly to the host-plant species that is most profitable in terms of host encounter rate (Geervliet *et al.*, 1998a,b). When given a three-oviposition experience on a plant species infested with a high host density, followed by a three-oviposition experience on another plant species with a low host density, *C. glomerata* females chose the plant species on which they had experienced the highest host density when given a choice between the two now equally infested host-plant species in a wind-tunnel (Geervliet *et al.*, 1998a). When the production of plant volatiles varies reliably with herbivore density, plant volatiles can be a suitable cue for predicting host density, an important characteristic of patch quality. Non-profitable plants are neglected and the area to be intensively searched is thus significantly reduced. This is especially beneficial to parasitoids that are time-limited, i.e. are likely to die without having laid their full complement of eggs.

Plant species vary significantly in the degree of specificity of volatile information they produce after they are damaged (for references, see Vet, 1999). Some plant species, such as broad bean plants, produce qualitatively different odours when attacked by different herbivore (aphid) species (Powell *et al.*, 1998). Life seems simple for the parasitoid *Cotesia kariyai*: maize plants produce different volatiles when suitable and unsuitable host stages feed on them. Takabayashi *et al.* (1995) showed that female wasps are attracted to maize plants infested by first- to fourth-instar caterpillars (suitable host instars), while plants infested by fifth- or sixth-instar larvae (non-suitable host instars) are not attractive. This high degree of specificity of plant volatiles can greatly help parasitoids to optimize plant choice decisions, i.e. to distinguish between plants infested with host and non-host species or suitable and unsuitable host stages. Other plant species, such as cabbage, merely produce a higher quantity of odours when attacked by any herbivore, demanding a different foraging strategy by parasitoids, such as the need to learn subtle quantitative differences in odour composition or the use of visual cues. When information from the plant is indistinct, parasitoids can be hampered in their efficiency to locate suitable patches and hosts. This was shown for the parasitoid *C. glomerata*, which attacks *Pieris* caterpillars on cabbage (Vos *et al.*, 2001). Parasitoids waste search time by being attracted to leaves containing non-host herbivores and by spending considerable amounts of time on such leaves. Since most plants in the field are likely to be infested by more than one herbivore species, Vos *et al.* (2001) modelled the implications of such time wasting at the community level. They showed that an increase in herbivore diversity on a plant initially promoted the persistence of parasitoid communities. However, when herbivore diversity was very high, parasitoids became extinct, due to insufficient parasitism rates. Hence, the informational value of plants determines the host-encounter rate of parasitoids and this can have far-reaching consequences for parasitoid–host interactions and community structure.

Visual cues

Although we may stress the importance of odour cues during foraging, a parasitoid's sensory world is not restricted to the chemical dimension. Individual substrates may also differ in respect of other sensory inform-ation. Plants or plant parts, for instance, may be recognized by visual cues (colour, shape or patterns of feeding damage), as well as other physical characteristics (surface structure, vibration conduction). The parasitoid's ability to use detailed visual information during foraging sets it apart from other parasites. In a parasitoid's natural habitat, there are several contexts in which visual cues could provide useful information. Parasitoids may employ colour signals to locate floral nectar, a source of food (see below), or to find herbivores feeding on conspicuously coloured plant structures. In those cases in which the coloration of plant structures is consistently and specifically altered due to herbivory (e.g. galls or leaf-mines), the coloration of infested plant structures could become a particularly reliable stimulus to host-seeking parasitoids. Irrespective of coloration, the shape of herbivore damage can be an apparent and reliable indicator of herbi-vore presence. Recent work has demonstrated that parasitoids use various visual parameters during the search for food and host sites, and learn to discriminate between sites on the basis of these visual characteristics (Wäckers and Lewis, 1999). In visual learning experiments, *Microplitis croceipes* performed equally well in shape discrimination as in colour learning. Parasitoids were conditioned in a flight chamber with plants containing two types of paper targets, differing either in colour (yellow/grey) or in shape (triangle/square). As only one visual alternative carried a *Spodoptera* larva, parasitoids could associate the visual information with an oviposition reward. After parasitizing six hosts, the parasitoid's response to targets was tested in a separate run in which no caterpillars were present. Parasitoids that had been conditioned to discriminate colours made 73% of their landings on the previously rewarded colour alternative. In the case of shape discrimination, this figure was 79%. This high relative rate of shape learning is remarkable as it is in strong contrast to the learning capacity of honey-bees. The latter have been shown to learn colour better than pattern and pattern better than shapes. The strong performance in shape-learning tasks by *M. croceipes* may be adaptive in dealing with the homochromatic but multishaped environment in which parasitoids have to locate their herbivorous hosts. Not only do plant species differ in their morphology, but individual plant structures are usually also distinguishable by their characteristic shape. However, the primary benefit of shape learning probably lies in the fact that it could allow parasitoids to recognize specific types of feeding damage.

Host Finding

The distinction between patch finding and the subsequent stage of host finding, where individual hosts are located within a patch, is not always clear-cut. Long-distance cues are used during long-range orientation. However, many of these cues may also be used at shorter range (e.g. herbivore-induced plant cues, visual host-damage cues). Within the gradual transition from long to short range, however, we often see a shift from indirect cues, such as plant cues, to more direct ones, such as contact chemicals directly derived from the host itself (e.g. frass, silk or other excretions). These contact chemicals are perceived by taste receptors on the antennae and tarsae, the stimulation of which generally elicits a strong and distinct behavioural response in the parasitoid, such as changes in walking speed and angles turned. The resulting intensified search of the restricted area where the cue is perceived enhances the chance of locating the host. As with patch finding, chemicals play a significant role, but physical cues (both visual and mechanosensory) can also be involved during the stage of host finding. Host-derived visual information used by parasitoids includes host colour (Powell *et al.*, 1998) and host movement.

Mechanosensory cues

Mechanosensory cues (sound and vibration) are commonly employed by arthropods as a means of communication. Parasitoids also employ mechanosensory communication as part of their courtship behaviour, but in addition they use mechanosensory cues during foraging. They may home in on sounds or substrate vibrations produced by their hosts during various types of activity (Meyhöfer and Casas, 1999). Tachinid flies represent a particularly well-evolved example of this type of orientation, as they are specifically tuned in to the mating calls of their cricket hosts (Ramsauer and Robert, 2000).

An entirely different category of vibrational orientation has been described in pupal parasitoids. Pupal parasitoids of the genus *Pimpla* have been shown to track down hidden hosts through a highly sophisticated mechanism called vibrational sounding (Wäckers *et al.*, 1998). During this type of orientation, parasitoids transmit self-produced vibrations on to a substrate. The resonance of the reverberating substrate provides parasitoids with information on the relative solidity of the substrate, allowing them to scan their habitat for hidden hosts. Similar to echo location, vibrational sounding is therefore an example of self-communication, in which the parasitoid is both sender and receiver at the same time. This form of sensory orientation is unique, as it is the only known example in which parasitoids employ self-produced signals to locate their hidden hosts, making them (at least to a certain extent) independent of host-derived stimuli. This is particularly crucial for pupal parasitoids, as pupae neither feed nor move and are often hidden in well-concealed pupation sites (Vet *et al.*, 1995). Parasitoid size is a crucial factor in the functioning of vibrational sounding, as larger females can produce 'louder' vibrations, allowing them to successfully locate hosts in a broader range of substrates (Otten *et al.*, 2001).

Multisensory orientation

We have seen that parasitoids possess a broad sensory arsenal, employing olfaction, vision and mechanoreception to locate their hosts. Rather than using the different types of information in isolation, the various sensory parameters are often combined in multisensory orientation (Wäckers and Lewis, 1994). The concurrent use of several sources of information enhances a parasitoid's flexibility in interacting with its environment in several ways. For one thing, it greatly increases sensory differentiation. Unlike our common laboratory approach, in which we attempt to expose the insect to a single sensory mode, objects in the field generally present a combination of sensory information to the foraging insect. The more of this information that is used, the better the insect will be able to differentiate between the range of objects encountered. When orientation is restricted to a single sensory modality, differentiation is limited to a single sensory dimension (e.g. odour). Each additional modality adds a new dimension to the differentiating power. The fact that multi-sensory information can enhance orientation could be demonstrated in experiments with free-flying *M. croceipes* (Wäckers and Lewis, 1994). When challenged to differentiate between two alternatives, parasitoids conditioned to a combination of olfactory and visual stimuli landed more accurately than individuals conditioned to only the visual image or odour, respectively.

Another reason why multisensory orientation enhances flexibility is because the specific advantages of the individual types of information can be combined. For instance, a parasitoid can be solely guided by odour in situations when the object is barred from vision, while visual orientation allows the parasitoid to stay on track when turbulence interrupts the odour plume.

Host Acceptance

When a host is found, the female parasitoid's decision-making continues. Will it accept the host for oviposition, will it use it for host feeding or will it reject the host? If it accepts the host for oviposition, the number of eggs and the sex of these eggs (which is under female control (see below)) need to be decided.

Host quality

Individual hosts differ in their suitability for offspring development or as a resource for the adult female, i.e. they differ in quality. Although some parasitoid species are limited to a single host species, many use a range of species. Upon encountering a host, the parasitoid needs to assess the species. For instance, some host species, such as the larvae of some species of *Drosophila* and some scale insects, are able to encapsulate the eggs after they have been parasitized (Blumberg and DeBach, 1981; Carton and Kitiano, 1981). This reduces the likelihood that an egg will success-fully develop to an adult, and hence the quality of the host. Within a host species, quality of hosts can vary with their age, size or stage. This is especially true for idiobionts, in which the host is killed at oviposition. Koinobionts, in which hosts continue to grow after being parasitized, can often develop on a much wider range of host sizes.

A final aspect of host quality is whether or not the host has previously been parasitized. It could either contain eggs from a different parasitoid species, from the same species or even from the female herself. Also, the number of eggs present can vary. A specific term is used for the ability to distinguish parasitized from unparasitized (sometimes called healthy) hosts: host discrimination. Intraspecific host discrimination is common, and there are species that are even able to discriminate between hosts parasitized by themselves and those parasitized by conspecifics (van Dijken *et al.*, 1992; Visser, 1993). Also, some species can distinguish between hosts with one and with two eggs ('counting') (Bakker *et al.*, 1990).

Accept or reject?

Assessing host quality is thus of crucial importance for parasitoids. However, even when parasitoids can assess host quality perfectly, low-quality hosts are not always rejected for oviposition. There are conditions, such as a very low density of high-quality hosts or under low survival probabilities, when accepting low-quality hosts will yield more offspring (higher fitness) than rejecting these hosts. There a vast body of literature, both theoretical and experimental, that deals with the question of host acceptance (the field of optimal foraging or optimal diet composition (Stephens and Krebs, 1986). As it falls well outside the scope of this chapter to give a comprehensive review of this field, we shall restrict ourselves to emphasizing the flexibility in host-acceptance behaviour of parasitoids.

Host acceptance is strongly affected by the number of eggs retained by the parasitoid (egg load) (Rosenheim and Rosen, 1991), but also by the expected number of hosts that will be encountered during the remainder of a parasitoid's life. Ideally, a parasitoid should die just after laying its last egg. When parasitoids are exposed to conditions that indicate low survival probabilities, they accept hosts more readily. Roitberg *et al.* (1992) mimicked the 'end of season' by rearing the parasitoid *L. heterotoma* under the autumn photoperiod and found that females accepted low-quality hosts (in this case, parasitized *Drosophila* larvae) more often than the control group, which were reared under the summer photoperiod.

Superparasitism

Many species of parasitoids exploit hosts that have a patchy distribution. When searching for hosts, such patches are depleted as more and more hosts will get parasitized. Females encounter these parasitized hosts and can either reject or superparasitize them. Theory predicts that this decision in solitary parasitoids strongly depends on the number of other females that are searching on the same patch (Van der Hoeven and Hemerik, 1990; Visser *et al.*, 1992a) as superparasitism becomes more profitable when more females search a patch. Not only does the likelihood that an encountered parasitized host has been parasitized by the female itself decrease with increasing competition, but also the likelihood that another female will superparasitize hosts the female itself has parasitized increases. It therefore starts to become profitable to 'defend' the hosts by self-superparasitism, even though only one parasitoid larva can complete its development in the host. And, indeed, experiments confirm the effect of competition on host-acceptance decisions. When two *L. heterotoma* females search a patch, more superparasitism is found than when a single females searches a patch (half the size and with half the number of hosts) (Visser *et al.*, 1990). Even more striking is the result that females that were kept together in a tube before being introduced singly on a patch with only unparasitized hosts self-superparasitized more frequently than females kept alone before the experiment (Visser *et al.*, 1990), thus 'anticipating' the likelihood of future superparasitism by conspecifics.

Clutch-size and sex-ratio decisions

When a host is accepted for oviposition, additional decisions need to be made: how many eggs to lay (clutch size) and what sex ratio to produce. Since parasitoids are haplodiploid (unfertilized, i.e. haploid, eggs result in males, while fertilized, i.e. diploid, eggs result in females), a female has perfect control over the sex of the eggs it lays, and it can produce precise sex ratios.

In solitary parasitoids, parasitoid larvae within a single host are aggressive to each other and generally only a single adult emerges. In gregarious parasitoids, the larvae share the host. Hence, solitary parasitoids usually lay a single egg (but not always (see above)) and gregarious species lay a number of eggs in a single host. This clutch size is strongly affected by the size of the host, even in koinobiont species. In idiobiont species, larger clutches in larger hosts are due to the trade-off between number and size of offspring from a host (with smaller offspring having a lower fitness (Visser, 1994)). In koinobionts, there may be an additional ultimate factor that determines clutch size. We may find smaller clutches in small hosts, because investing in smaller hosts is more risky in terms of lower survival, due to predation or starvation, of the host individual (Vet *et al.*, 1994).

The clutch size is also affected by the encounter rate with hosts. Within a searching bout, clutch size is rarely constant (Ikawa and Suzuki, 1982; Visser, 1996). Apparently, the female adjusts her clutch size as she balances the number of eggs she has available with the number of hosts she is likely to encounter during the remainder of her life. If she encounters more hosts than expected, the female's clutch size will decrease, while, if the opposite holds, it will increase. As most experiments are carried out at relatively high host densities, the former is more commonly found.

The sex ratio produced in a host is often biased. In species where mating takes place at the location where the offspring emerge from the host, sex ratios are often female-biased. This is because the sons are in competition with each other, rather than with unrelated males, for their sisters (local mate competition) (Hamilton, 1967). In that case, a mother should produce only just enough males to fertilize her daughters. With an increasing probability that other females will produce offspring at the same location, the optimal sex ratio approaches 50 : 50 rather quickly. When only a single egg is laid in a host, frequently sons are produced on small hosts and daughters on larger hosts. The reason for this is that host size determines offspring size and that being large is more important for daughters than it is for sons (Trivers and Willard, 1973). The sex of the egg laid is thus affected by the size of the host. However, parasitoids do not use absolute measures for large and small host sizes, but relative ones. Whether they lay a male offspring in a host of a certain size depends on the range of host sizes encountered (van der Assem, 1971; Charnov *et al.*, 1981). When a female *Lariophagus distinguendus* is presented with hosts of a single size (1.4 mm), it lays 15% male eggs. However, if it is given alternating large and small hosts and the 1.4 mm are the smallest, it will lay 30% males. If, in contrast, they are the largest, it lays 2%. Clearly, the parasitoids use a relative size rule, allowing them to optimally adjust their behaviour to varying circumstances.

Finally, although we have discussed the various host-acceptance decisions separately, we want to stress that it is unlikely that this is how parasitoids behave. Often, a simultaneous decision on host acceptance, clutch size and sex ratio may be needed. For example, autoparasitoids lay male eggs on hosts that already contain a conspecific larva but female eggs on unparasitized hosts (Viggiani, 1984). Hunter and Godfray (1995) showed that, under abundant host densities, when parasitoids are egg-limited, equal numbers of males and females are produced. Under low host densities, the sex ratio reflects the ratio of the two host types. Thus, clearly, the decision whether to accept or reject a host of a specific type depends on the sex ratio already produced on the patch. Hence, host-acceptance and sex-ratio decisions are not taken sequentially but simultaneously.

Decisions of developing parasitoids

Host acceptance may be the end of the mother parasitoid's decision trajectory in respect of a single host, but it is not the end of flexibility in parasitoid behaviour. After oviposition, decision-making is 'transferred to' the developing parasitoids. They now have to decide: (i) how to interact with siblings or other competitors (see Strand, Chapter 7, this volume); and (ii) when and how quickly to devour the host individual. Natural selection will act strongly on both processes and it goes beyond the scope of this chapter to discuss larval behaviour in detail. We want to mention only developmental timing, since this is a novel aspect for parasitoid behavioural ecology. The growth strategy of a parasitoid larva may have severe fitness consequences. The outcome will depend on the costs and benefits of a basic trade-off between developmental speed and adult wasp size at emergence. Early emergence can be highly advantageous in a growing population, while a large size at emergence can ensure greater survival, fecundity and foraging success. The developmental strategy has rarely been approached from an optimality point of view. Hemerik and Harvey (1999) show with a simulation model that developing parasitoids of *Venturia canescens* have to trade off their size as adults (which determines lifetime reproductive success) and development time. Especially in nutritionally suboptimal hosts, there is strong selection for longer development times, and a flexible growth strategy allows developing parasitoids to respond adaptively to such variation in host quality.

Patch Leaving

After having found a patch, a parasitoid exploits this patch by encountering hosts, which it might accept for oviposition or host feeding. Eventually, exploitation of a patch ends with a parasitoid leaving the spot. The decision of when to leave a patch is a classical question in behavioural ecology. Since efficient exploitation of patches determines the fitness of individual parasitoids and thus population dynamics, patch residence times have been the subject of many experimental and theoretical studies. Parasitoids that are able to react in a flexible way to varying conditions are supposed to realize more offspring than parasitoids using rigid rules. Patch residence times can be influenced by previous experience, host-related cues (kairomone concentration), egg load, encounters with (un)parasitized hosts, predation risk and the presence of conspecifics (by direct encounter, superparasitism or recognizing a mark).

Most of the early optimal foraging models consider animals as deterministically acting organisms. Only after advanced statistical tools became widely available did the viewpoint of researchers shift to how decision-making by parasitoids was affected by different cues. The fitness output of an individual parasitoid can depend heavily on its behaviour, irrespective of whether this has evolved because of different spatial distributions of hosts or because of encountering conspecifics. Since each parasitoid has to compete with its conspecifics for an often limited number of hosts, decision-making by parasitoids can, for a substantial part, be determined by the behaviour of one or more conspecifics. We shall give examples of early optimal foraging theory models, of factors affecting the behaviour of individual parasitoids, of behavioural strategies on different spatial distributions of hosts affecting lifetime fitness and of parasitoids reacting to the presence of conspecifics.

Early optimal foraging theory

In the early optimal foraging models, the parasitoids live in a static world and are assumed to be omniscient and to forage optimally, i.e. they try to produce as many offspring as possible. Charnov's (1976) marginal-value theorem shows elegantly how an omniscient predator or parasitoid should adapt its patch residence time in prey or host patches according to the expected encounter rates with prey or hosts in the full habitat. The simple rule that results from the analysis of the model is to leave a patch when fitness gain per time is higher outside than inside the patch. This rule predicts that, in poorer habitats, patches are exploited more fully than in richer ones. That this is the case for parasitoids has been shown by Visser *et al.* (1992b): *L. heterotoma* having experienced a poor environment on the previous day superparasitize more than those that experienced a rich environment beforehand. One of the main criticisms of the early models concerns omniscience of the forager and a static world. That parasitoids may sample their environment and thus gain experience on which to base their foraging decisions is simply not considered. Since the early models, researchers have tried to incorporate more reality. For instance, they model foraging animals in dynamic worlds (Houston and McNamara, 1999; Clark and Mangel, 2000). The outcome of these models consists of behavioural rules that are conditional on some characteristic(s) of the habitat and some state(s) of the individual, still reflecting only part of the total spectrum of parasitoids' behaviours in varying environments.

What factors affect leaving?

Patch-leaving decisions of parasitoids have frequently been studied in an evolutionary context, focusing on the underlying mechanisms of decision-making. For the parasitoid *L. clavipes,* for example, its tendency to leave a patch has been studied on an artificial patch (Hemerik *et al.,* 1993). This tendency is shown to decrease when kairomone is present. When the first encountered host is parasitized, the parasitoid's leaving tendency increases, whereas the opposite is true when it encounters a healthy one, showing the parasitoid's behavioural flexibility. For arriving at these conclusions, a technique stemming from survival analysis, namely, Cox's regression model, is used (Cox, 1972). This semi-parametric statistical model allows one to infer the effects of cues and the (complex) interplay between different factors.

With the same statistical technique, a sequence of patch visits by a parasitoid can be analysed. This was recently done for the parasitoid *C. glomerata* in a semi-field set-up allowing the parasitoids to visit more than one patch during the observation time of 1 h (Vos *et al.,* 1998). Here, we use this analysis to illustrate the great behaviour flexibility of generalist parasitoids. In Europe, the generalist parasitoid *C. glomerata* prefers parasitizing larvae of the large white butterfly, *Pieris brassicae,* a host with a clumped distribution (20–100 caterpillars on one leaf, but clumps are rare). It also attacks larvae of the small white butterfly, *Pieris rapae*, occurring in a more regular spatial distribution with low numbers per leaf (one to five caterpillars on a leaf, but infested leaves are not rare). Parasitoids of European *C. glomerata* were observed in three different environments: (i) with only larvae of the large white butterfly present; (ii) with only larvae of the small white butterfly present; or (iii) with larvae of both white butterflies present. This generalist parasitoid shows substantial behavioural flexibility. If only the clumped host *P. brassicae* can be found, it uses information on previously visited numbers of host-infested leaves and on ovipositions to determine when to leave. When foraging for *P. rapae*, these cues have no effect on the leaving tendency, whereas some (but not all) cues are used when both host species are present. Thus, this parasitoid's leaving rules are different for preferred versus non-preferred hosts (Vos *et al.*, 1998).

Host spatial distribution

The pay-off of different behavioural rules depends critically on the spatial distribution of hosts across patches (Iwasa *et al.*, 1981), exemplified by the work of Vos (2001: chapters 4 and 5) on the *C. glomerata–Pieris* system. More than 100 years ago, *C. glomerata* was introduced to the USA to combat *P. rapae*. Since *P. brassicae* is absent in the USA, *P. rapae* is the main host for US *C. glomerata*. The distribution of *P. rapae* in Europe and in early and late season in the USA can be described by a Poisson distribution, whereas a more clumped negative binomial distribution applies in the USA when host density is highest in midsummer. Vos (2001: chapter 4) showed that parasitoids of US and European origin behave differently in the same experimental set-up. The difference between foraging behaviour of US and European *C. glomerata* was not a simple change in one trait. All estimates for behavioural parameters, such as patch-arrival, patch-leaving and travel-time decisions, were considerably different between these populations. For instance, European parasitoids revisited infested patches more often than US parasitoids, they had larger travel times between patches and a higher tendency to leave after oviposition than US parasitoids. These behavioural data from semi-field environments are used to parameterize a simulation model of *C. glomerata* foraging behaviour in a large field with *Brassica* plants. Vos (2001: chapter 5) related the observed difference in foraging behaviour between US and European parasitoids through simulations with different spatial distributions of *P. rapae* to variation in lifetime reproductive success. Reproductive success is not different between the US and European strains when foraging on the (clumped) negative binomial distribution, whereas parasitoids with 'US' behaviour obtain higher lifetime fitness on a (more regular) Poisson distribution of hosts than 'European' parasitoids. All in all 'US' *C. glomerata* seem to have flexibly adapted their behaviour. This enables them to exploit a less clumped spatial distribution of hosts.

Conspecifics

The risk of superparasitism might lengthen the time spent by individual wasps on a patch, since they can reduce their offspring's mortality due to superparasitism by a conspecific parasitoid. In situations where host patches are scarce, many conspecifics are competing for limited resources and superparasitism pays only if the time between two eggs deposited in one host is short. In such cases, defending a patch against conspecifics may be a beneficial strategy. The egg parasitoid *Trissolcus basalis* frequently exhibits such defence behaviour (Field and Calbert, 1998). The defence behaviour of *T. basalis* is investigated with a combination of evolutionary game-theory models and experiments. When two competitors arrive simultaneously at the same patch, they first co-exploit the patch together for some time. The time to the start of a contest is determined by the size of the patch, the encounter rate with unparasitized hosts, the encounter rate with conspecifics and the previous investment (Field and Calbert, 1998). If the two competitors arrive at different times, the female that arrives first on the patch is more likely to retain the overall possession of the patch (Field and Calbert, 1999). The winner of the fight gets the ownership, after which the non-owner will retreat to the periphery of the patch. While the owner searches for hosts, the non-owner tries to return to the patch and gain an opportunity to superparasitize already parasitized hosts or to parasitize healthy ones. Finally, the owner ceases searching for hosts and starts guarding the patch to prevent the non-owner from superparasitism. This guarding is called 'the waiting game' (Field *et al.*, 1998). At first sight, this guarding seems a waste of valuable search time, but, by extending the period between the deposition of the first and second egg in a host, the guarding parasitoid guarantees a certain number of offspring.

Optimal foraging not only demands good host-finding ability (sensory abilities and patch-choice decisions), but also efficient decision rules for patch leaving. During all steps in the foraging cycle, parasitoids have to choose between different alternatives. The early theoretical models assuming parasitoids to behave in an optimal manner did not always predict what was found in experiments. Most of the time, there was qualitative agreement between model and data, but the average parasitoid is not always as omniscient as assumed by theoreticians. The combination of practical and theoretical approaches enhances our insight into how animals are able to make flexible decisions.

Feeding versus Reproduction

While the broad variation in parasitoid–host associations makes generalizations about parasitoid reproductive strategies difficult, the feeding associations of the adult stages are less diverse. Virtually all parasitoids require carbohydrates as a source of energy, especially for flight. Parasitoids cover their energetic needs by feeding on accessible sugar sources, such as (extra)floral nectar or honeydew. Carbohydrates can have a strong impact on several key fitness parameters. Sugar feeding is indispensable to parasitoid survival, a factor applying to both females and males. In addition, sugar feeding can also raise a female's fecundity, as well as her propensity to search for herbivorous hosts. While parasitoids as a group share a requirement for carbohydrates, hymenopteran parasitoids can be categorized on the basis of two fundamental feeding characteristics, each representing a distinct trade-off between reproduction and feeding.

Host-feeding versus oviposition

The issue of host-feeding constitutes a first divide. On the one hand, there are numerous parasitoid species that do not feed on the haemolymph of their host and as a result are entirely dependent on sugar-rich substrates for their nutrition. On the other hand, there are those species that do engage in host-feeding, in addition to feeding on separate carbohydrate sources. The two food sources usually cover separate requirements. Whereas nectar or honeydew feeding primarily provides carbohydrates to cover the parasitoid's energetic needs, insect haemolymph usually contains relatively low levels of carbohydrates (trehalose and glycogen). Instead, host-feeding constitutes a primary source of protein for physiological processes, such as egg maturation. As a result of the difference in their nutritional composition, the two food sources are only partly interchangeable.

In many parasitoid species, host-feeding and reproduction are mutually exclusive, as host-feeding leaves the host unsuitable for larval development. For host-feeding species, this may create conflict over whether to use a host for current (oviposition) or future reproductive success (host-feeding). The question of how parasitoids balance this dual exploitation of their host resources has been the topic of optimization models, as well as empirical studies (Ueno, 1999). While earlier models assumed equal host suitability to address the effect of varying host density, later work incorporated the effect of varying host quality. In the latter (more realistic) scenario, models predict that parasitoids should selectively use low-quality hosts for feeding and restrict oviposition to high-quality hosts (Kidd and Jervis, 1991).

Empirical studies have demonstrated that parasitoids do, indeed, selectively exploit their hosts according to various quality parameters. When given a choice between different host species, parasitoids tend to feed on the species that is the poorer host for parasitoid development. Within a single species, parasitoids can discriminate by size, using the smaller hosts for host-feeding (Rosenheim and Rosen, 1992). Parasitoids can also use information on host developmental stage (Kidd and Jervis 1991) or previous parasitization. In the latter case, parasitoids preferentially feed on hosts that contain offspring of conspecifics (Ueno, 1999) or heterospecifics, killing the resident parasitoid larvae.

Host search versus food foraging

In addition to the issue of host-feeding, parasitoids can also be divided according to the spatial association between host and carbohydrate sources. A first group includes those parasitoid species whose hosts are closely linked to carbohydrate-rich food sources. This applies to species whose hosts excrete suitable sugars, e.g. honeydew, or whose hosts occur on sugar-rich substrates, such as fruits or nectar-bearing plant structures. For these parasitoids, the task of locating hosts and carbohydrates is linked. Parasitoids from this group may show few specific adaptations to the exploitation of additional carbohydrate sources and little or no task differentiation between food foraging and host search. The second group is comprised of those parasitoids whose hosts are not reliably associated with a suitable carbohydrate source. These parasitoids have to alternate their search for hosts (reproduction) with bouts of food foraging, which requires a clear task differentiation. The latter group faces the issue of whether to stay in a host patch, thereby optimizing short-term reproductive success, or to leave the host patch in search of food sources, a strategy that could optimize reproduction in the long term.

Parasitoids are equipped with a number of mechanisms that enable them to deal with the dichotomy between searching for hosts (reproduction) and foraging for sugar sources (energy). They possess separate categories of innate responses, which are expressed relative to their physiological needs (Wäckers, 1994). Food-deprived parasitoids typically seek out stimuli that are associated with food, such as floral odours or colours. Following feeding, parasitoids lose interest in these food stimuli and start responding to host-associated cues (Wäckers, 1994). Associative learning of host- and food-associated information is also organized according to the parasitoid's physiological state (Takasu and Lewis, 1993). When parasitoids are conditioned using two distinct odours (vanilla or chocolate) in association with feeding and oviposition, respectively, they will learn both. When pitted against each other, they will choose between the two in accordance with their predominant physiological needs (Takasu and Lewis, 1993).

Epilogue

Our chapter has highlighted the behaviour of insect parasitoids in respect of patch finding, host finding, host acceptance, patch leaving and food ecology. We discussed the prominent role of olfaction and other sensory modes in foraging, placed behavioural variation in a functional context and looked at the connection between the experimental and theoretical approach. Throughout the chapter, we emphasized the flexibility of parasitoid foraging behaviour, i.e. variation in individual behaviour, which we think is one of the most intriguing characteristics of insect parasitoids. Especially through learning, parasitoids can alter their behaviour in an optimal way. They can enhance their response to temporarily important stimuli, alter their innate preferences and even acquire responses to novel stimuli that have proved to be reliable indicators of host presence. Stephens (1993) argues that learning is most beneficial when the environment is unpredictable between generations, but predictable within generations. For many insect parasitoids, this is a likely scenario. The primary source of information for foraging parasitoids is not the host itself, but the host's direct environment, such as its food. Mothers will often (have to) forage for different host food resources from those foraged for by their daughters, but within the lifetime of the individual there will certainly be short-term temporal regularities to detect. Hence learning can be expected as a likely adaptive mechanism for foraging parasitoids under natural conditions.

Parasitoids have been shown to be excellent model systems for answering evolutionary questions on behaviour, and we are convinced that much of the progress in theoretical behavioural ecology can be attributed to these small insects. Many behavioural ecologists have used parasitoids to couple theory with experiment and to address field-derived questions with fairly simple behavioural experiments in the laboratory. It is due to the direct link between foraging and fitness in insect parasitoids that this trajectory from field to laboratory experiment is relatively narrow and therefore somewhat justifiable. The great emphasis on laboratory experiments, rather than field observation, is also due to the fact that the behaviour of parasitoids is hard to study under field conditions. There are several reasons behind this. First of all, parasitoids are generally very small and lively insects and thus difficult to track visually. In addition, it is often hard to define a patch under field conditions. The chemical world that is perceived by the insect often remains ambiguous to the observer. But the most severe limitation is probably the great flexibility of parasitoid behaviour itself. Each parasitoid from a natural population has its own history, which, to a great extent, determines its behavioural decisions, as we have shown throughout our chapter. When observing a naturally foraging parasitoid, it will be impossible to know all the parameters that have influenced its behaviour. An example is host-acceptance behaviour, where parasitoids show an amazing flexibility. They take into account the suitability of the host (host quality), encounter rates with hosts, levels of competition, life expectancy and many more variables. This suggests that the underlying response mechanisms (which translate the external and internal cues the parasitoid receives into host-acceptance decisions) must be quite complex. Natural selection will shape these mechanisms such that, averaged over a large range of natural conditions, they result in adaptive decisions. It will be our challenge for the future to unravel these mechanisms (as presented in Fig. 3.2), at the sensory-physiological and neurobiological level, in combination with behavioural experimentation and a theoretical approach. At the same time, we need to study genetic variation in traits and the selective forces acting on them. Only this integration of proximate and ultimate studies will lead to a complete understanding of parasitoid foraging behaviour. Undoubtedly, technical developments will make it more feasible in the future to track parasitoids during their lifetime under field conditions, which will help us to further interconnect field and laboratory findings. Furthermore, molecular biology is now offering us the tools to study gene–environment interactions and to walk the path from genotype to phenotype. It will certainly open new and important avenues for gaining insight into the function and mechanism of behavioural variation in insect parasitoids.

Whatever our limitations, our goal remains to understand the real world in which parasitoids have evolved and function and how they manage to adapt to varying circumstances. As we have shown in this chapter, many other organisms, such as plants, non-hosts and

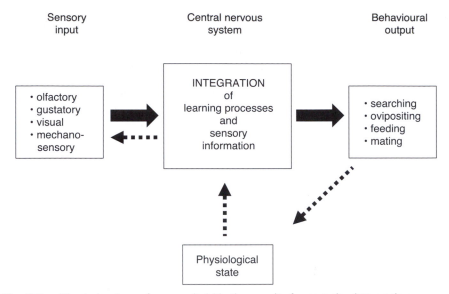

Fig. 3.2. The behaviour of a parasitoid is the result of a complex integration process in the central nervous system, with input from and feedback to the sensory level and the parasitoid's physiological state.

competitors, have a great influence on what parasitoids do and can achieve, and we need to incorporate these considerations in our behavioural studies and theoretical models. Simple optimal foraging studies under a specific set of conditions are becoming less popular, due to the increasing awareness of the importance of food-web ecology and complexity in species interactions. This is especially important when we want to assess the importance of behavioural variation for population and community processes. The behaviour-based approach, as highlighted in this chapter, is not only important for gaining insight into why individuals act as they do. The individual parasitoids together form the population and their individual success determines their individual number of offspring, which together will be the next generation. Thus behaviour of individuals determines the spatial distribution of parasitism over patches and ultimately the population dynamics of both host and parasitoid. Since insect parasitoids play a significant role in the combat of insect pests, we need to make a stronger effort to link the behavioural variation of parasitoids to parasitoid–host population dynamics. The study of the behaviour of insect parasitoids has a promising future. Its behavioural-ecological approach of blending theoretical and empirical aspects can probably serve as a model for behavioural studies with other parasite species.

References

Askew, R.R. (1971) *Parasitic Insects.* Heinemann, London, 316 pp.

Bakker, K., Peulet, P. and Visser, M.E. (1990) The ability to distinguish between hosts containing different numbers of parasitoid eggs by the solitary parasitoid *Leptopilina heterotoma* (Thompson) (Hym., Cynip.). *Netherlands Journal of Zoology* 40, 514–520.

Blumberg, D. and DeBach, P. (1981) Effects of temperature and host age upon the encapsulation of *Metaphycus helvolus* eggs by the brown soft scale *Coccus hesperidum. Journal of Invertebrate Pathology* 37, 73–79.

Carton, Y. and Kitano, H. (1981) Evolutionary relationships to parasitism by seven species of the *Drosophila melanogaster* subgroup. *Biological Journal of the Linnean Society* 16, 227–241.

Charnov, E.L. (1976) Optimal foraging: the marginal value theorem. *Theoretical Population Biology* 9, 129–136.

Charnov, E.L., Los-den Hartogh, R.L., Jones, W.T. and van der Assem, J. (1981) Sex ratio evolution in a variable environment. *Nature* 289, 27–33.

Clark, C.W. and Mangel, M. (2000) *Dynamic State Variable Models in Ecology: Methods and Applications.* Oxford University Press, Oxford, 289 pp.

Cox, D.R. (1972) Regression models and life tables. *Biometrics* 38, 67–77.

Dicke, M. and Vet, L.E.M. (1999) Plant–carnivore interactions: evolutionary and ecological consequences for plant, herbivore and carnivore. In: Olff, H., Brown, V.K. and Drent, R.H. (eds) *Herbivores Between Plants and Predators.* Blackwell Science, Malden, Massachusetts, pp. 483–520.

Field, S.A. and Calbert, G. (1998) Patch defence in the parasitoid wasp *Trissolcus basalis*: when to begin fighting. *Behaviour* 135, 629–642.

Field, S.A. and Calbert, G. (1999) Don't count your eggs before they're parasitized: contest resolution and the tradeoffs during patch defense in a parasitoid wasp. *Behavioral Ecology* 10, 122–127.

Field, S.A., Calbert, G. and Keller, M.A. (1998) Patch defence in the parasitoid wasp *Trissolcus basalis* (Insecta: Scelionidae): the time structure of pairwise contests, and the 'waiting game'. *Ethology* 104, 821–840.

Geervliet, J.B.F., Ariëns, S., Dicke, M. and Vet, L.E.M. (1998a) Long-distance assessment of patch profitability through volatile infochemicals by the parasitoids *Cotesia glomerata* and *C. rubecula* (Hymenoptera: Braconidae). *Biological Control* 11, 113–121.

Geervliet, J.B.F., Vreugdenhil, A.I., Dicke, M. and Vet, L.E.M. (1998b) Learning to discriminate between infochemicals from different plant–host complexes by the parasitoids *Cotesia glomerata* and *C. rubecula* (Hym: Braconidae). *Entomologia Experimentalis et Applicata* 86, 241–252.

Godfray, H.C.J. (1994) *Parasitoids, Behavioral and Evolutionary Ecology.* Princeton University Press, Princeton, New Jersey, 473 pp.

Hairston, N.G., Smith, F.E. and Slobodkin, L.B. (1960) Community structure, population control, and competition. *American Naturalist* 44, 421–425.

Hamilton, W.D. (1967) Extraordinary sex ratios. *Science* 156, 477–488.

Hemerik, L. and Harvey, J.A. (1999) Flexible larval development and the timing of destructive feeding by a solitary endoparasitoid: an optimal foraging problem in evolutionary perspective. *Ecological Entomology* 24, 308–315.

Hemerik, L., Driessen, G. and Haccou, P. (1993) Effects of intra-patch experiences on patch time, search time and searching efficiency of the parasitoid *Leptopilina clavipes* (Hartig). *Journal of Animal Ecology* 62, 33–44.

Houston, A. and McNamara, J.M. (1999) *Models of Adaptive Behaviour: an Approach Based on State.* Cambridge University Press, Cambridge, 378 pp.

Hunter, M.S. and Godfray, H.C.J. (1995) Ecological determinants of sex allocation in an autoparasitoid wasp. *Journal of Animal Ecology* 64, 95–106.

Ikawa, T. and Suzuki, Y. (1982) Ovipositional experience of the gregarious parasitoid *Apanteles glomeratus* L. (Hymenoptera: Braconidae), influencing her discrimination of host larvae, *Pieris rapae crucivora. Applied Entomology and Zoology* 17, 199–216.

Iwasa, Y., Higashi, M. and Yamamura, N. (1981) Prey distribution as a factor determining the choice of an optimal foraging strategy. *American Naturalist* 117, 710–723.

Kidd, N.A.C. and Jervis, M.A. (1991) Host feeding and oviposition strategies of parasitoids in relation to host stage. *Researches on Population Ecology* 33, 13–28.

LaSalle, J. and Gauld, I.D. (1993) *Hymenoptera and Biodiversity.* CAB International, Wallingford, UK.

Meyhöfer, R. and Casas, J. (1999) Vibratory stimuli in host location by parasitic wasps. *Journal of Insect Physiology* 45, 967–971.

Otten, H., Wäckers, F.L. and Dorn, S. (2001) Efficacy of vibrational sounding in the parasitoid *Pimpla turionellae* (Hymenoptera: Ichneumonidae) is affected by female size. *Animal Behaviour* 61, 671–677.

Papaj, D.R. and Vet, L.E.M. (1990) Odor learning and foraging success in the parasitoid, *Leptopilina heterotoma. Journal of Chemical Ecology* 16, 3137–3151.

Powell, W., Pennacchio, F., Poppy, G.M. and Tremblay, E. (1998) Strategies involved in the location of hosts by the parasitoid *Aphidius ervi* Haliday (Hymenoptera: Braconidae, Aphidiinae). *Biological Control* 11, 104–112.

Ramsauer, N. and Robert, D. (2000) Free-flight phonotaxis in a parasitoid fly: behavioral thresholds, relative attraction and susceptibility to noise. *Naturwissenschaften* 87, 315–319.

Roitberg, B.D., Mangel, M., Lalonde, R.G., Roitberg, C.A., Van Alphen, J.J.M. and Vet, L.E.M. (1992) Seasonal dynamic shifts in patch exploitation by a parasitic wasp. *Behavioral Ecology* 3, 156–165.

Rosenheim, J.A. and Rosen, D. (1991) Foraging and oviposition decisions in the parasitoid *Aphytis lingnanensis*: distinguishing the influences of egg load and experience. *Journal of Animal Ecology* 60, 873–894.

Rosenheim, J.A. and Rosen, D. (1992) Influence of egg load and host size on host feeding behaviour by the parasitoid *Aphytis lingnanensis. Ecological Entomology* 17, 263–272.

Sabelis, M.W., van Baalen, M., Bakker, F.M., Bruin, J., Drukker, B., Egas, M., Janssen, A.R.M., Lesna, I.K., Pels, B., van Rijn, P. and Scutareanu, P. (1999) The evolution of direct and indirect plant defence against herbivorous arthropods. In: Olff, H., Brown, V.K. and Drent, R.H. (eds) *Herbivores: Between Plants and Predators.* Blackwell Science, Oxford, pp. 109–166.

Simons, M.T.T.P., Suverkropp, B.P., Vet, L.E.M. and de Moed, G. (1992) Comparison of learning in related generalist and specialist eucoilid parasitoids. *Entomologia Experimentalis et Applicata* 64(2), 117–124.

Stephens, D.W. (1993) Learning and behavioural ecology: incomplete information and environmental predictability. In: Papaj, D.R. and Lewis, A.C. (eds) *Insect Learning, Ecological and Evolutionary Aspects.* Chapman & Hall, New York, pp. 195–218.

Stephens, D.W. and Krebs, J.R. (1986) *Foraging Theory*. Princeton University Press, Princeton, New Jersey, 247 pp.

Takabayashi, J., Takahashi S, Dicke, M. and Posthumus, M.A. (1995) Developmental stage of herbivore *Pseudaletia separata* affects production of herbivore-induced synomone by corn plants. *Journal of Chemical Ecology* 21, 273–287.

Takasu, K. and Lewis, W.J. (1993) Host- and food-foraging of the parasitoid *Microplitis croceipes*: learning and physiological state effects. *Biological Control* 3, 70–74.

Trivers, R.L. and Willard, D.E. (1973) Natural selection of parental ability to vary the sex ratio of offspring. *Science* 179, 90–92.

Turlings, T.C.J., Tumlinson, J.H. and Lewis, W.J. (1990) Exploitation of herbivore-induced plant odors by host-seeking parasitic wasps. *Science* 250, 1251–1253.

Turlings, T.C.J., Wäckers F.L., Vet, L.E.M., Lewis, W.J. and Tumlinson, J.H. (1993) Learning of host-finding cues by *Hymenopterous* parasitoids. In: Papaj, D.R. and Lewis, A.C. (eds) *Insect Learning: Ecological and Evolutionary Perspectives*. Chapman & Hall, New York, pp. 51–78.

Ueno, T. (1999) Multiparasitism and host feeding by solitary parasitoid wasps (Hymenoptera: Ichneumonidae) based on the pay-off from parasitized hosts. *Annals of Entomological Society of America* 92, 601–608.

van der Assem, J. (1971) Some experiments on sex ratio and sex regulation in the pteromalid *Lariophagus distinguendus*. *Netherlands Journal of Zoology* 21, 373–402.

Van der Hoeven, N. and Hemerik, L. (1990) Superparasitism as an ESS: to reject or not reject, that is the question. *Journal of Theoretical Biology* 146, 467–482.

van Dijken, M.J., van Stratum, P. and van Alphen, J.J.M. (1992) Recognition of individual-specific marked parasitized hosts by the solitary parasitoid *Epidinocarsis lopezi*. *Behavioral Ecology and Sociobiology* 30, 77–82.

Vet, L.E.M. (1983) Host-habitat location through olfactory cues by *Leptopilina clavipes* (Hartig) (Hym: Eucoilidae), a parasitoid of fungivorous *Drosophila*: the influence of conditioning. *Netherlands Journal of Zoology* 33, 225–248.

Vet, L.E.M. (1999) Evolutionary aspects of plant–carnivore interactions. In: Chadwick, D.J. and Goode, J. (eds) *Insect–Plant Interactions and Induced Plant Defence*. John Wiley & Sons, Chichester, UK, pp. 3–20.

Vet, L.E.M. and Dicke, M. (1992) Ecology of infochemical use by natural enemies in a tritrophic context. *Annual Review of Entomology* 37, 141–172.

Vet, L.E.M. and Papaj, D.R. (1992) Effects of experience on parasitoid movement in odour plumes. *Physiological Entomology* 17, 90–96.

Vet, L.E.M., Lewis, W.J., Papaj, D.R. and van Lenteren, J.C. (1990) A variable-response model for parasitoid foraging behaviour. *Journal of Insect Behavior* 3, 471–490.

Vet, L.E.M., Wäckers, F. and Dicke, M. (1991) How to hunt for hiding hosts: the reliability–detectability problem in foraging parasitoids. *Netherlands Journal of Zoology* 41, 202–213.

Vet, L.E.M., Datema, A., Janssen, A. and Snellen, H. (1994) Clutch size in a larval–pupal endoparasitoid: consequences for fitness. *Journal of Animal Ecology* 63, 807–815.

Vet, L.E.M., Lewis, W.J. and Cardé, R.T. (1995) Parasitoid foraging and learning. In: Cardé, R.T. and Bell, W.J. (eds) *Chemical Ecology of Insects*, 2nd edn. Chapman & Hall, New York.

Vet, L.E.M., de Jong, A.G., Franchi, E. and Papaj, D.R. (1998) The effect of complete versus incomplete information on odour discrimination in a parasitic wasp. *Animal Behavior* 55, 1271–1279.

Viggiani, G. (1984) Bionomics of the Aphelinidae. *Annual Review of Entomology* 29, 257–276.

Visser, M.E. (1993) Adaptive self- and conspecific superparasitism in the solitary parasitoid *Leptopilina heterotoma*. *Behavioral Ecology* 4, 22–28.

Visser, M.E. (1994) The importance of being large: the relationship between size and fitness in females of the parasitoid *Aphaereta minuta* (Hymenoptera: Braconidae). *Journal of Animal Ecology* 63, 963–987.

Visser, M.E. (1996) The influence of competition between foragers on clutch size decisions in an insect parasitoid with scramble larval competition. *Behavioral Ecology* 7, 109–114.

Visser, M.E., van Alphen, J.J.M. and Nell, H.W. (1990) Adaptive superparasitism and patch time allocation in solitary parasitoids; the influence of the number of parasitoids depleting a patch. *Behaviour* 114, 21–36.

Visser, M.E., van Alphen, J.J.M. and Hemerik, L. (1992a) Adaptive superparasitism and patch time allocation in solitary parasitoids: an ESS model. *Journal of Animal Ecology* 61, 93–101.

Visser, M.E., van Alphen, J.J.M. and Nell, H.W. (1992b) Adaptive superparasitism and patch time allocation in solitary parasitoids – the influence of prepatch experience. *Behavioral Ecology and Sociobiology* 31, 163–171.

Vos, M. (2001) Foraging under incomplete information: parasitoid behaviour and community dynamics. PhD thesis, Wageningen University, Wageningen, the Netherlands.

Vos, M., Hemerik, L. and Vet, L.E.M. (1998) Patch exploitation by the parasitoids *Cotesia rubecula* and *Cotesia glomerata* in multi-patch environments with different host distributions. *Journal of Animal Ecology* 67, 774–783.

Vos, M., Berrocal, S.M., Karamaouna, F., Hemerik, L. and Vet, L.E.M. (2001) Plant-mediated indirect effects and the persistence of parasitoid–herbivore communities. *Ecology Letters* 4, 38–45.

Wäckers, F.L. (1994) The effect of food deprivation on the innate visual and olfactory preferences in *Cotesia rubecula*. *Journal of Insect Physiology* 40, 641–649.

Wäckers, F.L. and Lewis, W.J. (1994) Olfactory and visual learning and their combined influence on host site location by the parasitoid *Microplitis croceipes*. *Biological Control* 4, 105–112.

Wäckers, F.L. and Lewis, W.J. (1999) A comparison between color-, shape- and pattern-learning by the hymenopteran parasitoid *Microplitis croceipes*. *Journal of Comparative Physiology A* 184, 387–393.

Wäckers, F.L., Mitter, E. and Dorn, S. (1998) Vibrational sounding by the pupal parasitoid *Pimpla* (*Coccygomimus*) *turionellae*: an additional solution to the reliability–detectability problem. *Biological Control* 11, 141–146.

Host Discrimination by Seed Parasites

4

Frank J. Messina

Department of Biology, Utah State University, 5305 Old Main Hill, Logan, UT 84322–5305, USA

Introduction

Angiosperm seeds are typically well provisioned with nutrients. Resources derived from the endosperm are needed for the successful germination, growth and establishment of seedlings (Parrish and Bazzaz, 1985; Härdling and Nilsson, 1999). Not surprisingly, many herbivores include nutrient-rich seeds as a major component of their diets (Mattson, 1980). We can define as seed parasites those herbivores that feed and develop wholly or mainly within host seeds (Price, 1997). Depending on the site and density of infestation, a seed parasite may kill the seed embryo, decrease the survival of the emerging seedling or have little effect on its host (Cipollini and Stiles, 1991; Jaeger *et al.*, 2001). In some parasites, juvenile stages cannot move between seeds and must complete development within a single 'natal' seed. These species thus exploit hosts in a way that makes them similar to many fruit parasites and insect parasitoids (see Vet *et al.*, Chapter 3, and Strand, Chapter 7, this volume). Insect taxa that commonly act as seed parasites include seed beetles (family Bruchidae), weevils (Curculionidae), grain borers (Bostrichidae), seed flies (Anthomyiidae and Tephritidae) and chalcid seed wasps (Torymidae and Eurytomidae).

Unlike free-ranging granivores, such as rodents and ants, seed parasites tend to have specialized diets, and many infest seeds before they disperse from the parent plant. Janzen (1980) found that 75% of the seed-beetle species in a Costa Rican forest were associated with a single plant species. Narrow host ranges of seed parasites may reflect fine-scale adaptation to the timing of the host reproduction, as well as to the specific chemical and physical defences of seeds. Plant investment in defence traits should depend in part on the fitness value of various tissues, and reproductive parts often show high concentrations of secondary metabolites (Hamilton *et al.*, 2001). Some legume seeds, for example, contain chemicals in the seed-coat or endosperm that are highly toxic to non-adapted herbivores, but they are frequently attacked by specialist parasites (Johnson, 1990; Povey and Holloway, 1992; Siemens *et al.*, 1992).

A second common feature of seed-parasite life histories is a high frequency of intra- and interspecific competition within seeds. Many hosts support the development of only one or two parasite larvae per seed (Messina, 1991a). Because of the narrow host ranges of most parasites, intraspecific competition is inevitable when parasite populations become large relative to local seed availability (Delgado *et al.*, 1997). The capacity of some seed parasites to infest nearly an entire seed crop has led to their importation for the biological control of exotic weeds (e.g. Briese, 2000; Radford *et al.*, 2001). Several behavioural, physiological and life-history traits of seed parasites have been interpreted as evolutionary responses to frequent intraspecific competition (Smith and Lessells, 1985). In this chapter, I focus on host-discrimination behaviour as a means by which egg-laying females reduce competition among their offspring within seeds.

In much of the parasite literature, host discrimination simply refers to a parasite's ability to recognize and exploit a host. A more restricted meaning, applied especially to insect parasitoids, is the tendency of an egg-laying female to avoid hosts that already bear conspecific eggs or larvae (Roitberg and Prokopy, 1987). Host discrimination can be expressed either as a lower rate of acceptance of infested hosts or as smaller clutch sizes on such hosts (Messina and Fox, 2001). By reducing the degree of competition experienced by offspring, discriminating females will under most circumstances have higher fitness than females that do not respond to previous infestation. Yet we might expect variation in the degree of host discrimination when there is concomitant variation in the associated gain in fitness (Nufio and Papaj, 2001). Host discrimination thus provides an opportunity to examine the adaptive modification of parasite behaviour in response to host traits.

I shall first review evidence for host discrimination and mechanisms by which females distinguish between occupied and unoccupied hosts. I then consider host-discrimination behaviour in the context of general foraging models. Essential to this objective is identifying the genetic and environmental causes of variation in the trait. Studies of seed beetles demonstrate how variation in seed size can shape the egg-laying behaviour of associated parasites. I conclude by examining host discrimination in a broader ecological context. A potential cost of host discrimination is that cues mediating the behaviour can be exploited by the seed parasite's own natural enemies. On the other hand, the action of natural enemies can make host discrimination advantageous even in the absence of intraspecific competition.

Evidence of Host Discrimination

Attempts to detect host discrimination among herbivorous insects have focused on species that are endoparasitic (feed within plant tissues) and infest small, discrete hosts, such as seeds, buds, flowers and fruits (Prokopy, 1972). Although host discrimination may be weak among external, leaf-feeding insects (Groeters *et al.*, 1992; Mappes and Mäkelä, 1993), these species have also received less attention (Schoonhoven, 1990; Poirier and Borden, 1991). Perhaps more relevant than the insect's feeding mode is the relative mobility of the insect larva (whether it can move to new hosts when resources become scarce) and the number of larvae that can be supported by one host individual (Vasconcellos-Neto and Monteiro, 1993). Benefits to host discrimination should also depend on whether a female can accurately 'survey' conspecific density on a potential host before she deposits her own eggs.

Three complementary lines of evidence have been used to determine whether seed parasites respond to conspecific eggs or larvae on potential hosts. The most common assay has been to provide females with occupied vs. unoccupied hosts in a laboratory choice test. A related technique is to offer a female a single host in a no-choice situation and to record her responses as a function of whether or not the host is already infested. Relevant variables include the probability of host acceptance, the time elapsed until a seed is accepted and the number of eggs laid. At the population level, host discrimination can be detected by exposing hosts to a single female or a group of females, and then measuring the dispersion of eggs among hosts. Whereas most natural insect populations show clumped or aggregated distributions, strong host discrimination should produce a uniform or regular distribution of eggs among host individuals.

Reliance on only one of these techniques may produce misleading results in respect of the presence or strength of host discrimination. Subtle, statistical preferences obtained in laboratory choice tests may not translate into effective host discrimination under natural conditions, where differences in intrinsic host quality or other factors may overwhelm any effects of previous infestation on host acceptance (Klijnstra and Schoonhoven, 1987). An even dispersion of juvenile stages in the field might indicate cannibalism or strong contest competition within seeds rather than host discrimination by ovipositing females. Conversely, a failure to observe an even dispersion of juvenile stages does not preclude host discrimination; periods of limited host availability may produce an aggregated or random dispersion of eggs in species for which host discrimination has been well documented (Averill and Prokopy, 1989; Fox and Mousseau, 1995).

Host discrimination has been detected in all major groups of seed parasites, including weevils (Ferguson *et al.*, 1999a,b), seed flies (Zimmerman, 1982; Pittara and Katsoyannos, 1990; Lalonde and Roitberg, 1992), cone flies (McClure *et al.*, 1998; Quiring *et al.*, 1998), seed-eating moths (Huth and Pellmyr, 1999) and seed wasps (Kouloussis and Katsoyannos, 1991). Yet some seed parasites fail to discriminate between occupied and unoccupied hosts, even though they use small hosts and their juvenile stages are unable to move between hosts (Povey and Sibly, 1992). One explanation is that there may be a non-linear relationship between larval density and fitness (sometimes called an Allée effect), such that larval performance is optimal at intermediate numbers of larvae per host. Allée effects can occur when the food source is improved by group feeding or when mortality from natural enemies is inversely density-dependent (Messina, 1998). Advantages to host discrimination also depend on the degree to which egg-laying females are constrained by time available for oviposition or resources available for egg production (see 'Foraging Considerations' below). Povey and Sibly (1992) have argued that host discrimination is absent or weak in the rice weevil (*Sitophilus oryzae*) because females are iteroparous (lay eggs in multiple bouts) and use adult feeding to replenish resources needed for continual egg production.

Host discrimination has been especially well studied among bruchid beetles. These insects have been intimately associated with legume seeds for much of the Cenozoic (Poinar, 1999). Throughout this chapter, I shall focus on pest bruchids (especially *Callosobruchus* spp.), which have infested human stores of grain legumes for thousands of years and are well suited to laboratory manipulations. Female *Callosobruchus* beetles lay eggs singly on the surfaces of legume seeds and pods. The hatching larva burrows into the underlying cotyledons, and must complete its development within a single host seed. The cowpea seed beetle, *Callosobruchus maculatus* (F.), has evolved multiple traits (including a dispersal polymorphism and a facultative diapause) that enable it to infest legume seeds both in human stores and in the field (Utida, 1972; Messina, 1987). Non-diapausing females infest hosts within hours after they emerge, so that populations build up rapidly in storage. As a consequence, traits that mediate intraspecific competition appear to be particularly important in this species (Smith and Lessells, 1985).

Among *Callosobruchus* beetles, host discrimination has been documented in several ways. Females have long been known to produce non-random, uniform distributions of eggs among seeds (Utida, 1943; Avidov *et al.*, 1965). The fitness benefits associated with uniform egg laying are easily quantified (Mitchell, 1975; Credland *et al.*, 1986); multigeneration studies have shown that competition within seeds sharply reduces the survival of larvae and the body size, egg size, fecundity and longevity of emerging adults (Messina, 1991a,b; Fox and Savalli, 1998). Paired choice tests have demonstrated that egg-laying females discriminate between seeds bearing different numbers of eggs, as well as between seeds with or without eggs (Messina and Renwick, 1985; Wilson, 1988; Horng, 1997). This quantitative response to egg density maintains uniform egg distributions even after all seeds bear several eggs. Finally, focal-animal observations have shown that the presence of an egg on a seed alters female behavioural sequences, as well as the probability of host acceptance (Messina and Dickinson, 1993; Parr *et al.*, 1996).

Proximate Cues

Few studies have identified the cues used by herbivorous insects to detect eggs or larvae on potential hosts (Hurter *et al.*, 1987). Most investigations have focused on chemical cues, which are called host-marking phero- mones or oviposition-deterring pheromones. These cues may reside in the eggs, larvae or larval faeces, but in many cases the egg-laying female shows a distinct behaviour that results in pheromone deposition immedi- ately after an egg is laid. For example, the female of the almond seed wasp, *Eurytoma amygdali* Enderlein, lays a single egg in the nucellar tissue of a developing seed and then drags the tip of her abdomen on the almond surface (Kouloussis and Katsoyannos, 1991). A series of experi- ments demonstrated that this dragging behaviour, which has been well studied in related fruit parasites, leads to the deposition of a water-soluble marking pheromone.

The distinctive marking behaviour of *E. amygdali* may originate as a means by which a female avoids reinfesting the same host. If egg-laying females search for hosts over a relatively small area, the probability of re-encountering a host can be high, and even a short-lived, water-soluble pheromone can be effectively used by a female to avoid competition among her own offspring. At the same time, a marking pheromone can inhibit egg laying by later-arriving, conspecific females, whose offspring would perform better in unoccupied hosts. Host-marking behaviour may therefore evolve in a similar way to and under the same constraints as other animal communication systems (Roitberg and Prokopy, 1987).

Marking pheromones deposited after oviposition are sometimes secreted by specialized glands (Quiring *et al.*, 1998; Ferguson *et al.*, 1999a). Most studies suggest that these compounds are perceived by contact chemoreception (gustation) or short-range olfaction, so that the female cannot detect whether a host is occupied without close inspection. Ablations of sense organs and electrophysiological recordings from particular sensillae have been used to identify the means of pheromone perception (Ferguson *et al.*, 1999b). Bioassays have shown that some marking pheromones are quite stable; others lose effectiveness after only a few hours. Theoretical analysis suggests that variation in pheromone persistence will depend on whether the pheromone signal is exploited by the parasite's natural enemies (Hoffmeister and Roitberg, 1998)

Interest in the active components of marking pheromones has been spurred by the potential use of these chemicals to control seed or fruit parasites that are economic pests. If marking pheromones deter ovi- position or cause females to disperse away from infested hosts, they might be used to manipulate parasite behaviour in a way that reduces crop damage. Unfortunately, identification of marking pheromones has been difficult, in part because chemicals that cause avoidance of occupied hosts (and uniform egg distributions) are hard to distinguish from a wider array of chemicals that also happen to deter host acceptance in laboratory choice tests.

Two examples illustrate potential problems. Although it is an external leaf feeder rather than a seed or fruit parasite, the cabbage butterfly, *Pieris brassicae* L., apparently avoids laying egg batches on leaves that have already received eggs (Rothschild and Schoonhoven, 1977). Novel alkaloids (miriamides) produced in the female's accessory glands are deposited with the eggs. Choice tests initially suggested that these compounds caused reduced acceptance of occupied leaves (Blaakmeer et al., 1994). Yet females also avoided leaves from which eggs are removed, even though such leaves do not contain detectable amounts of miriamides.

Subsequent experiments suggested that rejection of egg-bearing leaves might actually be mediated by volatile, plant-derived compounds that are induced by previous oviposition (Blaakmeer et al., 1994). Because miriamides are not volatile, plant-derived compounds may better explain the observation that females respond to egg-bearing leaves without contacting them. The release of volatiles by infested plants is well known to attract natural enemies of herbivorous insects, but volatiles may also be used by the herbivores themselves to avoid competition (De Moraes et al., 2001).

Similar confusion has surrounded attempts to identify the chemical means of host discrimination by *Callosobruchus* beetles. Early research suggested that ether-soluble compounds (especially fatty acids) act as bruchid marking pheromones (Oshima et al., 1973; Sakai et al., 1986). These compounds can be obtained from washes of egg-laden glass beads that serve as surrogate seeds. Yet washes of glass beads exposed to males or virgin females also deterred oviposition in choice tests, and it is difficult to imagine how these non-specific cues could mediate the quantitative assessment of egg density described above (Messina and Renwick, 1985).

Additional experiments suggested that *Callosobruchus* marking pheromone is perceived through the maxillary palps. Ablation of these mouth-parts eliminated uniform distributions of eggs; treated females actually tended to aggregate their eggs (Fig. 4.1; Messina et al., 1987). However, palpectomized females were still able to avoid seeds coated with putative marking pheromone in choice tests. Taken together, these results suggested a weak correspondence between the sensory receptors involved in the response to chemical extract and those needed for uniform egg laying. It is certainly possible that multiple chemical cues are used for detection of eggs on infested seeds (Credland and Wright, 1990), but another complication is that fatty acids can both stimulate and deter oviposition, depending on their concentrations (Parr et al., 1998).

Identification of chemicals responsible for host discrimination will probably be most successful if multiple behavioural assays (choice and

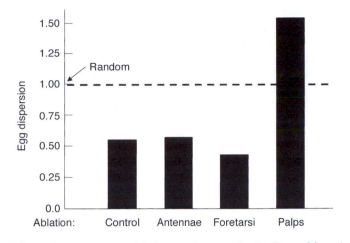

Fig. 4.1. Effect of sense-organ ablation on the egg distributions of females of *Callosobruchus maculatus.* Dispersion was measured as the variance-to-mean ratio: 1 = random, < 1 = uniform, and > 1 = aggregated.

no-choice tests) are combined with ablation experiments and electrophysiological recordings, so that both stimuli and receptors can be identified. Attempts to identify mechanisms of host discrimination should also not overlook the roles of non-chemical cues, including tactile or visual cues. In *Callosobruchus* beetles, females laid fewer eggs on seeds bearing egg-shaped models than they did on seeds bearing a flat dab of the material used to construct models (Messina and Renwick, 1985). Tactile cues are known to be important in the responses of *C. maculatus* females to different legume species and cultivars. A final consideration is that the use of gustatory cues after contact with a host does not preclude long-distance olfactory cues that could permit females to avoid entire patches of highly infested hosts (Ignacimuthu *et al.*, 2000).

Foraging Considerations

Female seed parasites must distribute their eggs among many scattered hosts. Because a search for suitable oviposition sites resembles a search for suitable food, foraging theory has been used to model the costs and benefits of host discrimination. These models predict circumstances under which a female should accept or reject an occupied host. For insects that lay clutches of eggs rather than single eggs, foraging theory has also been used to predict the optimal number of eggs to deposit on infested vs. uninfested hosts (Charnov and Skinner, 1984; Iwasa *et al.*, 1984).

At first glance, it might seem that a female should always reject a host that already bears conspecific eggs, especially when a seed can support the development of only a few larvae. The expenditure of time or eggs during oviposition on an infested host reduces a searching female's ability to exploit higher-quality hosts that she may encounter in the future (Rosenheim, 1999). However, the consequences of female egg-laying decisions depend on several factors, including a female's age, experience, egg load and survival probability. A particularly important factor is the shape of the larval competition curve, which describes the combined fitness of surviving larvae as a function of an increasing number of eggs on a host (Smith and Lessells, 1985).

In many cases, the combined fitness or total productivity from a host will increase with an increasing density of larvae, up to a maximum determined by host size and quality (Fig. 4.2). This value has been called the single-host maximum; any eggs beyond this number will only decrease the combined fitness of all offspring. A key point is that, if per capita fitness declines monotonically with each additional larva in a host, then the total productivity per host (estimated as the number of surviving larvae times per capita fitness) rises in a decelerating fashion up to the single-host maximum (Fig. 4.2). In the chestnut weevil (*Curculio elephas* (Gyll.)), for example, per capita fitness can be measured as a larva's survival probability times its potential fecundity (a function of its eventual adult weight). Per capita fitness is highest when one larva develops per seed, drops by 8% when two larvae share a host and drops by 15% when four larvae share a host (Desouhant *et al.*, 2000). Yet the highest total productivity (the single-host maximum) occurs when seven to eight larvae coexist in a single chestnut.

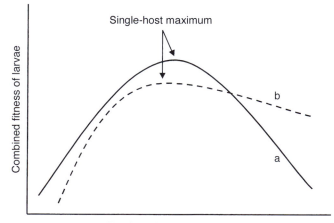

Fig. 4.2. Relationship between larval density and the combined fitness of all larvae in a seed (total productivity). Combined fitness is the product of larval density and per capita fitness, which in turn equals the product of offspring survival and fecundity. Curves a and b depict different 'penalties' associated with exceeding the most productive number of larvae (the single-host maximum).

Given a dome-shaped or concave-down relationship between larval density and the total productivity, Charnov and Skinner (1984) argued that host acceptance and clutch size should depend on the costs of searching. If hosts are plentiful (and search costs low), a female can maximize her rate of fitness gain by rejecting infested hosts and by depositing only a single egg on each unoccupied host. If hosts are scarce (and search costs high), females should become more likely to accept an occupied host and lay more eggs per clutch. Only when search costs are infinite (a female is not expected to encounter another host) should a female deposit as many eggs as the single-host maximum, and it is always maladaptive to lay more eggs than this value. In the chestnut weevil example, females typically lay only one or two eggs per host. Desouhant *et al.* (2000) suggested that clutch sizes in this species are well below the single-host maximum because the number of suitable chestnuts is usually high relative to the number of egg-laying weevils.

In general, a female should reject a host or stop laying additional eggs when the marginal rate of fitness gain drops to a point where she could gain fitness more rapidly by exploiting another host (Wajnberg *et al.*, 2000). Realistic models of egg-laying decisions need to incorporate effects of a female's physiological status, in addition to whether a host is already infested or not (Parker and Begon, 1986; Wilson and Lessells, 1994). Because some of these variables change continuously over a female's lifetime, dynamic optimization models have been used to show how the degree of host discrimination is expected to change according to a female's physiological state or the rate of encountering new hosts (Mangel, 1987).

Unlike seed predators (such as rodents), seed parasites do not quickly consume entire seeds. After a parasite female has deposited an egg, the seed remains available to other females. One might expect a female to lay fewer eggs if there is a risk of later-arriving females adding eggs to the same host. Ives (1989) demonstrated that the evolutionarily stable clutch size again depends on the shape of the larval competition curve. Small clutches are expected if per capita fitness declines linearly with an increasing number of larvae per host. In this case, there will be a steep decline in the total productivity once the number of larvae exceeds the single-host maximum (curve a in Fig. 4.2). If the first-arriving female lays a number of eggs that is close to the single-host maximum, any eggs added by subsequent females will greatly decrease offspring fitness. But, if increasing larval density causes a non-linear, decelerating drop in per capita fitness, then exceeding the single-host maximum causes only a gradual decrease in total productivity (curve b in Fig. 4.2). In this case, the optimal clutch size is largely independent of the number of females that oviposit on the same host.

The above discussion assumes that female egg-laying decisions depend on reducing competition among offspring. A different scenario occurs when seed parasites also act as facultative or obligate pollinators of their host plants. In the yucca moth (*Tegeticula yuccasella* (Riley)), females actively pollinate yucca flowers after ovipositing and thereby ensure a food supply for their seed-eating progeny. A flower that receives a high number of eggs is more likely to be selectively aborted, which results in the death of all eggs (Huth and Pellmyr, 1999) (differential abortion may be a means by which the plant ensures that only a fraction of its developing seeds is consumed). Thus, the plant's response will itself prevent competition for food among moth larvae, but it still 'pays' for a female not to overexploit a host (and risk destruction of her eggs).

Yucca moth females do lay fewer eggs on previously visited hosts, which they appear to recognize through a marking pheromone (Wilson and Addicott, 1998; Huth and Pellmyr, 1999). Some evidence suggests that females can even assess the density of conspecific eggs quantitatively, as noted before in the seed beetle *C. maculatus*. The yucca moth–yucca interaction is unusual because the progeny of early-arriving females cannot usurp resources from the progeny of late-arriving females; both early and late eggs will die as a consequence of floral abortion. Huth and Pellmyr (1999) have argued that these conditions favour effective host marking by the first female to deposit eggs. In seed-parasite/pollinator systems without selective abortion of flowers, plants use other means to protect seeds (Jaeger *et al.*, 2000, 2001), and host discrimination will probably reflect the usual avoidance of competition or cannibalism.

Variable Host Discrimination by Seed Beetles

Predictions from foraging models have rarely been tested with seed parasites; most examples of adaptive variation in host discrimination have involved parasitoids. Experiments using *Callosobruchus* seed beetles confirm the importance of the larval competition curve in determining the degree of host discrimination. These beetles also provide examples of adaptive plasticity in foraging behaviour as a result of a female's physiological state or experience.

Interfertile geographical populations of *C. maculatus* show striking differences in the mode and intensity of larval competition within seeds (Toquenaga, 1993; Messina, 1998; and references therein). In some populations, larvae engage in a strong 'contest' type of competition (including biting behaviour), so that small seeds virtually never yield two adults and even large seeds yield few adults per seed. Larvae from other populations display more of a scramble type of competition; larvae form burrows near the periphery of the seed and burrows are less likely to intersect. In these populations, small seeds often yield two adults and more than ten adults can emerge from a large seed, albeit with reduced size. Hybridization experiments revealed additive inheritance of differences in larval competitiveness within seeds (Messina, 1991b).

Variation in larval competitiveness produces very different larval competition curves, such as those depicted in Fig. 4.2. In populations with strongly competitive larvae, the single-host maximum is at most a few eggs and the 'penalty' for adding eggs to occupied seeds is much greater. This is particularly true because older larvae have a competitive advantage over younger larvae and only one larva will develop to adult emergence in a small seed (Messina, 1991b). We might therefore predict parallel differences in the strength of host discrimination among populations.

Geographical populations of *C. maculatus* do vary in the predicted direction (Messina and Mitchell, 1989; Mitchell, 1990). Females from an Asian strain with highly competitive larvae produce much more uniform egg distributions and show stronger preferences for egg-free seeds than do females from strains with less competitive larvae (Messina, 1991a; Messina *et al.*, 1991). Moreover, Asian females provided few seeds simply stop laying eggs after each seed has received two or three eggs, so that their lifetime realized fecundities are sharply reduced. Females from strains with less competitive larvae add several eggs per seed, and their realized fecundities do not drop substantially unless seeds are absent (Messina, 1999). Interpopulation differences in host discrimination depend both on differences in proximate cues and on differences in female responses to egg-laden seeds (Messina *et al.*, 1991). Reciprocal crosses of populations with divergent egg-laying behaviour suggested almost complete dominance towards strong host discrimination and very uniform egg laying (Messina, 1989).

Callosobruchus beetles have also been used to identify non-genetic sources of variation in host discrimination. Non-diapausing females emerge from seeds with about eight mature oocytes and, in the absence of suitable hosts, they will accumulate oocytes in their abdomens for about 2 days (Credland and Wright, 1989; Wilson and Hill, 1989). In one experiment, females from a population with strong host discrimination were deprived of hosts for 10 h and were compared with females with no host deprivation. Host deprivation caused the frequency of oviposition 'mistakes' to rise from 19% to 50%, where a mistake was defined as adding a second egg to a seed when at least one uninfested seed was still available (Messina *et al.*, 1992). In a strain with weak host discrimination (and less competitive larvae), a period of host deprivation had little effect on the frequency of egg-laying mistakes, which were already common among females with no deprivation.

There has been a long-standing interest in whether experience gained during a foraging bout modifies host discrimination by *C. maculatus* females. Mitchell (1975) proposed that learning was needed for females to maintain uniform egg distributions even after all seeds bear multiple eggs. In particular, he hypothesized that oviposition would be more likely if a currently visited seed has fewer eggs than the previous one. An alternative hypothesis is that a female simply has a lower probability of host acceptance with each increase in egg density. In this case, the specific probabilities of acceptance will depend on the female's physiological condition or the amount of time since the last egg was laid (Messina and Renwick, 1985). Wilson (1988) labelled these respective mechanisms as 'relative' and 'absolute' rules, and performed experiments that suggested female behaviour conformed to an absolute rule (no learning). However, simulations by Mitchell (1990) indicated that absolute rules could not account for females that are especially adept at distinguishing different egg densities on seeds.

The role of experience in *Callosobruchus* host discrimination was examined in detail by Horng (1997), who combined foraging models with empirical observations of female behaviour and larval competition. Females from populations with very competitive larvae did appear to use a relative rule in accepting or rejecting egg-laden seeds, but an absolute rule was adequate to explain the weaker discrimination by females from populations with less competitive larvae (Horng, 1997). The beetle–legume interaction may therefore serve as an example of genetic variation in the expression of parasite learning. A statistical analysis of behaviour sequences indicated that female behaviour is best described by an 'adjusted-threshold' model, in which thresholds for accepting hosts with different egg densities are modified by the female's current egg load and the amount of time since her last oviposition (Horng *et al.*, 1999).

Host Size and Parasite Evolution: a Selection Experiment

Genetic variation in host discrimination by female seed beetles appears to reflect differences in the intensity and type of larval competition within seeds. Yet this explanation begs the question as to why larvae from completely interfertile populations should compete in such different ways (Messina, 1991a,b). Several empirical and theoretical investigations of this question have converged on the idea that larval competitiveness is itself primarily an adaptation to seed size (Smith and Lessells, 1985; Toquenaga *et al.*, 1994; Tuda, 1998). The Asian strain discussed above was obtained from a region where most hosts are small and other local beetle populations exhibit contest-type competition within seeds (Mitchell, 1991).

Because optimal larval behaviour depends on the frequencies of possible phenotypes, Smith and Lessells (1985) used game theory to predict evolutionarily stable levels of aggression among seed-parasite larvae. They concluded that the cost of scramble or exploitative competition in a large seed is low enough for 'tolerant' behaviour (such as a tendency to form peripheral, non-intersecting burrows) to be advantageous, whereas an aggressive, 'contest' phenotype is superior in a small host (but see Colegrave, 1995). Both Toquenaga *et al.* (1994) and Tuda (1998) obtained data suggesting that transferring a population to a different-sized host alters larval competitiveness in the expected direction.

I performed a mass-selection experiment (F.J. Messina, unpublished data) to test directly the effects of host size on both larval competitiveness and host discrimination in *C. maculatus*. A response to selection may be expected in this system because earlier experiments confirmed between-population genetic variation for each trait (Messina, 1989), and host-discrimination behaviour was also found to be heritable within populations (Messina, 1993; Tanaka, 2000). The selection experiment used an Asian strain with highly uniform egg laying and contest competition. This base population was collected from a small host (mung bean) and was consistently maintained at large population sizes on this host (Messina, 1991a).

The original population was divided into six lines. Three replicate lines were maintained on the ancestral host, and three were transferred to a larger, novel host, cowpea (Fig. 4.3; the cowpeas had approximately three times more mass than mung beans). After more than 40 generations on each host, we examined larval competitiveness in each line by establishing either one or two larvae per seed in both mung bean and cowpea. We quantified host discrimination as the uniformity of egg distributions produced by females that were each given 40 seeds. We used the U score of Messina and Mitchell (1989; see also Horng, 1997), which is not as influenced as other dispersion indices by the number of eggs laid. This index typically varies between zero, which signifies random egg laying (no host discrimination), and one, which represents the most uniform distribution possible. To reduce possible host-related, non-genetic, maternal effects (Fox and Savalli, 1998, and references therein), we reared all lines on mung beans for two generations before the test generation. This step introduced a conservative bias because traits in cowpea-selected

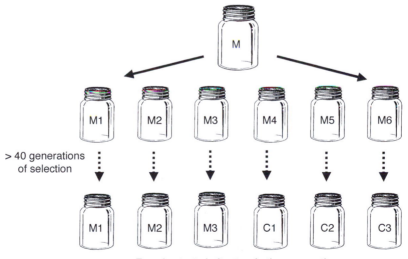

Females tested after two further generations
on mung bean

Fig. 4.3. Schematic diagram of a selection experiment in which an ancestral population of *Callosobruchus maculatus* on mung bean (M) was divided into replicate lines and maintained on the same host (M1–3) or transferred to cowpea (C1–3). All lines were returned to mung beans for two generations to avoid host-related maternal effects on the test generation.

lines could have partially reverted towards values in the ancestral line on mung beans.

All replicate cowpea-selected lines evolved to resemble populations that have long been associated with large hosts. Larval competitiveness decreased considerably. Mung beans receiving two, equal-aged larvae from cowpea-selected lines frequently yielded two emerging adults (> 60% of all seeds), but lines maintained on mung bean almost invariably yielded only one adult per seed (F.J. Messina, unpublished data). When the host was cowpea, approximately 90% of seeds with two cowpea-selected larvae yielded two adults, but only 13% of seeds with two mung bean-selected larvae did so. There was also a major effect of selection on the body size of emerging adults. Even in the absence of competition (only one larva per seed), adults from cowpea-selected lines were significantly smaller than those from either mung bean-selected lines or the ancestral population.

Cowpea-selected and mung bean-selected lines diverged in their degree of host discrimination. Egg distributions were measured for each line on each host in 12 treatment combinations. Statistical analysis indicated an effect of test host, as egg-laying females from all lines were better able to avoid occupied seeds (and produced higher U scores) on mung bean than on cowpea (Fig. 4.4). This result matches earlier studies showing more oviposition 'mistakes' on larger hosts (Mitchell, 1990). However, there was also a significant effect of selection host, as well as a selection-host × test-host interaction. When females were given mung

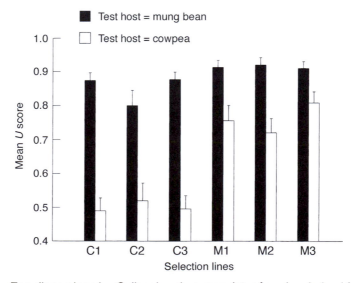

Fig. 4.4. Egg dispersions by *Callosobruchus maculatus* females derived from cowpea-selected (C1–3) or mung bean-selected (M1–3) lines and presented either cowpeas or mung beans. *U* scores vary between 0 (= random egg laying) and 1 (= completely uniform distributions) (see Messina and Mitchell, 1989).

beans, egg distributions of cowpea-selected females were slightly less uniform than those of mung bean-selected females. On cowpea seeds, cowpea-selected females were much 'sloppier' than mung bean-selected females in how they spread their eggs among seeds (Fig. 4.4).

By showing a decreased tendency to avoid occupied hosts, females from cowpea-selected lines converged towards the poorer host discrimination observed in cowpea-adapted populations. Selection for strong avoidance of occupied hosts may have been relaxed in the cowpea-selected lines because of the simultaneous evolution of less competitive larvae. The relatively quick decline in average U scores on cowpea (≤ 40 generations) also suggests that there may be a cost associated with a high degree of host discrimination, although the nature of this cost remains unclear (Messina, 1993). Taken together, the results of the selection experiment demonstrate how a host trait (size) can predictably modify both a juvenile trait (competitive ability) and an adult trait (host discrimination) in the parasite.

Effects of Parasite Enemies

Juvenile stages of seed parasites are of course susceptible to their own natural enemies as well as to competition or cannibalism. Several mechanisms have been identified by which natural enemies can modify the tendency for seed-parasite females to avoid occupied hosts. Risk of mortality from natural enemies also provides a separate explanation for why females may reject a potential host (or lay few eggs) when the current density on the host is well below the single-host maximum. Bruchid beetles have been useful subjects for addressing relationships between egg-laying behaviour by seed parasites and mortality risk from natural enemies. Eggs and larvae of most seed beetles are attacked by at least one species of parasitoid wasp.

Natural enemies could reduce advantages of host discrimination if cues that allow seed parasites to detect previous infestation are used by natural enemies to find parasite eggs or larvae. In a fruit parasite, marking pheromone deposited after oviposition acted as a kairomone to attract a specialist parasitoid (Prokopy and Webster, 1978). Use of marking pheromone as a kairomone causes a trade-off between minimizing competition and minimizing predation risk, and can therefore maintain different levels of host discrimination (and genetic variation) within a population (Roitberg and Lalonde, 1991; Hoffmeister and Roitberg, 1998).

A similar trade-off would occur if predation risk were inversely density-dependent, i.e. if single eggs or eggs in small clutches suffer greater mortality from enemies than do eggs in large clutches. Inverse density dependence could arise, for example, if a parasitoid tends to attack only one or two seed-parasite eggs before leaving the seed. Mitchell (1977) suggested that *Mimosestes* seed beetles often lay a stack of two eggs rather than a single egg because the presence of the top egg diminishes vulnerability to parasitoids for the bottom egg. When parasitoids are abundant, the average survival of a pair of eggs exceeds that of single eggs.

As yet, there is little evidence that marking pheromones of seed parasites are exploited by the parasite's natural enemies. In part, this is because so little is known about the composition of the marking pheromones themselves. A mixture of saturated hydrocarbons and diacylglycerols applied to a legume seed stimulated the oviposition or 'stinging' behaviour of a parasitic wasp, *Dinarmus basalis* (Rondani), that attacks the azuki seed beetle, *Callosobruchus chinensis* (L.) (Kumazaki *et al.*, 2000). The hydrocarbons are thought to be constituents of the marking pheromone in this bruchid, but neither the diacylglycerol nor other active glycerols are known to mediate avoidance of occupied seeds. Because eggs from different geographical populations of *C. maculatus* appear to differ in their tendency to deter further egg laying (Messina *et al.*, 1991), it would be interesting to determine whether there are parallel differences in prey location or oviposition stimulation of the parasitoid.

At least two studies of seed beetles have detected positive density dependence in the risk of mortality from parasitoids; eggs or larvae on heavily infested seeds are more likely to be attacked than those on lightly infested seeds. In a study of *Callosobruchus chinensis* and its parasitoid, positive density dependence was caused by a substantial increase in the parasitoid's searching efficiency when seeds bore four beetle eggs instead of one (Ryoo and Chun, 1993). There is also strong competition among *C. chinensis* larvae within seeds. The net reproductive rate of an emerging female shows a negative linear relationship with the number of larvae per seed. Thus, both risk of natural enemies and intraspecific competition favour females that lay only one egg per seed and avoid seeds that already bear eggs. Ryoo and Chun (1993) could not disentangle the relative importance of competition vs. enemies in determining *C. chinensis* behaviour.

Siemens and Johnson (1992) suggested that positively density-dependent mortality from natural enemies was more important than intraspecific competition in explaining why females of the seed beetle *Stator limbatus* (Horn) lay few eggs per seed. Highest beetle productivity from a Palo Verde seed (the single-host maximum) occurred when 13–15 larvae shared a seed, but females typically laid one or a few eggs on an individual seed. Siemens and Johnson (1992) argued that parasitoid-induced mortality was responsible for this behaviour and that the action of the parasitoid would itself reduce intraspecific competition, at least in a large host seed. However, they also showed that competition decreased the per capita fitness of larvae with each increase in larval density, so that total productivity increased in a decelerating fashion, as depicted in Fig. 4.2. For reasons discussed earlier, the combination of a dome-shaped larval competition curve and high seed availability (low search costs) favours small egg clutches, even without density-dependent attack from parasitoids. Females supplied with an abundance of seeds in the laboratory did in fact lay only one egg per seed (also Fox *et al.*, 1996).

A final potential influence of natural enemies is via effects on larval competitiveness, which, as we have seen, tends to coevolve with host discrimination in *Callosobruchus* beetles. After transferring *C. chinensis* beetles from a large-seeded legume host to a smaller one, Tuda (1998) noted an evolutionary shift from scramble-type competition among larvae to contest-type competition. Interestingly, beetle populations exposed to parasitoids evolved in this direction more slowly than did populations without parasitoids. This delay in the evolution of highly competitive larvae was predicted from population models in which parasitoids contribute an additional source of mortality that lessens the severity of competition (Tuda and Iwasa, 1998). By modifying larval competition within seeds, chronic attack by parasitoids can affect which type of egg-laying behaviour is advantageous. It is worth noting, however, that transfer to a small host eventually increased the frequency of contest competition even in the presence of parasitoids (Tuda, 1998).

Conclusion

Herbivorous insects obviously use multiple cues and a variety of sensory inputs in the hierarchical process of locating and consuming plants (Bernays and Chapman, 1994). Seed parasites, which typically have sessile larvae and narrow host ranges, illustrate the sometimes close relationship between female egg-laying behaviour and offspring fitness. We therefore expect parasite females to display a suite of non-random behaviours that enhance the survival and reproduction of their progeny. In one species that attacks its host in the flowering stage, the parasite female may actually manipulate the host to ensure greater fruit set and hence the amount of food available to larvae (Brody and Morita, 2000).

Host discrimination is a common behaviour of seed parasites because of the importance of minimizing competition among larvae within seeds. Yet we cannot ignore other determinants of host use, such as intrinsic plant quality or mortality from natural enemies. Some of these factors may in fact generate trade-offs between behaviour that reduces larval competition and other components of offspring or maternal fitness. Differences in local host availability, host size and larval aggression are all predicted to modify the strength of host discrimination. Variation in host discrimination can be expressed either as genetic differences at the population level or as behavioural plasticity within the lifetime of an individual forager.

Seed beetles and host legumes have been particularly useful for testing these predictions, in part because they are quite amenable to laboratory manipulations and in part because several species are major economic pests. Population crosses and breeding designs have revealed abundant genetic variation in host-discrimination behaviour, and selection experiments have helped identify causal mechanisms underlying this variation. Focal-animal observations and host-deprivation experiments have allowed female behaviour to be compared with expectations from optimal foraging models. Further information is needed regarding the proximate mechanisms of host discrimination and the nature of trade-offs between this trait and others. Costs of host discrimination are of particular interest because the behaviour appears to decline quickly when larval competition is reduced and selection is relaxed. Although this chapter has focused on traits that affect intraspecific competition, seed parasites and their hosts are similarly well suited for examining other aspects of parasite–host interactions.

References

Averill, A.L. and Prokopy, R.J. (1989) Distribution patterns of *Rhagoletis pomonella* (Diptera: Tephritidae) eggs in hawthorn. *Annals of the Entomological Society of America* 82, 38–44.

Avidov, Z., Applebaum, S.W. and Berlinger, M.J. (1965) Physiological aspects of host specificity in the Bruchidae: II – ovipositional preference and behaviour of *Callosobruchus chinensis* L. *Entomologia Experimentalis et Applicata* 8, 96–106.

Bernays, E.A. and Chapman, R.F. (1994) *Host-plant Selection by Phytophagous Insects.* Chapman & Hall, New York, 312 pp.

Blaakmeer, A., Hagenbeek, D., van Beek, T.A., de Groot, A., Schoonhoven, L.M. and van Loon, J.J.A. (1994) Plant response to eggs vs. host marking pheromone as factors inhibiting oviposition by *Pieris brassicae. Journal of Chemical Ecology* 20, 1657–1665.

Briese, D.T. (2000) Impact of *Onopordum capitulum* weevil *Larinus latus* on seed production by its host-plant. *Journal of Applied Ecology* 37, 238–246.

Brody, A.K. and Morita, S.I. (2000) A positive association between oviposition and fruit set: female choice or manipulation? *Oecologia* 124, 418–425.

Charnov, E.L. and Skinner, S.W. (1984) Evolution of host selection and clutch size in parasitoid wasps. *Florida Entomologist* 67, 5–21.

Cipollini, M.L. and Stiles, E.W. (1991) Seed predation by the bean weevil *Acanthoscelides obtectus* on *Phaseolus* species: consequences for seed size, early growth and reproduction. *Oikos* 60, 205–214.

Colegrave, N. (1995) The cost of exploitation competition in *Callosobruchus* beetles. *Functional Ecology* 9, 191–196.

Credland, P.F. and Wright, A.W. (1989) Factors affecting female fecundity in the cowpea seed beetle, *Callosobruchus maculatus* (Coleoptera: Bruchidae). *Journal of Stored Products Research* 25, 125–136.

Credland, P.F. and Wright, A.W. (1990) Oviposition deterrents of *Callosobruchus maculatus* (Coleoptera: Bruchidae). *Physiological Entomology* 15, 285–298.

Credland, P.F., Dick, K.M. and Wright, A.W. (1986) Relationships between larval density, adult size and egg production in the cowpea seed beetle, *Callosobruchus maculatus. Ecological Entomology* 11, 41–50.

Delgado, C., Couturier, G. and Delobel, A. (1997) Oviposition of seed-beetle *Caryoborus serripes* (Sturm) (Coleoptera: Bruchidae) on palm (*Astrocaryum chambira*) fruits under natural conditions in Peru. *Annals of the Entomological Society of France* 33, 405–409.

De Moraes, C.M., Mescher, M.C. and Tumlinson, J.H. (2001) Caterpillar-induced nocturnal plant volatiles repel conspecific females. *Nature* 410, 577–580.

Desouhant, E., Debouzie, D., Ploye, H. and Menu, F. (2000) Clutch size manipulations in the chestnut weevil, *Curculio elephas*: fitness of oviposition strategies. *Oecologia* 122, 493–499.

Ferguson, A.W., Solinas, M., Ziesmann, J., Isadoro, N., Williams, I.H., Scubla, P., Mudd, A., Clark, S.J. and Wadhams, L.J. (1999a) Identification of the gland secreting oviposition-deterring pheromone in the cabbage seed weevil, *Ceutorhynchus assimilis*, and the mechanism of pheromone deposition. *Journal of Insect Physiology* 45, 687–699.

Ferguson, A.W., Ziesmann, J., Blight, M.M., Williams, I.H., Wadhams, L.J., Clark, S.J., Woodcock, C.M. and Mudd, A. (1999b) Perception of oviposition-deterring pheromone by cabbage seed weevil (*Ceutorhynchus assimilis*). *Journal of Chemical Ecology* 25, 1655–1670.

Fox, C.W. and Mousseau, T.A. (1995) Determinants of clutch size and seed preference in a seed beetle, *Stator beali* (Coleoptera: Bruchidae). *Environmental Entomology* 24, 1557–1561.

Fox, C.W. and Savalli, U.M. (1998) Inheritance of environmental variation in body size: superparasitism of seeds affects progeny and grandprogeny size via a nongenetic maternal effect. *Evolution* 52, 172–182.

Fox, C.W., Martin, J.D., Thakar, M.S. and Mousseau, T.A. (1996) Clutch size manipulations in two seed beetles: consequences for progeny fitness. *Oecologia* 108, 88–94.

Groeters, F.R., Tabashnik, B.E., Finson, N. and Johnson, M.W. (1992) Oviposition preference of the diamondback moth (*Plutella xylostella*) unaffected by the presence of conspecific eggs or *Bacillus thuringiensis*. *Journal of Chemical Ecology* 18, 2353–2362.

Hamilton, J.G., Zangerl, A.R., DeLucia, E.H. and Berenbaum, M.R. (2001) The carbon-nutrient balance hypothesis: its rise and fall. *Ecology Letters* 4, 86–95.

Härdling, R. and Nilsson, P. (1999) Parent–offspring and sexual conflicts in the evolution of angiosperm seeds. *Oikos* 84, 27–34.

Hoffmeister, T.S. and Roitberg, B.D. (1998) Evolution of signal persistence under predator exploitation. *Ecoscience* 5, 312–320.

Horng, S. (1997) Larval competition and egg-laying decisions by the bean weevil *Callosobruchus maculatus*. *Animal Behaviour* 53, 1–12.

Horng, S., Lin, H., Wu, W. and Godfray, H.C.J. (1999) Behavioral processes and egg-laying decisions of the bean weevil, *Callosobruchus maculatus*. *Researches in Population Ecology* 41, 283–290.

Hurter, J., Boller, E.F., Städler, E., Blattmann, B., Buser, H.-R., Bosshard, N.U., Damm, L., Kozlowski, M.W., Schöni, R., Raschdorf, F., Dahinden, R., Schlumpf, E., Fritz, H., Richter, W.J. and Schreiber, J. (1987) Oviposition-deterring pheromone in *Rhagoletis cerasi* L.: purification and determination of the chemical constitution. *Experientia* 43, 157–163.

Huth, C.J. and Pellmyr, O. (1999) Yucca moth oviposition and pollinator behavior is affected by past flower visitors: evidence for a host-marking pheromone. *Oecologia* 119, 593–599.

Ignacimuthu, S., Wäckers, F.L. and Dorn, S. (2000) The role of chemical cues in host finding and acceptance by *Callosobruchus chinensis*. *Entomologia Experimentalis et Applicata* 96, 213–219.

Ives, A.R. (1989) The optimal clutch size of insects when many females oviposit per patch. *American Naturalist* 133, 671–687.

Iwasa, Y., Suzuki, Y. and Matsuda, H. (1984) Theory of oviposition strategy of parasitoids. I. Effect of mortality and limited egg number. *Theoretical Population Biology* 26, 205–227.

Jaeger, N., Till-Bottraud, I. and Deprés, L. (2000) Evolutionary conflict between *Trollius europaeus* and its seed-parasite pollinators *Chiastocheta* flies. *Evolutionary Ecology Research* 2, 885–896.

Jaeger, N., Pompanon, F. and Deprés, L. (2001) Variation in predation costs with *Chiastocheta* egg number on *Trollius europaeus*: how many seeds to pay for pollination? *Ecological Entomology* 26, 56–62.

Janzen, D.H. (1980) Specificity of seed-attacking beetles in a Costa Rican deciduous forest. *Journal of Ecology* 68, 929–952.

Johnson, C.D. (1990) Coevolution of Bruchidae and their hosts: evidence, conjecture, and conclusions. In: Fujii, K., Gatehouse, A.M.R., Johnson, C.D., Mitchel, R. and Yoshida, T. (eds) *Bruchids and Legumes: Economics, Ecology and Coevolution.* Kluwer Academic, Dordrecht, the Netherlands, pp. 181–188.

Klijnstra, J.W. and Schoonhoven, L.M. (1987) Effectiveness and persistence of the oviposition deterring pheromone of *Pieris brassicae* in the field. *Entomologia Experimentalis et Applicata* 45, 227–235.

Kouloussis, N.A. and Katsoyannos, B.I. (1991) Host discrimination and evidence for a host marking pheromone in the almond seed wasp, *Eurytoma amygdali. Entomologia Experimentalis et Applicata* 58, 165–174.

Kumazaki, M., Matsuyama, S., Suzuki, T., Kuwahara, Y. and Fujii, K. (2000) Parasitic wasp, *Dinarmus basalis*, utilizes oviposition-marking pheromone of host azuki bean weevils as host-recognizing kairomone. *Journal of Chemical Ecology* 26, 2677–2695.

Lalonde, R.G. and Roitberg, B.D. (1992) Host selection behavior of a thistle-feeding fly: choices and consequences. *Oecologia* 90, 534–539.

McClure, M., Quiring, D.T. and Turgeon, J.J. (1998) Proximate and ultimate factors influencing oviposition site selection by endoparasites of conifer seed cones: two sympatric dipteran species on larch. *Entomologia Experimentalis et Applicata* 87, 1–13.

Mangel, M. (1987) Oviposition site selection and clutch size in insects. *Journal of Mathematical Biology* 25, 1–22.

Mappes, J. and Mäkelä, I. (1993) Egg and larval load assessment and its influence on oviposition behaviour of the leaf beetle *Galerucella nymphaeae. Oecologia* 93, 38–41.

Mattson, W.J. (1980) Herbivory in relation to plant nitrogen content. *Annual Review of Ecology and Systematics* 11, 119–161.

Messina, F.J. (1987) Genetic contribution to the dispersal polymorphism of the cowpea weevil (Coleoptera: Bruchidae). *Annals of the Entomological Society of America* 80, 12–16.

Messina, F.J. (1989) Genetic basis of variable oviposition behavior in *Callosobruchus maculatus* (Coleoptera: Bruchidae). *Annals of the Entomological Society of America* 82, 792–796.

Messina, F.J. (1991a) Life-history variation in a seed beetle: adult egg-laying vs. larval competitive ability. *Oecologia* 85, 447–455.

Messina, F.J. (1991b) Competitive interactions between larvae from divergent strains of the cowpea weevil (Coleoptera: Bruchidae). *Environmental Entomology* 20, 1438–1443.

Messina, F.J. (1993) Heritability and 'evolvability' of fitness components in *Callosobruchus maculatus*. *Heredity* 71, 623–629.

Messina, F.J. (1998) Maternal influences on larval competition in insects. In: Mousseau, T.A. and Fox, C.W. (eds) *Maternal Effects as Adaptations*. Oxford University Press, New York, pp. 227–243.

Messina, F.J. (1999) Expression of a life-history trade-off in a seed beetle depends on environmental context. *Physiological Entomology* 24, 358–363.

Messina, F.J. and Dickinson, J.A. (1993) Egg-laying behavior in divergent strains of the cowpea weevil (Coleoptera: Bruchidae): time budgets and transition matrices. *Annals of the Entomological Society of America* 86, 207–214.

Messina, F.J. and Fox, C.W. (2001) Offspring size and number. In: Fox, C.W., Roff, D. and Fairbairn, D.J. (eds) *Evolutionary Ecology: Concepts and Case Studies*. Oxford University Press, New York, pp. 113–127.

Messina, F.J., and Mitchell, R. (1989) Intraspecific variation in the egg-spacing behavior of the seed beetle *Callosobruchus maculatus*. *Journal of Insect Behavior* 2, 727–741.

Messina, F.J. and Renwick, J.A.A. (1985) Ability of ovipositing seed beetles to discriminate between seeds with differing egg loads. *Ecological Entomology* 10, 225–230.

Messina, F.J., Barmore, J.L. and Renwick, J.A.A. (1987) Oviposition deterrent from eggs of *Callosobruchus maculatus*: spacing mechanism or artifact? *Journal of Chemical Ecology* 13, 219–226.

Messina, F.J., Gardner, S.L. and Morse, G.E. (1991) Host discrimination by egg-laying seed beetles: causes of population differences. *Animal Behaviour* 41, 773–780.

Messina, F.J., Kemp, J.L. and Dickinson, J.A. (1992) Plasticity in the egg-spacing behavior of a seed beetle: effects of host deprivation and seed patchiness (Coleoptera: Bruchidae). *Journal of Insect Behavior* 5, 609–621.

Mitchell, R. (1975) The evolution of oviposition tactics in the bean weevil, *Callosobruchus maculatus* (F.). *Ecology* 56, 696–702.

Mitchell, R. (1977) Bruchid beetles and seed packaging by Palo Verde. *Ecology* 58, 644–651.

Mitchell, R. (1990) Behavioral ecology of *Callosobruchus maculatus*. In: Fujii, K., Gatehouse, A.M.R., Johnson, C.D., Mitchel, R. and Yoshida, T. (eds) *Bruchids and Legumes: Economics, Ecology and Coevolution*. Kluwer Academic, Dordrecht, the Netherlands, pp. 317–330.

Mitchell, R. (1991) The traits of a biotype of *Callosobruchus maculatus* (F.) (Coleoptera: Bruchidae) from South India. *Journal of Stored Products Research* 27, 221–224.

Nufio, C.R. and Papaj, D.R. (2001) Host marking behavior in phytophagous insects and parasitoids. *Entomologia Experimentalis et Applicata* 99, 273–293.

Oshima, K., Honda, H. and Yamamoto, I. (1973) Isolation of an oviposition marker from azuki bean weevil, *Callosobruchus chinensis* (L.). *Agricultural and Biological Chemistry* 37, 2679–2680.

Parker, G.A. and Begon, M. (1986) Optimal egg size and clutch size: effects of environment and maternal phenotype. *American Naturalist* 128, 573–592.

Parr, M.J., Tran, B.M.D., Simmonds, M.S.J. and Credland, P.F. (1996) Oviposition behaviour of the cowpea seed beetle, *Callosobruchus maculatus*. *Physiological Entomology* 21, 107–117.

Parr, M.J., Tran, B.M.D., Simmonds, M.S.J., Kite, G.C. and Credland, P.F. (1998) Influence of some fatty acids on oviposition by the bruchid beetle, *Callosobruchus maculatus*. *Journal of Chemical Ecology* 24, 1577–1593.

Parrish, J.A.D. and Bazzaz, F.A. (1985) Nutrient content of *Abutilon theophrasti* seeds and the competitive ability of the resulting plants. *Oecologia* 65, 247–251.

Pittara, I.S. and Katsoyannos, B.I. (1990) Evidence for a host-marking pheromone in *Chaetorellia australis*. *Entomologia Experimentalis et Applicata* 54, 287–295.

Poinar, G.O. (1999) A fossil palm bruchid, *Caryobruchus dominicanus* sp. n. (Pachymerini: Bruchidae) in Dominican amber. *Entomologica Scandinavica* 30, 219–224.

Poirier, L.M. and Borden, J.H. (1991) Recognition and avoidance of previously laid egg masses by the oblique-banded leafroller (Lepidopterist: Tortricidae). *Journal of Insect Behavior* 4, 501–508.

Povey, S.R. and Holloway, G.J. (1992) The effect of energy trade-offs on life history and fitness in the rice weevil, *Sitophilus oryzae*. *Oikos* 64, 441–450.

Povey, S.R. and Sibly, R.M. (1992) No oviposition plasticity in *Sitophilus oryzae* (L.) (Coleoptera: Curculionidae). *Journal of Stored Products Research* 28, 11–14.

Price, P.W. (1997) *Insect Ecology*, 3rd edn. John Wiley & Sons, New York, 874 pp.

Prokopy, R.J. (1972) Evidence for a marking pheromone deterring repeated oviposition in apple maggot flies. *Environmental Entomology* 1, 326–332.

Prokopy, R.J. and Webster, R.P. (1978) Oviposition deterring pheromone of *Rhagoletis pomonella*: a kairomone for its parasitoid *Opius lectus*. *Journal of Chemical Ecology* 4, 481–494.

Quiring, D.T., Sweeney, J.W. and Bennett, R.G. (1998) Evidence for a host-marking pheromone in white spruce cone fly, *Strobilomyia neanthracina*. *Journal of Chemical Ecology* 24, 709–721.

Radford, I.J., Nicholas, D.M. and Brown, J.R. (2001) Assessment of the biological control impact of seed predators on the invasive shrub *Acacia nilotica* (prickly acacia) in Australia. *Biological Control* 20, 261–268.

Roitberg, B.D. and Lalonde, R.G. (1991) Host marking enhances parasitism risk for a fruit-infesting fly *Rhagoletis basiola*. *Oikos* 61, 389–393.

Roitberg, B.D. and Prokopy, R.J. (1987) Insects that mark host plants. *Bioscience* 37, 400–406.

Rosenheim, J.A. (1999) The relative contributions of time and eggs to the cost of reproduction. *Evolution* 53, 376–385.

Rothschild, M. and Schoonhoven, L.M. (1977) Assessment of egg load by *Pieris brassicae* (Lepidoptera: Pieridae). *Nature* 266, 352–355.

Ryoo, M.I. and Chun, Y.S. (1993) Oviposition behavior of *Callosobruchus chinensis* (Coleoptera: Bruchidae) and weevil population growth: effects of larval parasitism and competition. *Environmental Entomology* 22, 1009–1015.

Sakai, A., Honda, H., Oshima, K. and Yamamoto, I. (1986) Oviposition marking pheromone of two bean weevils, *Callosobruchus chinensis* and *Callosobruchus maculatus*. *Journal of Pesticide Science* 11, 163–168.

Schoonhoven, L.M. (1990) Host-marking pheromones in Lepidoptera, with special reference to two *Pieris* spp. *Journal of Chemical Ecology* 16, 3043–3052.

Siemens, D.H. and Johnson, C.D. (1992) Density-dependent egg parasitism as a determinant of clutch size in bruchid beetles (Coleoptera: Bruchidae). *Environmental Entomology* 21, 610–619.

Siemens, D.H., Johnson, C.D. and Ribardo, K.J. (1992) Alternative seed defense mechanisms in congeneric plants. *Ecology* 73, 2152–2166.

Smith, R.H. and Lessells, C.M. (1985) Oviposition, ovicide and larval competition in granivorous insects. In: Sibly, R.M. and Smith, R.H. (eds) *Behavioral Ecology*. Blackwell Scientific, Oxford, pp. 423–448.

Tanaka, Y. (2000) Realized heritability of behavioral responsiveness to an oviposition deterring pheromone in the azuki bean weevil *Callosobruchus chinensis*. *Entomologia Experimentalis et Applicata* 96, 239–243.

Toquenaga, Y. (1993) Contest and scramble competitions in *Callosobruchus maculatus* (Coleoptera: Bruchidae) II. Larval competition and interference mechanisms. *Researches in Population Ecology* 35, 57–68.

Toquenaga, Y., Ichinose, M., Hoshino, T. and Fujii, K. (1994) Contest and scramble competitions in an artificial world: genetic analysis with genetic algorithms. In: Langton, C.G. (ed.) *Artificial Life III, SFI Studies in the Sciences of Complexity*. Addison-Wesley, Reading, Massachusetts, pp. 177–199.

Tuda, M. (1998) Evolutionary character changes and population responses in an insect host–parasitoid experimental system. *Researches in Population Ecology* 40, 293–299.

Tuda, M. and Iwasa, Y. (1998) Evolution of contest competition and its effect on host–parasitoid dynamics. *Evolutionary Ecology* 12, 855–870.

Utida, S. (1943) Studies on the experimental population of the azuki bean weevil, *Callosobruchus chinensis* (L.). VIII. Statistical analysis of the frequency distribution of the emerging weevils on beans. *Memoirs of the College of Agriculture of Kyoto Imperial University* 54, 1–22.

Utida, S. (1972) Density dependent polymorphism in the adult of *Callosobruchus maculatus* (Coleoptera, Bruchidae). *Journal of Stored Products Research* 8, 111–126.

Vasconcellos-Neto, J. and Monteiro, R.F. (1993) Inspection and evaluation of host plant by the butterfly *Mechanitis lysimnia* (Nymph., Ithomiinae) before laying eggs: a mechanism to reduce intraspecific competition. *Oecologia* 95, 431–438.

Wajnberg, E., Fauvergue, X. and Pons, O. (2000) Patch leaving decision rules and the marginal value theorem: an experimental analysis and a simulation model. *Behavioral Ecology* 11, 577–586.

Wilson, K. (1988) Egg laying decisions by the bean weevil *Callosobruchus maculatus*. *Ecological Entomology* 13, 107–118.

Wilson, K. and Hill, L. (1989) Factors affecting egg maturation in the bean weevil *Callosobruchus maculatus*. *Physiological Entomology* 14, 115–126.

Wilson, K. and Lessells, C.M. (1994) Evolution of clutch size in insects. I. A review of static optimality models. *Journal of Evolutionary Biology* 7, 339–363.

Wilson, R.D. and Addicott, J.F. (1998) Regulation of mutualism between yuccas and yucca moths: is oviposition behavior responsive to selective abscission of flowers? *Oikos* 81, 109–118.

Zimmerman, M. (1982) Facultative deposition of an oviposition-deterring pheromone by *Hylemya*. *Environmental Entomology* 11, 519–522.

Soil and Plant Interactions' Impact on Plant-parasitic Nematode Host Finding and Recognition

5

A. Forest Robinson

USDA-ARS, Southern Crops Research Laboratory, 2765 F&B Road,
College Station, TX 77845, USA

Introduction

Our understanding of host finding and recognition by plant-parasitic nematodes is limited. Since plant-parasitic nematodes share morphology, ancestry, niches and behaviour with other nematode species, it is important to examine what is known regarding them as well.

I shall focus this review on nematode locomotion. Croll and Sukhdeo (1981), however, list more than 20 kinds of nematode movement in addition to locomotion, ranging from isolated to coordinated repetitive motions and complex sequences of actions. Control of various internal or supporting muscles attached to the oesophagus, stoma and reproductive apparatus is involved during feeding, copulation, oviposition, defecation and perforation of barriers during hatching, skin penetration and tissue migrations.

Definitions

The behavioural terminology adopted here is that proposed by Burr (1984) and adopted by Dusenbery (1992). The most important concepts for nematodes are migration, taxis and kinesis. Migration is the net movement of an individual or population in response to a stimulus gradient and can be accomplished by taxis or kinesis. Taxis is migration achieved by directed turns, which orientate the body axis relative to the gradient, whereas kinesis is migration achieved by undirected responses, such as by altering the rate of movement (orthokinesis) or incidence of random turning (klinokinesis) when going up or down a gradient. Taxis and kinesis are root words to which prefixes are often added. Prefixes that have been used for operative stimuli affecting nematodes include chemo-

(chemical), galvano- (electric field), geo- (gravity), magneto- (magnetic field), photo- (light), rheo- (fluid flow), thermo- (temperature) and thigmo-(touch). Other environmental variables shown or proposed to influence directional movement by nematodes include electric current, osmotic pressure, pH, redox potential, soil moisture, surface tension and soil texture.

General characteristics of nematode movement

The function of sensory structures in plant-parasitic nematodes (Perry, 1996) has received less investigation than homologous structures in the bacteriophagous nematode *Caenorhabditis elegans*. The primary sensors are the amphids, located on either side of the head. Through laser ablation and mutant behavioural experiments, specific nerve endings in *C. elegans* amphids have been assigned roles in perception of volatiles, aqueous solutes and temperature (Bargmann and Mori, 1997). The sensory structures needed to respond to volatiles, solutes, temperature and touch are also present in parasitic nematodes. Ocelli, needed for phototaxis, are absent in plant-parasitic nematodes (Burr and Babinszki, 1990; Robinson *et al.*, 1990).

From a behavioural perspective, plant-parasitic nematodes fall into two broad groups: those that infect roots and those that infect leaves, stems and flowers. Both must be able to move through soil and through plant tissue during part of the life cycle. For those that infect plant parts above ground, one or more developmental stages must also be able to move quickly within thin and often transient films of water on foliar surfaces. Infective stages of plant-parasitic nematodes typically exhibit uninterrupted spontaneous movements, which are several times more rapid in foliar than in root parasites. Infective juveniles of certain vertebrate parasites may remain inactive until stimulated to move by heat (Croll, 1971), vibration (Wicks and Rankin, 1997) or light (Robinson *et al.*, 1990).

Most plant-parasitic nematodes are vermiform in post-egg stages. Exceptions include the sedentary, saccate females of about a dozen genera in the order Tylenchida (*Cactodera*, *Globodera*, *Heterodera*, *Meloidodera*, *Meloidogyne*, *Nacobbus*, *Rotylenchulus*, *Sphaeronema*, *Trophotylenchulus*, *Tylenchulus* and *Verutus*). Vermiform plant-parasitic nematodes usually move forwards in water and on solid surfaces by sinusoidal dorsoventral waves of the entire body, which are rhythmically propagated backwards from the anterior end (Wallace, 1968b, 1969), a type of undulatory propulsion (Gray, 1953). At less frequent intervals, determined by the species, substrate and other factors, forward waves that propel the nematode backwards can be propagated from the posterior end. Forward waves may result when the anterior end encounters an obstacle (Croll, 1976; Wicks and Rankin, 1997) or can occur spontaneously at more or less regular intervals. The typical pattern seen on agar is a quick withdrawal for one or more body waves, hesitation and resumed forward movement in a new direction. A kink, considered to be a refractory state by Croll and Sukhdeo (1981), can be formed when backward and forward waves meet.

Special problems in using agar

Most plant-parasitic nematodes occupy habitats that are optically opaque. Most experiments to study directed movement *in vitro* have been on the surface of water agar where nematodes and their tracks can be seen. These studies suffer the deficiency noted by Croll (1970): bilaterally symmetrical nematodes typically undulate dorsoventrally on their sides on a water agar surface, which precludes lateral steering. If the bilaterally positioned amphids are in fact separated sufficiently in plant-parasitic nematodes to resolve edaphically realistic chemical and thermal gradients, as argued by Dusenbery (1988a), and if they normally serve to guide lateral steering when nematodes move through soil, the primary orientation mechanism may be seriously handicapped, if not entirely incapacitated, on agar. In the case where nematodes can burrow through the agar and rotate on the body axis, gradients to which they respond must occur within the agar, not just along its surface. Interestingly, the only nematode that appears to have been critically examined for lateral steering, the phototaxing adult female of the orthopteran parasite *Mermis nigrescens*, shows the response very clearly. When phototaxing, the anterior end of this large nematode sweeps the air laterally, while dorsoventral body waves lift and lower the ventral surface from a horizontal moist felt-cloth substrate (Gans and Burr, 1994).

Nematode species vary in the way they move on agar (Robinson, 2000). The robust 650 and 1100 µm long infective juveniles of *Ditylenchus dipsaci* and *Steinernema glaseri* burrow easily through 0.75% agar and become trapped in water films less frequently than do the 350–400 µm long infectives of *Meloidogyne incognita*, *Rotylenchus reniformis* and *Tylenchulus semipenetrans* (A.F. Robinson, personal observation). Some species (*S. glaseri* and *Ditylenchus phyllobius*, for example) readily and others never ascend the walls of plastic Petri dishes. Some nematodes produce curved tracks on agar, exhibiting slew caused by dorsoventral asymmetry of the laterally positioned body (Croll and Sukhdeo, 1981; Green, 1977). Slew is particularly strong in males of plant-parasitic cyst nematodes with ventrally curved tails. Simulation models have shown that changes in slew curvature in response to changes in stimulus intensity within a stimulus field can help males to find females (Green, 1977). Slew does not seem to account for the circular tracks made by thermotaxing wild-type *C. elegans* on radial temperature gradients on agar (Hedgecock and Russell, 1975; Mori and Ohshima, 1995; Bargmann and Mori, 1997).

Some studies have examined horizontal movement in shallow containers of Sephadex beads, sand or agar coated with a thin or partial layer of sand to provide heterogeneity, if not a third dimension (Robinson, 2000). A sand layer on agar stimulates nictation by some *Steinernema* spp. (Campbell and Kaya, 1999) and alters the pattern of movement by *C. elegans* (Anderson *et al.*, 1997a,b).

Soil texture and moisture

Most plant-parasitic nematodes require thin moisture films for movement, lack the strength to dislodge soil particles and easily become trapped in water films (Wallace, 1959c). Consequently, their movement is markedly influenced by the porosity and moisture of soil. Most of what we know about the influence of soil texture and moisture on nematode movement came from the classical experiments of Wallace (1958a,b,c, 1959a,b,c, 1960, 1968a). These experiments are best understood in the context of a major advance in plant physiology that resulted in the 1950s from the realization that water movement through plants and soil could be explained best in terms of the Gibbs free energy of water (the water potential) at the leaf–air interface, within leaf cells, in roots and in the soil (Milburn, 1979; Papendick and Campbell, 1981; Kramer, 1983). The water status of plants was found to be directly affected not by the quantity of water in soil, but rather by the energy required to extract water, due primarily to the strong attraction of soil particles for water molecules (the matric potential). In comparison, the osmotic pressure of soil water was found to be too small to be of physiological significance to plants in most cases. Ultimately, concepts and notation from electrical engineering were incorporated into plant physiology to explain and predict the direction and rate of movement of water in soil and plants. This notation partitioned the total Gibbs free energy, or water potential, into matric, osmotic, gravitational and turgor potential. Today, the water potential in soil, air and plant tissues is usually expressed as a negative pressure given in bars or pascals (Pa) (1 bar = 0.1 MPa = 10^6 dyn cm^{-2} = *c.* 1 atm).

Wallace built devices to control and measure soil water potential and monitor nematode migration in three dimensions. When he compared movement by nematode species and stages of 15 different mean body lengths, ranging from 186 to 2000 μm, in four soil fractions, with particle size ranging from 75 to 1000 μm, he found that, at equivalent water potential, large nematodes require soil with larger particles than small nematodes do and the optimum particle size is linearly related to nematode length (Wallace, 1958c). This relationship is not apparent if soil water content rather than matric potential is kept constant, because finely textured soils can hold several times as much water as coarsely textured soils at the same matric potential. A second result was that, although the optimum particle size for movement by a nematode of a given length is constant in sand, when complex soil containing significant amounts of colloidal clay and silt is dried and sieved into crumb-size classes and then rewetted, it is crumb size, not soil particle size, that is critical, i.e. in most natural soils where fine particles can stick together to form much larger crumbs, soil structure is the primary determinant.

A perhaps more important finding of Wallace's was that movement by nematodes is fastest not in water-saturated soil, but rather when soil water content is near field capacity (approximately –0.05 bar = 50 cm water), i.e. when soil pores have drained just to the extent achievable by gravity. In fact, when movement was measured at various water potentials in a series of soils with particle sizes ranging from fine to coarse, the optimal water potential for movement was the same (–0.05 bar) regardless of soil texture or moisture content. This effect was shown with infective juveniles of both *Heterodera schachtii* and the much larger *D. dipsaci* (Wallace, 1956, 1958a,b). Finally, in any soil at soil matric potentials drier than *c.* –0.5 bar, nematode movement essentially stopped.

Osmotic pressure and salts

Nematodes have a large surface-to-volume ratio and a body covering permeable to water, and require body turgor for normal movement. Thus, osmoregulation is particularly important to behaviour. Nematodes within soil touch the strongly hygroscopic, electrically charged surfaces of clay particles, and nematodes within plant tissue feed on, perforate and often find themselves wedged between plant cells with high internal turgor. The internal osmotic pressures measured for plant, fresh water and soil nematodes (Wright and Newall, 1976; Wright, 1998) are 50–100 mM NaCl (equivalent to –2.2 to –4.2 bar), and thus several times greater than typical ionic concentrations in soil water (0 to –0.5 bar). These nematode species are generally able to regulate their body water contents well in hypo-osmotic solutions, consistent with the low osmotic pressure of water in soil (*c.* 22 mosmol kg^{-1}).

In glucose, mannitol and polyethylene glycol solutions, however, vermiform stages of various plant-parasitic nematodes remain motile and suffer no obvious volume loss at 110 mosmol kg^{-1} (–2.5 bar) (Viglierchio *et al.*, 1969; Wyss, 1970; Castro and Thomason, 1973; Robinson *et al.*, 1984a,b; Robinson and Carter, 1986). Wallace's observation that migration in soil ceases at about –0.5 bar matric potential, therefore, would indicate that the primary factor limiting nematode movement in soil is physical rather than physiological. Blake (1961) elegantly demonstrated this for *D. dipsaci* by showing that the relationship between matric potential and the speed of nematode movement in soil was the same whether soil was wetted with water or a subtoxic solution of urea. On the other hand, nematodes within living plant tissues may be exposed to water potentials near –15 bar, and osmoregulation in these nematodes, which is virtually unstudied, may be quite important.

Given the high permeability of nematodes to ions (Marks *et al.*, 1968; Castro and Thomason, 1973; Robinson and Carter, 1986) and the profound effects that unbalanced salts have on membrane-bound ion pumps, nerves and muscles in many organisms, the specific ion composition of water bathing a nematode can influence water balance and motility independently of osmotic pressure. Extreme effects from unbalanced salts were seen in the experiments of Robinson *et al.* (1984b), where monocationic solutions of Mg^{2+} or Na^+, with Cl^- or NO_3^- present as the anion, stopped all movement by the normally highly spontaneously motile J4 of the foliar parasite *D. phyllobius* in water at 100 mequiv l^{-1}. Under the same conditions, sucrose, mannitol and a synthetic soil solution at equivalent osmolality had no obvious effect on motility or water content. No volume loss was observed at toxic concentrations of Mg^{2+} or Na^+ and nematodes fully recovered from Na^+ but not Mg^{2+} exposure when returned to a synthetic soil solution. Thus, Wright (1998) justifiably criticized the use of single salt solutions for studying nematode water and ion regulation. The same warning should be extended to the study of plant-parasitic nematode behaviour in soil and plant tissues.

Temperature

Like most organisms, plant-parasitic nematodes appear to be thermally adapted to their habitats. Croll (1975) noted that motility optima for juvenile stages of a stem parasite, a root parasite and two animal parasites that ascend foliage were around 20–25°C, whereas optima for active penetrators of warm-blooded hosts were near 40°C. Robinson (1989) found that, in southern Texas, infectives of two foliar parasites, *D. dipsaci* and *D. phyllobius*, had optima for motility 10°C cooler than those of root parasites, *R. reniformis* and *T. semipenetrans*, from the same locale. This observation is consistent with the ecological need for the foliar parasites to be maximally active on foliar surfaces during the cool rainy periods in the summer when the foliar moisture films required for infection of foliar buds are present. A similar pattern is apparent among entomopathogenic nematodes, which, like plant-parasitic nematodes, reproduce within poikilothermic hosts. Optimal temperature ranges for host infection, establishment within the host and reproduction for 12 species and strains of entomopathogenic nematodes from different latitudes around the world were broadly similar to the climatic conditions of their origin (Grewal *et al.*, 1994). Further information on thermal adaptation is available in Trudgill's (1995a,b) analyses of base temperatures and thermal constants for embryogenesis and development in relation to niche temperature and reproductive strategy for more than 20 plant-parasitic, animal-parasitic and free-living nematodes.

Many nematodes are attracted to heat or repelled by it, and at least nine species exhibit a preferred temperature towards which they migrate. The latter include the root parasites *Globodera rostochiensis* (Rode, 1969a,b), *M. incognita* (Diez and Dusenbery, 1989a), *R. reniformis* (Robinson, 1989; Robinson and Heald, 1989, 1993) and *T. semipenetrans* (Robinson, 1989), the foliar parasites *D. dipsaci* (Wallace, 1961; Croll,

1967; Robinson, 1989) and *D. phyllobius* (Robinson, 1989), the entomopathogenic *Steinernema carpocapsae* (Burman and Pye, 1980), the
bacteriophagous *C. elegans* (Hedgecock and Russell, 1975) and the seal
and fish parasite *Tervanova decipiens* (Ronald, 1960). For plant-parasitic
nematodes within the soil, temperature is probably the most consistent
vertical cue in a world in which most life-limiting factors (root architecture, oxygen, carbon dioxide, moisture) are to be found by moving
vertically. Several studies have suggested that plant-parasitic nematodes
utilize diurnally fluctuating vertical gradients that extend through the
root zone to locate optimum depths for root finding and survival. Models
proposed by Dusenbery (1988a,c, 1989) and tested by Robinson (1994)
have shown that, when nematodes are exposed to the vertically propagated heat waves that occur naturally in soil, unexpected movement
towards the surface, towards great depths or towards specific depths can
result from interactions between the rate of nematode movement and the
rate of thermal adaptation. The possible role of metabolic heat from roots
as a host-recognition cue is discussed below.

Directed Movement

Movement towards carbon dioxide

Carbon dioxide may be the most common and potent nematode attractant
in nature. It is released abundantly by living and decaying plant and
animal tissues, providing an obvious cue to the possible presence of food.
It has also been suggested that carbon dioxide serves as a collimating
stimulus in soil, providing a directional reference for other responses
(Pline and Dusenbery, 1987). Because of their probable importance,
responses to CO_2 are discussed in detail as a prelude to a general
discussion of the literature on host finding by plant-parasitic nematodes.

Plant-parasitic nematode attraction to known sources of CO_2 *in vitro*
was first observed (or refuted) about 40 years ago (Bird, 1959, 1960;
Klingler, 1959; Rohde, 1960; Johnson and Viglierchio, 1961). CO_2 attracts
nematodes from a wide range of trophic groups, including bacterial
feeders (Balan and Gerber, 1972; Dusenbery, 1985; Viglierchio, 1990),
insect parasites (Gaugler *et al.*, 1980), root parasites (Bird, 1960; Johnson
and Viglierchio, 1961; Pline and Dusenbery, 1987; Robinson and Heald,
1991; Robinson, 1995), tree-trunk parasites (Miyazaki *et al.*, 1978a,b),
foliar parasites (Klingler, 1959, 1961, 1970, 1972), free-living marine
nematodes (Riemann and Schragge, 1988) and vertebrate parasites
(Granzer and Haas, 1991).

Interpreting the effects of CO_2 can be difficult. Most nematodes
are partly or completely anaesthetized in water equilibrated against air
containing more than 5% CO_2 (v/v) and, in at least some published
studies, nematodes have been subjected to flow rates of pure CO_2 that
probably produced anaesthetic concentrations in water agar (Robinson,

2000). These effects undoubtedly contributed to debates among early investigators regarding the importance of CO_2-induced changes in redox potential, pH, carbonic acid, bicarbonate and carbonate in the soil. In unbuffered, enzyme-free solutions at standard atmospheric pressure, dissolved CO_2 reacts slowly with water to form carbonic acid, which in turn dissociates ($pK = 6.1$) to yield bicarbonate and a proton, lowering the pH. Buffers, enzymes, atmospheric pressure and temperature influence the ratios of CO_2, H_2CO_3, HCO_3^-, CO_3^{2-} and H^+ present. The current consensus is that CO_2 can attract many nematodes at subtoxic concentrations and either dissolved CO_2 or carbonic acid is the nematode-attractive component.

Repulsion by ammonia may modulate attraction to CO_2. The root-knot nematode *M. incognita* is repelled by ammonia and several nitrogenous salts *in vitro* (Castro *et al.*, 1990, 1991). When entomopathogenic nematodes are released into the soil in large numbers, they accumulate around roots and root invasion by plant-parasitic nematodes is suppressed. Grewal *et al.* (1999) suggested that the plant-parasitic nematodes in this case are repelled by ammonia released by the entomopathogenic nematodes' symbiotic *Xenorhabdus* bacteria.

Pline and Dusenbery (1987) analysed responses of *M. incognita* on agar exposed to horizontal bilaminar flow of air from two parallel gas jets emitting different concentrations of CO_2. This produced a different atmosphere over each half of the agar plate. They found that the threshold gradient varied with the ambient concentration, i.e. the nematodes became more sensitive as the ambient concentration dropped and thus could detect about the same relative change at any ambient concentration. The threshold for *M. incognita* (Pline and Dusenbery, 1987) corresponded to a relative change of about 3% cm^{-1}, which was subsequently calculated (Dusenbery, 1987) to allow detection of roots from at least 5 cm and perhaps as far away as 500 cm. This effect is very important ecologically because it shows that nematodes can detect gradients at far greater distances from the source than would be possible with a fixed concentration differential threshold. As emphasized by Dusenbery, these predictions contrast sharply with Prot's (1980) conclusion that CO_2 only attracts nematodes within 1 or 2 cm of the source.

Klingler (1963) positioned a capillary CO_2 delivery tube within a thin layer of air between agar and a glass cover to attract *D. dipsaci*. The minimum effective gradient in the air, as determined by gas chromatography, was approximately 1 mmol CO_2 mol^{-1} cm^{-1}. In a more recent study, gradients were established in cylinders of moist sand by enclosing the cylinder in a plastic tube and equilibrating the two ends against air masses containing controlled CO_2 concentrations. *M. incognita*, *R. reniformis* and *S. glaseri* migrated up-gradient in response to a change of 0.2% cm^{-1} (2 mmol mol^{-1} cm^{-1}) at a mean CO_2 concentration of 1.2% (12 mmol mol^{-1}), which is a relative change of about 16% cm^{-1} or 1% per nematode body length (Robinson, 1995; Robinson and Jaffee, 1996). For comparison, first-instar western maize root worms can distinguish

1.12 mmol mol^{-1} from 0.99 mmol mol^{-1}, i.e. a 12% relative difference in concentration (Bernklau and Bjostad, 1998a,b). In the Robinson (1995) study, effective release rates for attracting *M. incognita* and *R. reniformis* to a point source in sand ranged from 6 to 35 µl min^{-1}, with an optimal flow of 15 µl min^{-1}. Intervals of 40 and 29 h, respectively, were required to achieve maximal attraction of the two species from a distance of 52 mm. Enough gas to achieve the same level of attraction was theoretically achievable with a germinating sunflower seed.

Movement towards plants

Host finding by plant-parasitic nematodes is examined in over 100 published studies. Contemporary general reviews include Perry (1996, 1997) and Perry and Aumann (1998). All root-parasitic and most foliar-parasitic nematodes have one or more infective stages that occur in the soil. These stages typically must find roots or find stems emerging from the soil to ascend foliage, in order to complete the life cycle.

Since roots respire, extract water, differentially take up salts and release various organic compounds, they can significantly modify local soil chemistry. As roots are approached, soil moisture decreases, Na$^+$ increases, K$^+$ and NO$_3^-$ decrease, O$_2$ decreases, CO$_2$ increases and amino acids, sugars and secondary metabolites increase. Very near roots, temperature increases slightly. Soil gases (ammonia, ethylene, methane), pH, redox potential and electrical potential may increase or decrease. Since the diffusivity of gases in the soil decreases by several orders of magnitude as soil interstices fill with water (Campbell, 1985), soil moisture content can markedly influence gradients of volatiles. Release and diffusion of behaviourally active chemicals are influenced by temperature (Pervez and Bilgrami, 2000).

Although Steiner (1925) postulated that nematodes are attracted to roots, the well-known nematode accumulations near roots on agar plates (Linford, 1939; Wieser, 1955, 1956; Widdowson *et al.*, 1958) were still thought by some researchers in the early 1960s to result only from chance encounters and physical trapping (Kühn, 1959; Sandstedt *et al.*, 1961; Sandstedt and Schuster, 1962). Numerous studies demonstrated directed movement towards roots in soil, towards living but not dead roots on agar plates and towards soil without roots but in which roots had been grown (Prot and VanGundy, 1981).

By the mid-1960s, contradictory results were obtained regarding nematode responses under controlled conditions to most substances known to occur as gradients around roots (Klingler, 1965; Prot, 1980). Although some studies showed that secondary metabolites correlated with host specificity, most evidence pointed to general attractants. The only consistent directions of movement were towards CO$_2$ and the wet end of a soil-moisture gradient (Wallace, 1960). Movement towards moisture vertically would help nematodes avoid desiccation but horizontally would take them away from roots. Therefore, CO_2 seemed the prime candidate as a root signal.

Subsequent investigations of responses to salts, usually unbalanced salts, in or on gels, sand and soil (Prot, 1978a,c, 1979a,b; Bilgrami and Jairajpuri, 1984; Riddle and Bird, 1985; Castro et al., 1990, 1991; Abou-Setta and Duncan, 1998), made salt gradients appear increasingly important. Experiments on long-distance movement by many species (Rode, 1962; Prot, 1980; Thomas, 1981), including the notoriously sluggish juveniles of root-knot nematodes (Johnson and McKeen, 1973; Prot, 1975, 1976, 1978b; Prot and Netscher, 1978; Prot and VanGundy, 1981; Dickson and Hewlett, 1986; Pinkerton et al., 1987), showed that nematodes frequently moved more than 15 cm and sometimes 1 m in less than 1 month. Movement by the root-knot nematode Meloidogyne javanica towards moisture in tubes of sand was reversed when the concentration of Hoagland's plant nutrient solution and other salts was highest at the wet end (Prot, 1979b). Salts affected different species some-what differently, however, and, while two root-knot nematodes were repelled by a wide range of salts, Heterodera oryzae was repelled only by sodium, and Scutellonema cavenessi was unaffected. Movement away from salts vertically would usually lead nematodes to deeper regions of higher water content. The role of salts as root-finding cues, however, was not clear, because sodium, which occurs in highest concentrations near roots, was repellent.

During the late 1980s and 1990s, temperature and CO_2 gradients were re-examined under more stringent conditions than in the 1960s. The extreme sensitivity of M. incognita (Dusenbery, 1988b; Pline et al., 1988) to small temperature changes (0.001°C) rekindled the hypothesis that metabolic heat from roots could attract nematodes (El-Sherif and Mai, 1969). Laboratory simulations of the heat waves that move through natural soil every day as a result of surface heating and cooling confirmed that heat waves can greatly alter the vertical distribution of plant nematodes within the root zone within hours, and at least two species responded by moving in opposite directions (Robinson, 1994), consistent with computer models of nematode movement (Dusenbery, 1988a,c). Nematodes were attracted to minute sources of CO_2 in the soil and behaviourally relevant release rates of the gas from a point source were shown to be achievable by roots and other biological sources (Robinson, 1995). In pot studies, ammonium was shown to repel nematodes from tomato plants (Castro et al., 1991).

Perry (1997) suggested that signals affecting nematodes in the soil are likely to be water-soluble, facilitating the establishment of concentration gradients in soil water. They are also likely to be volatile, and many substances are both. As noted by Campbell (1985), the fact that diffusivities in the liquid phase are four orders of magnitude smaller than those in the gas phase indicates that respiratory gas exchange in a soil profile without a continuous air phase is, for all practical purposes, zero, and gas transport in unsaturated soil profiles can be assumed to occur only through the gas phase. Gradients of volatile substances dissolved in the soil water surrounding nematodes, like those of respiratory gases, can also result not from diffusion through the soil water, but rather through equilibrium between the air and aqueous phase within the soil interstices, with movement of volatile molecules significant distances through the soil occurring primarily through the air phase.

The consensus today is that most responses of nematodes to plants are to general cues and lead to similar rates of root penetration in good and poor hosts. Thus, general stimuli are probably commonly used, because the host cannot easily avoid generating them. Gradients of CO_2 and salts probably play the most important roles, but many other factors affect movement. Holistic comparative studies in natural soil profiles are still badly needed to sort out the relative importance of various stimuli to different nematode species.

Some plant species repel or fail to attract nematodes. Nematode responses, however, often do not correlate well with host specificity. Roots of cucumber plants carrying the *Bi* (bitter) locus for triterpenoid curcurbitacins, for example, are reported to attract root-knot nematodes less than roots of other cucumber plants (Kaplan and Keen, 1980), but, none the less, plants with the *Bi* gene can become highly infected and form extensive galls or swellings in response to nematode feeding. The resistant grass *Aegilops variabilis* was less attractive to juveniles of *Meloidogyne naasi* than susceptible barley (*Hordeum vulgare*) and wheat (*Triticum aestivum*) (Balhadere and Evans, 1994), but susceptible and resistant barley cultivars were similarly attractive. Griffin (1969) observed that a nematode-susceptible lucerne genotype attracted two very different nematodes, *D. dipsaci* and *M. incognita*, less than a resistant genotype did, but did not show this effect for additional resistant and susceptible cultivars. Roots of two very different plants, cabbage and carrot, perhaps not surprisingly, attracted *Hirschmanniella oryzae* more than roots of a third very different plant, onion (Bilgrami *et al.*, 1985). Repellents may cause some of these differences. In a search for both volatile and non-volatile attractants of *M. incognita* to host roots, the only attractive chemical found was CO_2, although the presence of complex unidentified repellent chemicals was demonstrated (Diez and Dusenbery, 1989b; McCallum and Dusenbery, 1992).

In some cases, a high degree of host specificity seems irrefutable. Potato root exudate increases the motility of the obligate potato parasite, *G. rostochiensis*, in sand (Clarke and Hennessy, 1984) and triggers electrophysiological activity in the anterior end of the nematode (Rolfe *et al.*, 2000). Lee and Evans (1973) observed a strong positive correlation between nematode fecundity and the attractiveness of 16 genotypes of rice for *Aphelenchoides besseyi*. The foliar nematode, *D. phyllobius*, has a restricted host range, limited to foliage of certain *Solanum* spp. (Robinson *et al.*, 1978), and the nematode is strongly attracted to some unknown compound that is apparently unique to *Solanum* spp. (Robinson *et al.*, 1979). The latter case has been studied in some detail. The type host, *Solanum elaeagnifolium*, accumulates attractant in all foliar tissues throughout development, and the attractant can be collected by rinsing foliage with water (Robinson, 1992). The infective fourth-stage juveniles in soil can utilize gradients of the compound around the bases of stems during rainy weather to locate stems, which they then ascend, invading foliar buds up to 60 cm above the soil surface. Attractant activity is retained during freeze-drying and is freely soluble in water (Robinson and Saldaña, 1989), and so seems well suited for this purpose.

Movement through plant tissues

The tissue migrations, feeding sites and reproduction of animal and plant parasites have been studied in some cases for more than 100 years. The last two decades have seen increased interest in the evolution of parasitic behaviours, particularly in relation to host finding, site finding and feeding-site establishment. A recent conceptual advance has been to explain differential migrations in terms of a small number of fixed action patterns elicited by different stimuli in different nematode species or at different times during development (Sukhdeo, 1997). Examples include the characteristic sequences of activities exhibited by plant-parasitic nematodes during hatching and root invasion, the resumption of spontaneous activity by *S. carpocapsae* and *Ancylostoma caninum* when vibrated or by *Trichonema* sp. and *Agamermis catedecaudata* when illuminated, nictating by *S. carpocapsae* and *A. caninum* in response to CO_2, conversion from tortuous to straight locomotion by males of *Panagrellus redivivus* on agar when exposed to sex attractant (Samoiloff *et al.*, 1973) and accelerated movement by animal parasites in response to bile (Sukhdeo, 1997).

VonMende (1997) reviewed movement of juvenile stages of sedentary root parasites through roots, emphasizing work with the model plant, *Arabidopsis thaliana*. Wyss (1997) examined literature for feeding-site establishment in roots across a continuum of parasitic specialization among 14 illustrated types of feeding patterns or nematode-induced feeding sites. Scheres *et al.* (1997) have discussed nematode-elicited cellular and tissue modifications in plants in relation to contemporary genetic analysis of cell determination in roots.

Based on interpretations of tissue migrations by animal parasites, it seems plausible that a small group of fixed action patterns evoked in plant-parasitic nematodes at key points during plant tissue development and invasion may guide nematodes through tissue and regulate their involvement in feeding-site establishment. In 11 species of Tylenchida, contact with roots led to a predictable sequence of actions referred to as local exploration (lip rubbing and stylet probing), followed by a cell-wall cutting cycle (body immobilization and rhythmic stylet thrusting to cut a slit in the cell against which the lips are appressed) and then cell penetration (Doncaster and Seymour, 1973). Once inside the root and a cell to be fed on was encountered, a different sequence occurred. Feeding by *Aphelenchus avenae* (Fisher, 1975) in liquid culture showed that membrane penetration is not necessary for feeding, but several activities are: cessation of body movement, followed by bending of the head, stylet thrusting and protracted stylet exsertion during pumping of the oesophageal metacorpus.

A given species typically exhibits the same migration and feeding pattern in a wide range of plant hosts, and a small set of stimuli and responses probably guide these activities. It has been suggested that CO_2 mediates the acropetal migration of *Meloidogyne* spp. intercellularly through the root cortex following root entry in cotton (McClure and Robertson, 1973). After reaching the meristematic tissue just behind the growing tip, these nematodes turn basipetally to rest and feed on one or several cells, which are induced to develop into enlarged hypermetabolic nurse cells, usually called giant cells. The same migration pattern is seen in many hosts (Wyss *et al.*, 1992; VonMende, 1997). As one of many possible contrasting examples, root entry by the infective, vermiform females of the reniform nematode (*R. reniformis*) is largely intracellular, rather than intercellular through the cortex, and perpendicular, rather than parallel to the root axis. In a wide range of hosts, the infective female usually comes to rest and feeds permanently on an endodermal cell, eliciting the formation of a nurse syncytium comprised of several contiguous pericycle cells, without the nematode ever perforating the pericycle. This same feeding-site establishment behaviour, like that of root-knot nematodes, is exhibited in dozens of distantly related plant hosts (Robinson *et al.*, 1997).

Crop-protection opportunities

Only a few studies have explored interference with plant-parasitic nematode behaviour as a crop-protection tool. These studies have examined co-formulation of the sex attractant, vanillic acid, with biological control agents in pots and in the field (Meyer and Huettel, 1996; Meyer *et al.*, 1997) to manage the soybean cyst nematode *Heterodera glycines*, release of CO_2 as a bait (Bernklau and Bjostad 1998a,b) to increase the efficacy of pelletized mycophagous fungi (Robinson and Jaffee, 1996), manipulation of CO_2 and temperature gradients to increase Baermann funnel extraction efficiency (Robinson and Heald, 1989, 1991), application of lectins to block chemosensory function (Marbán-Mendoza *et al.*, 1987), shielding of plant roots with ammonium nitrate and other repellent fertilizer salts (Castro *et al.*, 1991) and application of tannic acid attractants to the soil as confusants (Hewlett *et al.*, 1997).

References

Abou-Setta, M.M. and Duncan, L.W. (1998) Attraction of *Tylenchulus semipenetrans* and *Meloidogyne javanica* to salts *in vitro*. *Nematropica* 28, 49–59.

Anderson, A.R.A., Young, I.M., Sleeman, B.D., Griffiths, B.S. and Robertson, W.M. (1997a) Nematode movement along a chemical gradient in a structurally heterogeneous environment. 1. Experiment. *Fundamental and Applied Nematology* 20, 157–163.

Anderson, A.R.A., Sleeman, B.D., Young, I.M. and Griffiths, B.S. (1997b) Nematode movement along a chemical gradient in a structurally heterogeneous environment. 2. Theory. *Fundamental and Applied Nematology* 20, 165–172.

Balan, J. and Gerber, N.N. (1972) Attraction and killing of the nematode *Panagrellus redivivus* by the predaceous fungus *Arthrobotrys dactyloides*. *Nematologica* 18, 163–173.

Balhadere, P. and Evans, A.A.F. (1994) Characterization of attractiveness of excised root tips of resistant and susceptible plants for *Meloidogyne naasi*. *Fundamental and Applied Nematology* 17, 527–536.

Bargmann, C.I. and Mori, I. (1997) Chemotaxis and thermotaxis. In: Riddle, D.L., Blumenthal, T., Meyer, B.J. and Priess, J.R. (eds) *C. elegans II*. Cold Spring Harbor Laboratory Press, Plainview, New York, pp. 717–737.

Bernklau, E.J. and Bjostad, L.B. (1998a) Behavioral responses of first-instar western corn rootworm (Coleoptera: Chrysomelidae) to carbon dioxide in a glass bead bioassay. *Ecology and Behavior* 91, 444–456.

Bernklau, E.J. and Bjostad, L.B. (1998b) Reinvestigation of host location by western corn rootworm larvae (Coleoptera: Chrysomelidae): CO_2 is the only volatile attractant. *Ecology and Behavior* 91, 1331–1340.

Bilgrami, A.L. and Jairajpuri, M.S. (1984) The responses of *Mononchus aquaticus* to chemicals and pH. *Indian Journal of Nematology* 14, 171–174.

Bilgrami, A.L., Ahmad, I. and Jairajpuri, M.S. (1985) Responses of adult *Hirschmanniella oryzae* towards different plant roots. *Revue de Nématologie* 8, 265–272.

Bird, A.F. (1959) The attractiveness of roots to the plant parasitic nematodes *Meloidogyne javanica* and *M. hapla*. *Nematologica* 4, 322–335.

Bird, A.F. (1960) Additional notes on the attractiveness of roots to plant parasitic nematodes. *Nematologica* 5, 217.

Blake, C.D. (1961) Importance of osmotic potential as a component of the soil water on the movement of nematodes. *Nature* 191, 144–145.

Burman, M. and Pye, A.E. (1980) *Neoaplectana carpocapsae*: movements of nematode populations on a thermal gradient. *Experimental Parasitology* 49, 258–265.

Burr, A.H. (1984) Photomovement behavior in simple invertebrates. In: Ali, M.A. (ed.) *Photoreception and Vision in Invertebrates*. Plenum Press, New York, pp. 179–215.

Burr, A.H.J. and Babinszki, C.P.F. (1990) Scanning motion, ocellar morphology and orientation mechanisms in the phototaxis of the nematode *Mermis nigrescens*. *Journal of Comparative Physiology A* 167, 257–268.

Campbell, G.S. (1985) *Soil Physics with Basic, Transport Models for Soil–Plant Systems*. Elsevier, New York, 150 pp.

Campbell, J.F. and Kaya, H.K. (1999) How and why a parasitic nematode jumps. *Nature* 397, 485–486.

Castro, C.E. and Thomason, I.J. (1973) Permeation dynamics and osmo-regulation in *Aphelenchus avenae*. *Nematologica* 19, 100–108.

Castro, C.E., Belser, N.O., McKinney, H.E. and Thomason, I.J. (1990) Strong repellency of the root knot nematode, *Meloidogyne incognita* by specific inorganic ions. *Journal of Chemical Ecology* 16, 1199–1205.

Castro, C.E., McKinney, H.E. and Lux, S. (1991) Plant protection with inorganic ions. *Journal of Nematology* 23, 409–413.

Clarke, A.J. and Hennessy, J. (1984) Movement of *Globodera rostochiensis* (Wollenweber) juveniles stimulated by potato-root exudate. *Nematologica* 30, 206–212.

Croll, N.A. (1967) Acclimatization in the eccritic thermal response of *Ditylenchus dipsaci*. *Nematologica* 13, 385–389.

Croll, N.A. (1970) *The Behaviour of Nematodes: Their Activity, Senses, and Responses.* St Martin's Press, New York, 117 pp.

Croll, N.A. (1971) Movement patterns and photosensitivity of *Trichonema* spp. infective larvae in non-directional light. *Parasitology* 62, 467–478.

Croll, N.A. (1975) Behavioural analysis of nematode movement. In: Dawes, B. (ed.) *Advances in Parasitology.* Academic Press, New York, pp. 71–122.

Croll, N.A. (1976) When *Caenorhabditis elegans* (Nematoda: Rhabditidae) bumps into a bead. *Canadian Journal of Zoology* 54, 566–570.

Croll, N.A. and Sukhdeo, M.V.K. (1981) Hierarchies in nematode behavior. In: Zuckerman, B.M. and Rohde, R.A. (eds) *Plant Parasitic Nematodes*, Vol. III. Academic Press, New York, pp. 227–251.

Dickson, D.W. and Hewlett, T.E. (1986) Vertical migration of *Meloidogyne arenaria* in soil columns in the field. *Revue de Nématologie* 9, 295.

Diez, J.A. and Dusenbery, D.B. (1989a) Preferred temperature of *Meloidogyne incognita*. *Journal of Nematology* 21, 99–104.

Diez, J.A. and Dusenbery, D.B. (1989b) Repellent of root-knot nematodes from exudate of host roots. *Journal of Chemical Ecology* 15, 2445–2455.

Doncaster, C.C. and Seymour, M.K. (1973) Exploration and selection of penetration site by Tylenchida. *Nematologica* 19, 137–145.

Dusenbery, D.B. (1985) Video camera–computer tracking of nematode *Caenorhabditis elegans* to record behavioral responses. *Journal of Chemical Ecology* 11, 1239–1247.

Dusenbery, D.B. (1987) Theoretical range over which bacteria and nematodes locate plant roots using carbon dioxide. *Journal of Chemical Ecology* 13, 1617–1625.

Dusenbery, D.B. (1988a) Avoided temperature leads to the surface: computer modeling of slime mold and nematode thermotaxis. *Behavioral Ecology and Sociobiology* 22, 219–223.

Dusenbery, D.B. (1988b) Behavioral responses of *Meloidogyne incognita* to small temperature changes. *Journal of Nematology* 20, 351–355.

Dusenbery, D.B. (1988c) Limits of thermal sensation. *Journal of Theoretical Biology* 131, 263–271.

Dusenbery, D.B. (1989) A simple animal can use a complex stimulus pattern to find a location: nematode thermotaxis in soil. *Biological Cybernetics* 61, 431–437.

Dusenbery, D.B. (1992) *Sensory Ecology: How Organisms Acquire and Respond to Information.* W.H. Freeman, New York, 558 pp.

El-Sherif, M. and Mai, W.F. (1969) Thermotactic response of some plant parasitic nematodes. *Journal of Nematology* 1, 43–48.

Fisher, J.M. (1975) Chemical stimuli for feeding by *Aphelenchus avenae*. *Nematologica* 21, 358–364.

Gans, C. and Burr, A.H.J. (1994) Unique locomotory mechanism of *Mermis nigrescens*, a large nematode that crawls over soil and climbs through vegetation. *Journal of Morphology* 222, 133–148.

Gaugler, R., LeBeck, L., Nakagaki, B. and Boush, G.M. (1980) Orientation of the entomogenous nematode *Neoaplectana carpocapsae* to carbon dioxide. *Environmental Entomology* 9, 649–652.

Granzer, M. and Haas, W. (1991) Host-finding and host recognition of infective *Ancylostoma caninum* larvae. *International Journal for Parasitology* 21, 429–440.

Gray, J. (1953) Undulatory propulsion. *Quarterly Journal of Microscopical Science* 94, 551–578.

Green, C.D. (1977) Simulation of nematode attraction to a point in a flat field. *Behaviour* 61, 130–146.

Grewal, P.S., Selvan, S. and Gaugler, R. (1994) Thermal adaptation of entomopathogenic nematodes: niche breadth for infection, establishment, and reproduction. *Journal of Thermal Biology* 19, 245–253.

Grewal, P.S., Lewis, D.E. and Venkatachari, S. (1999) A possible mechanism of suppression of plant-parasitic nematodes by entomopathogenic nematodes. *Nematology* 1, 735–743.

Griffin, G.D. (1969) Attractiveness of resistant and susceptible alfalfa to stem and root-knot nematodes. *Journal of Nematology* 1, 9.

Hedgecock, E.M. and Russell, R.L. (1975) Normal and mutant thermotaxis in the nematode *Caenorhabditis elegans*. *Proceedings of the National Academy of Sciences,USA* 72, 4061–4065.

Hewlett, T.E., Hewlett, E.M. and Dickson, D.W. (1997) Response of *Meloidogyne* spp., *Heterodera glycines*, and *Radopholus similis* to tannic acid. *Journal of Nematology* 29(4S), 737–741.

Johnson, P.W. and McKeen, C.D. (1973) Vertical movement and distribution of *Meloidogyne incognita* (Nematodea) under tomato in a sandy loam greenhouse soil. *Canadian Journal of Plant Science* 53, 837–841.

Johnson, R.N. and Viglierchio, D.R. (1961) The accumulation of plant parasitic nematode larvae around carbon dioxide and oxygen. *Proceedings of the Helminthological Society of Washington* 28, 171–174.

Kaplan, D.T. and Keen, N.T. (1980) Mechanisms conferring plant incompatibility to nematodes. *Revue de Nématologie* 3, 123–134.

Klingler, J. (1959) Anziehung von Collembolen und Nematoden durch Kohlendioxydquellen. *Mitteilungen der Schweizerischen Entomologischen Gesellschaft/Bulletin de la Société Entomologique Suisse* 2, 311–316.

Klingler, J. (1961) Anziehungsversuche mit *Ditylenchus dipsaci* unter Berucksichtigung der Wirkung des Kohlendioxyds, des Redoxpotentials und anderer Faktoren. *Nematologica* 6, 69–84.

Klingler, J. (1963) Die Orientierung von *Ditylenchus dipsaci* in Gemessenen kunstilichen und biologischen CO_2-Gradienten. *Nematologica* 9, 185–199.

Klingler, J. (1965) On the orientation of plant nematodes and of some other soil animals. *Nematologica* 11, 4–18.

Klingler, J. (1970) The reaction of *Aphelenchoides fragariae* to slit-like micro-openings and to stomatal diffusion gases. *Nematologica* 16, 417–422.

Klingler, J. (1972) The effect of single and combined heat and CO_2 stimuli at different ambient temperatures on the behavior of two plant-parasitic nematodes. *Journal of Nematology* 4, 95–100.

Kramer, P.J. (1983) *Water Relations of Plants.* Academic Press, Orlando, Florida, 489 pp.

Kühn, H. (1959) Zum Problem der Wirtsfindung phytopathogener Nematoden. *Nematologica* 4, 165–171.

Lee, Y.-B. and Evans, A.A.F. (1973) Correlation between attractions and suscepti-bilities of rice varieties to *Aphelenchoides besseyi* Christie, 1942. *Korean Journal of Plant Protection* 12, 147–151.

Linford, M.B. (1939) Attractiveness of roots and excised shoot tissues to certain nematodes. *Proceedings of the Helminthological Society of Washington* 6, 11–18.

McCallum, M.E. and Dusenbery, D.B. (1992) Computer tracking as a behavioral GC detector: nematode responses to vapor of host roots. *Journal of Chemical Ecology* 18, 585–592.

McClure, M.A. and Robertson, J. (1973) Infection of cotton seedlings by *Meloidogyne incognita* and a method of producing uniformly infected root segments. *Nematologica* 19, 428–434.

Marbán-Mendoza, N., Jeyaprakash, A., Jansson, H.-B., Damon, H.A. and Zuckerman, B.M. (1987) Control of root-knot nematode on tomato by lectins. *Journal of Nematology* 19, 331–335.

Marks, C.F., Thomason, I.J. and Castro, C.E. (1968) Dynamics of the permeation of nematodes by water, nematocides and other substances. *Experimental Parasitology* 22, 321–337.

Meyer, S.L.F. and Huettel, R.N. (1996) Application of a sex pheromone, pheromone analogs, and *Verticillium lecanii* for management of *Heterodera glycines*. *Journal of Nematology* 28, 36–42.

Meyer, S.L.F., Johnson, G., Dimock, M., Fahey, J.W. and Huettel, R.N. (1997) Field efficacy of *Verticillium lecanii*, sex pheromone, and pheromone analogs as potential management agents for soybean cyst nematode. *Journal of Nematology* 29, 282–288.

Milburn, J.A. (1979) *Water Flow in Plants.* Longman, London, 225 pp.

Miyazaki, M., Yamaguchi, A. and Oda, K. (1978a) Behaviour of *Bursaphelenchus lignicolus* in response to carbon dioxide. *Journal of the Japanese Forestry Society* 60, 203–208.

Miyazaki, M., Yamaguchi, A. and Oda, K. (1978b) Behaviour of *Bursaphelenchus lignicolus* in response to carbon dioxide released by respiration of *Mono-chamus alternatus* pupa. *Journal of the Japanese Forestry Society* 60, 249–254.

Mori, I. and Ohshima, Y. (1995) Neural regulation of thermotaxis in *Caenorhabditis elegans*. *Nature* 376, 344–348.

Papendick, R.I. and Campbell, G.S. (1981) Theory and measurement of water potential. In: Kral, D.M. and Cousin, M.K. (eds) *Water Potential Relations in Soil Microbiology*. Soil Science Society of America, Madison, Wisconsin, pp. 1–22.

Perry, R.N. (1996) Chemoreception in plant parasitic nematodes. *Annual Review of Phytopathology* 34, 181–199.

Perry, R.N. (1997) Plant signals in nematode hatching and attraction. In: Fenoll, C., Grundler, F.M.W. and Ohl, S.A. (eds) *Cellular and Molecular Aspects of Plant–Nematode Interactions*. Kluwer Academic Publishers, Dordrecht, pp. 38–50.

Perry, R.N. and Aumann, J. (1998) Behaviour and sensory responses. In: Perry, R.N. and Wright, D.J. (eds) *The Physiology and Biochemistry of Free-living and Plant-parasitic Nematodes.* CAB International, Wallingford, pp. 75–102.

Pervez, R. and Bilgrami, A.L. (2000) Some factors influencing chemoattraction behaviour of dorylaim predators, *Laimydorus baldus* and *Discolaimus major*, towards prey kairomones. *International Journal of Nematology* 10, 41–48.

Pinkerton, J.N., Mojtahedi, H., Santo, G.S. and O'Bannon, J.H. (1987) Vertical migration of *Meloidogyne chitwoodi* and *Meloidogyne hapla* under controlled temperature. *Journal of Nematology* 19, 152–157.

Pline, M. and Dusenbery, D.B. (1987) Responses of plant-parasitic nematode *Meloidogyne incognita* to carbon dioxide determined by video camera–computer tracking. *Journal of Chemical Ecology* 13, 873–888.

Pline, M., Diez, J.A. and Dusenbery, D.B. (1988) Extremely sensitive thermotaxis of the nematode *Meloidogyne incognita. Journal of Nematology* 20, 605–608.

Prot, J.-C. (1975) Recherches concernant le déplacement des juveniles de *Meloidogyne* spécies vers les racines. *Cahiers ORSTOM série Biologie* 3, 251–262.

Prot, J.-C. (1976) Amplitude et cinétique des migrations du nématode *Meloidogyne javanica* sous l'influence d'un plant de tomate. *Cahiers ORSTOM série Biologie* 11, 157–166.

Prot, J.-C. (1978a) Influence of concentration gradients of salts on the movement of second stage juveniles of *Meloidogyne javanica. Revue de Nématologie* 1, 21–26.

Prot, J.-C. (1978b) Vertical migration of four natural populations of *Meloidogyne. Revue de Nématologie* 1, 109–112.

Prot, J.-C. (1978c) Behaviour of juveniles of *Meloidogyne javanica* in salt gradients. *Revue de Nématologie* 1, 135–142.

Prot, J.-C. (1979a) Influence of concentration gradients of salts on the behaviour of four plant parasitic nematodes. *Revue de Nématologie* 2, 11–16.

Prot, J.-C. (1979b) Horizontal migrations of second-stage juveniles of *Meloidogyne javanica* in sand in concentration gradients of salts and in a moisture gradient. *Revue de Nématologie* 2, 17–21.

Prot, J.-C. (1980) Migration of plant-parasitic nematodes towards plant roots. *Revue de Nématologie* 3, 305–318.

Prot, J.-C. and Netscher, C. (1978) Improved detection of low population densities of *Meloidogyne. Nematologica* 24, 129–132.

Prot, J.-C. and VanGundy, S.D. (1981) Effect of soil texture and the clay component on migration of *Meloidogyne incognita* second-stage juveniles. *Journal of Nematology* 13, 213–217.

Riddle, D.L. and Bird, A.F. (1985) Responses of the plant parasitic nematodes *Rotylenchulus reniformis, Anguina agrostis* and *Meloidogyne javanica* to chemical attractants. *Parasitology* 91, 185–195.

Riemann, F. and Schragge, M. (1988) Carbon dioxide as an attractant for the free living marine nematode *Aconcholaimus thalassophygas. Marine Biology* 98, 81–85.

Robinson, A.F. (1989) Thermotactic adaptation in two foliar and two root-parasitic nematodes. *Revue de Nématologie* 12, 125–131.

Robinson, A.F. (1992) Cambios en la atracción de *Ditylenchus phyllobius* hacia tejidos de *Solanum elaeagnifolium* durante el desarrollo de la planta. *Nematropica* 22, 37–45.

Robinson, A.F. (1994) Movement of five nematode species through sand subjected to natural temperature gradient fluctuations. *Journal of Nematology* 26, 46–58.

Robinson, A.F. (1995) Optimal release rates for attracting *Meloidogyne incognita, Rotylenchulus reniformis*, and other nematodes to carbon dioxide in sand. *Journal of Nematology* 27, 42–50.

Robinson, A.F. (2000) Techniques for studying nematode movement and behavior on physical and chemical gradients. In: Ecology Committee, Society of Nematologists (eds) *Methods of Nematode Ecology*. Society of Nematologists, Hyattsville, Maryland, pp. 1–20.

Robinson, A.F. and Carter, W.W. (1986) Effects of cyanide ion and hypoxia on the volumes of second-stage juveniles of *Meloidogyne incognita* in polyethylene glycol solutions. *Journal of Nematology* 18, 563–570.

Robinson, A.F. and Heald, C.M. (1989) Accelerated movement of nematodes from soil in Baermann funnels with temperature gradients. *Journal of Nematology* 21, 370–378.

Robinson, A.F. and Heald, C.M. (1991) Carbon dioxide and temperature gradients in Baermann funnel extraction of *Rotylenchulus reniformis*. *Journal of Nematology* 23, 28–38.

Robinson, A.F. and Heald, C.M. (1993) Movement of *Rotylenchulus reniformis* through sand and agar in response to temperature, and some observations on vertical descent. *Nematologica* 39, 92–103.

Robinson, A.F. and Jaffee, B.A. (1996) Repulsion of *Meloidogyne incognita* by alginate pellets containing hyphae of *Monacrosporium cionopagum, M. ellipsosporum*, or *Hirsutella rhossiliensis*. *Journal of Nematology* 28, 133–147.

Robinson, A.F. and Saldaña, G. (1989) Characterization and partial purification of attractants for nematode *Orrina phyllobia* from foliage of *Solanum elaeagnifolium*. *Journal of Chemical Ecology* 15, 481–495.

Robinson, A.F., Orr, C.C. and Abernathy, J.R. (1978) Distribution of *Nothanguina phyllobia* and its potential as a biological control agent for silver-leaf nightshade. *Journal of Nematology* 10, 362–366.

Robinson, A.F., Orr, C.C. and Abernathy, J.R. (1979) Behavioral responses of *Nothanguina phyllobia* to selected plant species. *Journal of Nematology* 11, 73–77.

Robinson, A.F., Orr, C.C. and Heintz, C.E. (1984a) Activity and survival of *Orrina phyllobia*: preliminary investigations on the effects of solutes. *Journal of Nematology* 16, 26–30.

Robinson, A.F., Orr, C.C. and Heintz, C.E. (1984b) Effects of the ionic composition and water potential of aqueous solution on the activity and survival of *Orrina phyllobia*. *Journal of Nematology* 16, 30–37.

Robinson, A.F., Baker, G.L. and Heald, C.M. (1990) Transverse phototaxis by infective juveniles of *Agamermis* sp. and *Hexamermis* sp. *Journal of Parasitology* 76, 147–152.

Robinson, A.F., Inserra, R.N., Caswell-Chen, E.P., Vovlas, N. and Troccoli, A. (1997) *Rotylenchulus* species: identification, distribution, host ranges, and crop plant resistance. *Nematropica* 27, 127–180.

Rode, H. (1962) Untersuchungen über das Wandervermogen von Larven des Kartoffelnematoden (*Heterodera rostochiensis* Woll.) in Modellversuchen mit verschiedenen Bodenarten. *Nematologica* 7, 74–82.

Rode, H. (1969a) Über das Thermopräferendum von Larven des Kartoffelnematoden (*Heterodera rostochiensis* Woll.) unter Berücksichtigung ihrer Wanderaktivität im Reizgefälle. *Pedobiologia* 9, 405–425.

Rode, H. (1969b) Über Verhalten und Reaktionsempfindlichkeit von Larven des Kartoffelnematoden gegenüber thermischen Reizgefällen im überoptimalen Temperaturbereich. *Nematologica* 15, 510–524.

Rohde, R.A. (1960) The influence of carbon dioxide on respiration of certain plant-parasitic nematodes. *Proceedings of the Helminthological Society of Washington* 27, 160–164.

Rolfe, R.N., Barrett, J. and Perry, R.N. (2000) Analysis of chemosensory responses of second stage juveniles of *Globodera rostochiensis* using electro-physiological techniques. *Nematology* 2, 523–533.

Ronald, K. (1960) The effects of physical stimuli on the larval stage of *Terranova decipiens* (Krabbe, 1878) (Nematoda: Anisakidae). I. Temperature. *Canadian Journal of Zoology* 38, 623–642.

Samoiloff, M.R., Balakanich, S. and Pertovich, M. (1973) Evidence for the two-state model of nematode behavior. *Nature* 247, 73–74.

Sandstedt, R. and Schuster, M.L. (1962) Liquid trapping of *Meloidogyne incognita incognita* about roots in agar medium. *Phytopathology* 52, 174–175.

Sandstedt, R., Sullivan, T. and Schuster, M.L. (1961) Nematode tracks in the study of movement of *Meloidogyne incognita incognita*. *Nematologica* 6, 261–265.

Scheres, B., Sijmons, P.C., Van den Berg, C., DeVrieze, G., Willemsen, V. and Wolkenfelt, H. (1997) Root anatomy and development, the basis for nematode parasitism. In: Fenoll, C., Grundler, F.M.W. and Ohl, S.A. (eds) *Cellular and Molecular Aspects of Plant–Nematode Interactions*. Kluwer Academic Publishers, Dordrecht, pp. 25–37.

Steiner, G. (1925) The problem of host selection and host specialization of certain plant-infesting nemas and its application in the study of nemic pests. *Phytopathology* 15, 499–534.

Sukhdeo, M.V.K. (1997) Earth's third environment: the worm's eye view. *Bioscience* 47, 141–149.

Thomas, P.R. (1981) Migration of *Longidorus elongatus*, *Xiphinema diversicaudatum* and *Ditylenchus dipsaci* in soil. *Nematologia Mediterranea* 9, 75–81.

Trudgill, D.L. (1995a) An assessment of the relevance of thermal time relationships to nematology. *Fundamental and Applied Nematology* 18, 407–417.

Trudgill, D.L. (1995b) Why do tropical poikilothermic organisms tend to have higher threshold temperatures for development than temperate ones? *Functional Ecology* 9, 136–137.

Viglierchio, C.R., Croll, N.A. and Gortz, J.H. (1969) The physiological response of nematodes to osmotic stress and an osmotic treatment for separating nematodes. *Nematologica* 15, 15–21.

Viglierchio, D.R. (1990) Carbon dioxide sensing by *Panagrellus silusiae* and *Ditylenchus dipsaci*. *Revue de Nématologie* 13, 425–432.

VonMende, N. (1997) Invasion and migration behaviour of sedentary nematodes. In: Fenoll, C., Grundler, F.M.W. and Ohl, S.A. (eds) *Cellular and Molecular Aspects of Plant–Nematode Interactions*. Kluwer Academic Publishers, Dordrecht, pp. 51–64.

Wallace, H.R. (1956) Migration of nematodes. *Nature* 177, 287–288.

Wallace, H.R. (1958a) Movement of eelworms I. The influence of pore size and moisture content of the soil on the migration of larvae of the beet eelworm, *Heterodera schachtii* Schmidt. *Annals of Applied Biology* 46, 74–85.

Wallace, H.R. (1958b) Movement of eelworms II. A comparative study of the movement in soil of *Heterodera schachtii* Schmidt and of *Ditylenchus dipsaci* (Kuhn) Filipjev. *Annals of Applied Biology* 46, 86–94.

Wallace, H.R. (1958c) Movement of eelworms III. The relationship between eelworm length, activity and mobility. *Annals of Applied Biology* 46, 662–668.

Wallace, H.R. (1959a) Movement of eelworms. IV. The influence of water percolation. *Annals of Applied Biology* 47, 131–139.

Wallace, H.R. (1959b) Movement of eelworms. V. Observations of *Aphelenchoides ritzemabosi* (Schwartz, 1912) Steiner, 1932 on florists' chrysanthemums. *Annals of Applied Biology* 47, 350–360.

Wallace, H.R. (1959c) The movement of eelworms in water films. *Annals of Applied Biology* 47, 366–370.

Wallace, H.R. (1960) Movement of eelworms. VI. The influences of soil type, moisture gradients and host plant roots on the migration of the potato-root eelworm *Heterodera rostochiensis* Wollenweber. *Annals of Applied Biology* 48, 107–120.

Wallace, H.R. (1961) The orientation of *Ditylenchus dipsaci* to physical stimuli. *Nematologica* 6, 222–236.

Wallace, H.R. (1968a) The dynamics of nematode movement. *Annual Review of Phytopathology* 6, 91–114.

Wallace, H.R. (1968b) Undulatory locomotion of the plant parasitic nematode, *Meloidogyne javanica*. *Parasitology* 58, 377–391.

Wallace, H.R. (1969) Wave formation by infective larvae of the plant parasitic nematode *Meloidogyne javanica*. *Nematologica* 15, 65–75.

Wicks, S.R. and Rankin, C.H. (1997) Effects of tap withdrawal response habituation on other withdrawal behaviors: the localization of habituation in the nematode *Caenorhabditis elegans*. *Behavioral Neuroscience* 111, 342–353.

Widdowson, E., Doncaster, C.C. and Fenwick, D.W. (1958) Observations on the development of *Heterodera rostochiensis* in sterile root cultures. *Nematologica* 3, 308–314.

Wieser, W. (1955) The attractiveness of plants to larvae of root-knot nematodes. I. The effect of tomato seedlings and excised roots on *Meloidogyne hapla* Chitwood. *Proceedings of the Helminthological Society* 22, 106–112.

Wieser, W. (1956) The attractiveness of plants to larvae of root-knot nematodes. II. The effect of excised bean, eggplant, and soybean roots on *Meloidogyne hapla* Chitwood. *Proceedings of the Helminthological Society* 23, 59–64.

Wright, D.J. (1998) Respiratory physiology, nitrogen excretion and osmotic and ionic regulation. In: Perry, R.N. and Wright, D.J. (eds) *The Physiology and Biochemistry of Free-living and Plant-parasitic Nematodes*. CAB International, Wallingford, UK, pp. 103–131.

Wright, D.J. and Newall, D.R. (1976) Nitrogen excretion, osmotic and ionic regulation in nematodes. In: Croll, N.A. (ed.) *The Organization of Nematodes*. Academic Press, London, pp. 163–210.

Wyss, U. (1970) Zur Toleranz wandernder Wurzelnematoden gegenüber zunehmender Austrocknung des Bodens und hohen osmotischen Drücken. *Nematologica* 16, 63–73.

Wyss, U. (1997) Root parasitic nematodes: an overview. In: Fenoll, C., Grundler, F.M.W. and Ohl, S.A. (eds) *Cellular and Molecular Aspects of Plant–Nematode Interactions.* Kluwer Academic Publishers, Dordrecht, pp. 5–22.

Wyss, U., Grundler, F.M.W. and Munch, A. (1992) The parasitic behaviour of second-stage juveniles of *Meloidogyne incognita* in roots of *Arabidopsis thaliana. Nematologica* 38, 98–111.

Environmental Control of 6
Nematode Life Cycles

M.E. Viney

School of Biological Sciences, University of Bristol,
Woodland Road, Bristol BS8 1UG, UK

Parasitic nematodes have complex life cycles. Many aspects of these life cycles, including their development, are affected and controlled by environmental conditions within and without the host. The genus *Strongyloides* has a particularly complex life cycle, which is unique among nematode parasites of vertebrates. Many factors act and interact to control this life cycle in a remarkably complex and sophisticated way.

The Life Cycle of *Strongyloides* spp.

The parasitic stage of the *Strongyloides* life cycle consists of parasitic female worms only, which lie embedded in the mucosa of the host's small intestine. In *Strongyloides ratti,* these female parasites reproduce by a mitotic parthenogenesis, such that the progeny of any one female worm are genetically identical to each other and to their mother (Viney, 1994). The female parasite lays eggs and these or newly hatched first-stage larvae (L1s) pass out of the host with the faeces, where the extensive free-living phase of the life cycle occurs (Fig. 6.1).

In this free-living cycle, there are two types of development, direct and indirect. In direct (also known as homogonic) development L1s moult via an L2 stage into infective third-stage larvae (iL3s). These iL3s are able to infect new hosts by skin penetration, after which they migrate through the host body, developing via an L4 stage into adult females. *S. ratti* larvae have been shown to migrate through the nasal–frontal part of the head *en route* to the gut (Tindall and Wilson, 1988).

In indirect (also known as heterogonic) development, L1 progeny of parasitic females moult via three larval stages (L2, L3, L4) into free-living adult male and female worms. These worms mate and reproduce by conventional sexual reproduction (Viney *et al.*, 1993). The free-living

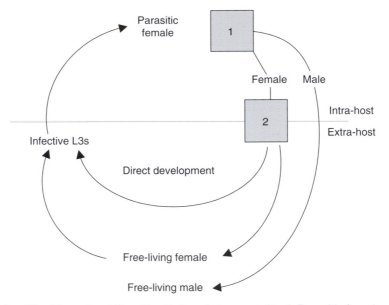

Fig. 6.1. The life cycle of *S. ratti*, with larval stages omitted. Parasitic females lay eggs and these and/or first-stage larvae (L1s) pass out of the host. Sex is determined by this stage (Box 1). Female L1s can moult though one larval stage (L2) into infective L3s. Alternatively, L1s moult through three larval stages (L2, L3, L4) into free-living females. The developmental choice of female larvae is controlled by environmental conditions (Box 2). Male larvae moult through four larval stages into free-living males only. Free-living adults reproduce by sexual reproduction, the female lays eggs and larvae moult into infective L3s. Infective L3s infect hosts by skin penetration, and migrate to the intestine, moulting through a fourth larval stage into parasitic females. (From Harvey and Viney, 2001.)

female worms lay eggs that hatch into L1 progeny, which moult via an L2 stage into iL3s, as in direct development. In *S. ratti* and *Strongyloides stercoralis*, only one round of indirect development can occur; however, up to nine have been observed with *Strongyloides planiceps*, a parasite of the cat (Yamada *et al.*, 1991).

Therefore, in the *Strongyloides* life cycle, young larvae appear to have a number of developmental potentials. The control of the development in this life cycle has been investigated in various ways for a number of species of *Strongyloides* (Schad, 1989). This background and more recent work has now developed a good understanding of the control of the *Strongyloides* life cycle, which is most thoroughly and best understood for the parasite of rats, *S. ratti*. This has shown that there are two separate levels of control in this life cycle: a male/female sex-determination event and a female-only developmental choice (Fig. 6.1).

Sex Determination

Cytological studies of a number of species of *Strongyloides* suggested that sex may be determined chromosomally (Nigon and Roman, 1952). This has been confirmed molecularly in *S. ratti*, where it has been shown that there is an XX/XO, female/male mechanism of sex determination (Harvey and Viney, 2001; Box 1 in Fig. 6.1). This mechanism of sex determination is fairly common among nematodes and is used by the model free-living nematode *Caenorhabditis elegans*. Extensive genetic analysis in *C. elegans* has shown that sex is actually determined by the ratio of the number of X chromosomes to the number of autosomes, rather than by the number of X chromosomes *per se* (Hodgkin, 1988), but whether this in the case in *S. ratti* is not known.

In the *S. ratti* life cycle, XX parasitic females produce genetically male (XO) and female (XX) larvae. The cytological details of how this occurs are not known, though it may be similar to a system in aphids where a modified form of mitotic parthenogenesis occurs, in which the X chromosome undergoes a specialized behaviour (Blackman, 1980). *S. ratti* male larvae are only able to develop into free-living adult males. However, female larvae have a further developmental choice (below) and can develop directly into iL3s or indirectly into free-living adult females. When the free-living adults mate, all the progeny inherit the one paternal X chromosome, thereby ensuring that the iL3 progeny of this mating have a female, XX, genotype. For this to occur, it requires either: (i) that all the male's sperm carry an X chromosome; or (ii) that any zygotes which do not have an XX genotype do not survive (Harvey and Viney, 2001). Which of these possibilities is correct remains to be determined. However, it would seem more efficient, and hence more likely, for spermatogenesis to ensure that all sperm carry an X chromosome, and, indeed, this occurs in aphids (Blackman, 1974).

Female Development

Chromosomally female L1 progeny of parasitic females have a developmental choice between direct development into iL3s and indirect development into free-living adult females (Box 2 in Fig. 6.1). The molecular and genetic basis of these different developmental fates is not known. However, it is likely to be similar to that which controls a similar developmental choice in the *C. elegans* life cycle (Riddle and Albert, 1997; see below).

The Effect of the Host Immune Response on Development in the *S. ratti* Life Cycle

Analysis of the proportion of larvae that develop by these two developmental routes (Boxes 1 and 2 in Fig. 6.1) shows that as an *S. ratti* infection progresses two changes occur. First, the proportion of larvae that develop into free-living males increases, i.e. the sex ratio of the progeny of a parasitic female becomes increasingly male-biased (Harvey *et al.*, 2000). Secondly, the proportion of female larvae produced by an infection that develop by the indirect route of development into free-living females increases (Harvey *et al.*, 2000). As an infection progresses two things are happening: parasitic female worms are becoming older and the host is developing an anti-*S. ratti* immune response. This immune response will eventually completely eliminate the infection, after which rats remain strongly immune to reinfection (Kimura *et al.*, 1999). To determine which (if any) of these factors alters the sex ratio and/or the development of female larvae, development via these routes was measured in infections of hosts of different immune status. The sex ratio of larvae passed from rats previously immunized with *S. ratti* were more male-biased compared with infections in naïve, control rats (Harvey *et al.*, 2000). Furthermore, the sex ratio of larvae passed from immune-compromised rats (which cannot mount an anti-*S. ratti* immune response) did not change as the infection progressed. Combined, these observations show that the host immune response, and not the age of the parasitic female, affects the sex ratio of the progeny of the parasitic females, with a change in the ratio towards males occurring in hosts mounting an anti-*S. ratti* immune response (Harvey *et al.*, 2000). By the same analysis, it was found that the host immune response also affects the development of female larvae, with development by the indirect route into free-living females favoured by larvae passed from rats with an anti-*S. ratti* immune response (Harvey *et al.*, 2000). The host immune response therefore has a coordinated effect on the development of the free-living phase of the *S. ratti* life cycle, namely, that development into both free-living adult males and females is favoured by larvae passed from hosts mounting an immune response.

It is not clear how the host immune response brings about this effect on *S. ratti* sex determination. It is almost certain that it occurs before eggs are laid, since newly laid eggs contain fully developed L1s and it is difficult to envisage that sex can be changed at this late stage of development. This suggests that the negative effects which the host immune response has on the fitness of parasitic females (see below; Kimura *et al.*, 1999) also affects the behaviour of the X chromosomes during the reproduction of the parasitic female. In *C. elegans*, the sex ratio is usually very heavily biased towards hermaphrodites (XX), with males (XO) comprising less than 0.5% of the population (Hodgkin *et al.*, 1979). These XO males are produced as a result of rare, spontaneous non-disjunction of the X chromosomes during meiosis (Hodgkin *et al.*, 1979). The rate of non-disjunction and hence the proportion of the progeny which are males increase with the age of the parent hermaphrodite (Goldstein and Curtis, 1987). The sex ratio can also be changed towards the production of males by environmental stresses, such as heat shock (Sulston and Hodgkin, 1988). However, precisely how and why the temperature experienced by a hermaphrodite or its age affects the cytology of the X chromosomes during meiosis are not known. Importantly, though, the observations in *C. elegans* and *S. ratti* are analogous in that the sex ratio of progeny of worms can facultatively change in response to apparent stresses on parent worms.

Other Environmental Effects on Development in the *S. ratti* Life Cycle

Many observations have been made on the effect of environmental conditions external to the host on the development of the free-living phase of the *S. ratti* life cycle. Generally, this has involved placing freshly passed faeces containing young larvae into different environmental conditions and assessing the effect of these treatments on the proportion of larvae that develop by the different developmental routes. Examples of such treatments include the incubation of faecal cultures in different gaseous environments (Taylor and Weinstein, 1990) and experiments in which various concentrations of a range of fatty acids were added to faecal cultures (Minematsu *et al.*, 1989). However, one of the most thoroughly investigated environmental manipulations is the environmental temperature external to the host (Viney, 1996). This has shown that temperature only affects the female developmental choice, with direct development to iL3s favoured at lower temperatures and indirect development into free-living adult females favoured at higher temperatures (Viney, 1996; Harvey *et al.*, 2000). Temperature pulse-chase experiments have shown that these larvae retain a developmental response to environmental temperature until the end of the L2 stage (Viney, 1996). Temperature does not affect the sex ratio of the progeny of a parasitic female.

The Interaction of the Host Immune Response and Temperature

The host immune response and the environmental temperature interact in their effect on the development of female larvae of *S. ratti*. The temperature sensitivity of the developmental choice of female larvae can be expressed as the difference in the proportion of female larvae that develop into free-living females between two temperatures (19 and 25°C). The greater this difference in proportion, the greater the developmental sensitivity of larvae to environmental temperature. This difference in proportion increases as an infection progresses, i.e. during an infection the developmental choice of female larvae becomes increasingly sensitive to environmental temperature. This change in developmental sensitivity to temperature is affected by the host anti-*S. ratti* immune response. Thus, the difference in the proportion of female larvae that develop into free-living females: (i) is greater in infections in rats previously immunized with *S. ratti*, compared with infections in naïve, control rats; and (ii) does not change as the infection progresses in infections in immuno-compromised rats. There is no such effect on the male/female sex determination switch, since this is not temperature-sensitive (see above).

The integration of these signals in controlling development is remarkable, since it integrates environmental signals that probably operate at different times in the life of developing larvae. Thus, the host immune response can directly affect the parasitic female and her eggs and L1s in the host intestine. Larvae are exposed to the environmental temperature only from when they are passed from a host. This would suggest that, for the host immune response and environmental temperature to interact, larvae developing outside the host must have a 'memory' of the immune status of the host from which they were passed.

There are several ways in which we may speculate how this happens. Since the parasitic female is also subject to the host immune response, when producing her eggs she may, in some way, preprogramme their future development in respect of the state of the host immune response. Alternatively, the host immune response may affect the physiological conditions of the gut and hence faeces, such that larvae developing in faeces are able to determine the immune status of the host from which they came. This latter possibility seems unlikely to be correct, since experiments in which larvae passed from newly infected (and thus not mounting an anti-*S. ratti* immune response) hosts which were transferred to faeces collected from rats immune to *S. ratti* did not alter their development (M. Crook and M.E. Viney, unpublished observations).

Transgenerational 'memory' of environmental conditions has been investigated in the water-flea, *Daphnia pulex*. Its growth, development and mode of reproduction of an individual are affected by the food available in its environment. However, a mismatch (usually down) in the quantity of food available in the environment of the flea compared with that of its mother also results in a change in the development and, importantly, reproduction of the progeny (LaMontagne and McCauley, 2001). The demonstration of this phenomenon elegantly shows how a 'memory' of conditions affecting one generation can affect the life of a subsequent generation. It remains to be seen whether such mechanisms operate in the generational 'memory' of *Strongyloides* and other nematodes.

Neuronal Control of *Strongyloides* Development

In addition to these analyses of factors that affect development in the *Strongyloides* life cycle, there has been investigation of its neuronal basis. All nematodes have bilateral, anterior sensory-rich structures, known as amphids, which are principally involved in chemo- and thermosensation. The ultrastructure of the amphidial neurons of *S. stercoralis* have recently been reconstructed from serial electron-microscope sections (Ashton and Schad, 1996). Their role in larval development has been investigated by ablating neurons using a laser microbeam and assessing the effect of this on worm development. This has shown that in *S. stercoralis* the developmental decision of female L1s to develop into free-living females requires the presence of two (known as ASF and ASI) amphidial neurons (Ashton *et al.*, 1998). Thus, larvae in which these pairs of neurons were killed develop directly into infective L3s, whereas untreated larvae or those in which only one pair of neurons were ablated develop into free-living females (Ashton *et al.*, 1998). This also suggests that the default development of *S. stercoralis* female L1s is direct development into iL3s, and that this can be altered by environmental signals so that larvae develop into free-living females. It is probable that young larvae of *S. ratti* sense their environmental conditions, including temperature, via similar amphidial mechanisms. Temperature sensation in *C. elegans* is known to occur via amphidial neurons (Bargmann and Mori, 1997). Parasitic females of *Strongyloides* spp. also have amphids and it is tempting to speculate that these are used for the sensation of the host immune response, which is used to affect the developmental potential of the progeny of a parasitic female.

This control of *S. stercoralis* development is analogous to the control of a developmental switch in the life cycle of many free-living nematodes, including *C. elegans. C. elegans* larvae have a developmental choice: when there is a limited supply of food and a high conspecific population density, larvae develop into long-lived, arrested third-stage dauer larvae; when there is a plentiful supply of food and a low conspecific population density, 'normal', non-dauer development continues (Riddle and Albert, 1997). The environmental signals (food availability and conspecific population density) are sensed by the amphids and this signal is transduced via a number of signalling pathways, which unite to bring about the physical and physiological development into dauer larvae, rather than into 'normal', non-dauer L3s. Experimental ablation of amphidial neurons has identified those which participate in this signalling in *C. elegans* (Riddle and Albert, 1997). The hypothesized analogy of the control of the developmental switches of these two species (Riddle and Albert, 1997; Viney, 1999) is supported by the demonstration that apparently analogous amphidial neurons of *S. stercoralis* and *C. elegans* control the respective developmental switches. However, it is noteworthy that in *C. elegans* both hermaphrodites and males can develop into dauer larvae. Thus, the environmental control of development is not restricted to one sex, as it is in *S. ratti*. This suggests that, if during evolution there has been conservation of the molecular basis of the *C. elegans* and *S. ratti* developmental switches, then the point in the life cycle at which this can operate has changed during the evolution of these species.

Variation in *S. ratti* Development

The development of different isolates of *S. ratti* varies. *S. ratti* isofemale lines (a line of parasites derived from a single iL3) of isolates from different geographical locations vary in the proportion of their larvae that develop by the indirect and direct routes of development (Viney *et al.*, 1992). Thus, some lines develop almost exclusively by the direct route of development during an infection, with virtually no larvae developing into free-living adults. Other lines show mixed direct and indirect development throughout the infection. The development of isofemale lines can be altered by selection. Thus, lines of parasites have been selected for direct and indirect development (from a line that underwent mixed development) by the repeated passage of iL3s that developed by these routes, respectively. There is an effective response to this selection, which can be very rapid (Viney, 1996). It is likely that the selection has acted on both levels of control (sex determination and the developmental choice of female larvae) in the life cycle, but the extent to which this occurred is not known. However, from our understanding of the control of the life cycle, selection for lines that develop by direct development must have at least occurred by selection for a heavily female-biased sex ratio.

In view of the neuronal control of the developmental switches of *Strongyloides*, there are other possibilities for the basis of variation between isofemale lines in their development. Thus, lines may vary in their sensitivity to environmental conditions *per se* and/or in the transduction of the information in the environmental signal to the developmental response. The observed response of lines of *S. ratti* to artificial selection for developmental propensity may therefore also be due to selection for different sensitivities to environmental conditions and/or variation in the transduction of that sensory information.

Different wild isolates of *C. elegans* also vary in their developmental response to environmental conditions. In laboratory assays in which eggs are placed in conditions likely to induce dauer larvae (limited food and dauer pheromone (Golden and Riddle, 1984)), the proportion of larvae that develop into dauer larvae, rather than into non-dauer L3s and L4s, varies between different isolates (M.E. Viney, J.A. Jackson and M.P. Gardner, unpublished observations). The developmental and/or sensory basis of this is not known, but is ripe for investigation and is more easily tractable in *C. elegans* compared with many other species, including *S. ratti*.

Why is *Strongyloides* Development Affected by Environmental Conditions?

Overall, these findings show a remarkable sophistication in the control of development in the *Strongyloides* life cycle. The variation in the development of isofemale lines shows that different developmental strategies are appropriate in different geographical settings. It is relatively easy to envisage that the developmental response of female larvae to temperature is an adaptation in which one can hypothesize that it is used as a seasonal cue, etc., which in turn may predict, for example, the probability of survival and reproduction as a free-living female or the probability of an iL3 encountering a new host.

A principal difference between indirect and direct development is that indirect development includes sexual reproduction, alternating with the asexual reproduction of the parasitic females, making the development of this life cycle effectively cyclically parthenogenetic. In direct development, only asexual, parthenogenetic reproduction occurs. In other words, sexual reproduction is facultative in the *Strongyloides* life cycle, and its occurrence varies in response to conditions within and outside the host. The selective advantages of sexual reproduction in nature are a continuing area of debate and investigation (Barton and Charlesworth, 1998). However, its widespread occurrence has been taken to show its general selective advantage. This advantage, whatever it is, presumably applies equally to *Strongyloides* spp. It is notable that many organisms that have facultative sexual reproduction generally indulge in it at times of stress. For example, aphids reproduce parthenogenetically during the summer, which results in rapid population increases. However, when environmental conditions deteriorate and food supplies become limiting, sexual reproduction occurs, the resulting progeny of which are resistant, overwintering stages (Hughes, 1989). Similarly, temperate populations of *Daphnia* spp. alternate between sexual and parthenogenetic reproduction, with sexual reproduction favoured at times of environmental stress (Herbert, 1987). With this perspective, the favouring of the development of free-living adults of *S. ratti* and the consequent sexual reproduction is consistent with the general trend of sexual reproduction occurring at times of environmental stress. It is notable with *S. ratti* that the environmental stress of the host immune response acts against the stages within the host, whereas the forms that undergo sexual reproduction are in the external environment. This may suggest that, in this respect, the host immune response is being used as a predictor of the state of a host environment in which future iL3s may find themselves.

In *C. elegans* all reproduction is sexual, the only variation being inbreeding during hermaphrodite sex and potential outcrossing during male/hermaphrodite sex. Therefore *C. elegans* has no sexual response to environmental stress. Instead, its developmental response to stress is to form dauer larvae. These are long-lived, environmentally resistant stages which, biologically, are presumably the dispersive stage of the life cycle. Thus, here, the strategic response to stress is to wait until the unfavourable conditions have passed.

Effects of the Host Immune Response on Nematode Infections

Strongyloides infections

The host immune response has significant effects on the development of the *S. ratti* life cycle. In addition, it has other, apparently direct, effects on the parasitic females of *Strongyloides* spp., and similar effects are seen in a range of other parasitic nematodes. For *S. ratti*, these effects can be seen in changes that occur at different times through an infection, which correlate with the development of an anti-*S. ratti* immune response. Thus, as an infection progresses, worms become shorter and their fecundity (number of eggs *in utero*) is reduced (Moqbel and McLaren, 1980); nematode fecundity is closely related to adult female body size (Skorping *et al.*, 1991). Concurrently, the worms move posteriorly from the small intestine to the caecum and colon (Kimura *et al.*, 1999). Ultrastructural observations of these worms show degenerate changes in worm tissues. It is not known whether this damage is a direct effect of the host immune response (e.g. the action of host molecules or effector cells) or an indirect result of other, more general (e.g. metabolic stress), effects of the immune response. As worms move posteriorly in the gut, they appear to be functionally starved, due to the presence of oral plugs, presumably of host origin (Moqbel and McLaren, 1980). Thus, here, it is hypothesized that the reduced ability of worms to feed and the consequent reduction in available energy result in the observed damage and that this is the basis of the reduction in fecundity (Moqbel and McLaren, 1980). That these effects are due to immune responses is demonstrated by the observations that: (i) the transfer of worms from hosts mounting an immune response to naïve hosts results in an increase in fecundity (Schad *et al.*, 1997); and (ii) immune suppression of hosts mounting an immune response results in an increase in parasite fecundity (Grove *et al.*, 1983; Schad *et al.*, 1984). Thus, overall, broad aspects of the fitness of nematode infections are affected by the host immune response. These are usually considered as manifestations of stresses imposed on the parasites by the host immune response. However, which particular nematode organ systems, metabolic pathways or processes are stressed or how these effects are brought about is not known. Similarly, as discussed above, how the effect of the host immune response is transduced to affect the sex-determination mechanism and the female-only developmental choice of *S. ratti* is not known.

Intuitively, it is envisaged that the host immune response affects parasitic worms in general, either directly by physically damaging them or indirectly by interfering with their feeding and nutrient supply or perhaps by otherwise making their environment generally less hospitable. These effects are, in turn, manifest as the reduced fitness of infections (reduced size and fecundity of worms) and, for *S. ratti*, the change in the sex ratio and in the developmental choice of female larvae. However, for *S. ratti*, it is difficult to see how 'damage' to a parasitic female brings about a change in the sex ratio of her progeny, since this would imply that suboptimal or defective operation of this process is the basis for the control of the developmental response of this life cycle to environmental conditions. This seems unlikely, given the observed variation in developmental potentials of isofemale lines (see above), which is the product of natural selection, because it is unlikely that natural selection has operated on a 'mechanism' controlling development where that 'mechanism' is the suboptimal or defective operation of a cellular process. Rather, it is more likely that the change in development of the *S. ratti* life cycle is an accurate and controlled response to environmental conditions, in this case the anti-*S. ratti* host immune response. In this respect, it is worth remembering that the life-cycle changes of aphids that occur at times of environmental stress are not induced or measured by environmental 'stress' *per se*. Rather, since 'stress' occurs typically and repeatedly in autumn and this season correlates with changes in day length, aphids use day length as the proximate measure of autumn, and hence environmental stress, in their environment (Hughes, 1989).

Thus, by analogy, what cue may *Strongyloides* use to assess the host immune response? One option is that parasitic females use the concentration of particular components of the host immune response as a measure of an immune response acting against them. However, for this to work, the cue would have to be a sufficiently reliable cue of an immune response that has an anti-*S. ratti* effect, rather than any immune response with a non- or lesser anti-*S. ratti* effect. Experimental observations suggest that this is unlikely to happen, since the effect of the host immune response on the development of *S. ratti* larvae appears to depend on the specificity of that response. Thus, challenge infections of *S. ratti* in rats previously immunized with nematodes other than *Strongyloides* or with non-nematode antigens show less change in their developmental route, compared with infections in rats previously immunized with *S. ratti* (West *et al.*, 2001). Indeed, the greater the effect of the immunization on the reduction in worm output (a measure of parasite fitness and thus of the effectiveness of the anti-*S. ratti* immune response) of the challenge infection, the greater the proportion of larvae that developed into sexual adults (West *et al.*, 2001). Therefore, while it is possible that *S. ratti* measures the host immune response directly, it is difficult to envisage what components of the immune response are only produced during an anti-*S. ratti* immune response and not during an immune response against any other infectious organisms or antigen sources.

There are alternatives. Parasitic females may measure other aspects of the host physiology that are an accurate indicator of an anti-*S. ratti* response that is occurring. However, it is again difficult to see what components would be specific to an anti-*S. ratti* immune response. A third alternative is that the host immune response may bring about a change in the internal physiology of the parasitic worms, and this change may be sensed internally and used as a cue. This would seem to have the advantage that this would be an accurate and specific measure of the effectiveness of an anti-*S. ratti* immune response.

In considering how *S. ratti* senses its immune environment, it is worth noting the details of the environmental control of the *C. elegans* life cycle. Here the arrested third-stage dauer larvae are formed at times of environmental 'stress'. *C. elegans* integrates three specific cues in its decision between dauer and non-dauer development. These are the concentration of available food (measured as the concentration of neutral carbohydrate-like substances produced by *Escherichia coli*), the concentration of conspecific worms (measured as the concentration of a pheromone produced constitutively by worms) and temperature (Riddle and Albert, 1997). Intuitively, the integration of food availability and conspecific density is a good measure of the quality of a worm's environment. However, this also shows that *C. elegans* determines environmental stress from the measurement of specific external factors. Furthermore, the use of these specific cues may be a better and more reliable indicator of the environment that the progeny of, in this case, *C. elegans* hermaphrodites will have to survive in than other measures of 'stress'. Thus, for example, internal measurement of quality or stress (such as hermaphrodite food stores) may be related to environmental conditions, though these conditions: (i) are also subject to other factors, independent of environmental conditions *per se*; and (ii) are likely to reflect past (though they may possibly reflect recent) conditions. This would therefore argue by analogy that *S. ratti* may not use its internal physiology as a measure of the anti-*S. ratti* immune response.

Other nematodes

Other nematodes have specific responses to the host immune response. *Nippostrongylus brasiliensis* reacts to immune pressure by secreting large quantities of acetylcholinesterase (Selkirk *et al.*, 2001). The reasons for this are not clear, though it has been hypothesized to play a role in enhancing nematode persistence by various means, including regulation of lymphoid cell functions and modification of wider host physiological processes that affect parasite persistence. This behaviour seems to be regulated by the host immune response, rather than by presence within a host, since the transfer of worms to immunologically naïve animals results in reduced acetylcholinesterase production (Selkirk *et al.*, 2001). However, by analogy with the argument above, it is also conceivable that some other aspect of host physiology that correlates with anti-*N. brasiliensis* immunity or some internal measure of parasite physiology that is an accurate measure of the detrimental effect of the immune response could also be used as cues for acetylcholinesterase production.

Many nematode species modulate their host's immune response as a means of enhancing their own survival (Maizels *et al.*, 1993). This can be seen, for example, in *Heligmosomoides polygyrus* infections. Immunization of mice with dead *H. polygyrus* larvae generates a high level of resistance to challenge infections. However, mice inoculated with both live and dead larvae do not exhibit this resistance to subsequent challenge. Thus, here, the live *H. polygyrus* immunomodulates the host immune response and thus suppresses an immune response directed against a challenge infection (Behnke *et al.*, 1983). Host immuno-modulation has also been described for species of filarial nematodes (Maizels *et al.*, 1993). It is probable that species which immunomodulate their hosts do so in response to their presence in a host, rather than in response to an immune response acting against them. Induction of immunomodulation automatically on entry into a host would seem to be a reliable mechanism for ensuring survival, on the assumption that an infection will induce an antiparasite immune response. However, if immunomodulation is energetically expensive and/or the host population varies widely in the 'strength' of antiparasite immune responses that may occur, it would seem more efficient for a parasite only to immuno-modulate their hosts' immune response specifically in response to a specific immune response acting against them.

Arrested development

Other nematodes have facultative developmental decisions in their life cycles, which are affected by various environmental conditions. Thus, for some species of nematodes (e.g. *Ostertagia, Trichostrongylus*) infective larvae can arrest their development after having entered a host. This phenomenon of arrested development has been most thoroughly investigated in nematode parasites of cattle. In these instances, L3s are ingested by hosts and the larvae moult to an L4 stage, at which point they can arrest their development for several months (Gibbs, 1986). This may be an adaptation to ensure that subsequent fecundity is most likely to coincide with newly available hosts, later in the hosts' reproductive year.

There has been much debate about the factors that induce arrest. It appears that host immune effects can induce arrest, at least for some species of nematodes (Michel *et al.*, 1979; Gibbs, 1986). However, the growth and arrested development of *Haemonchus contortus* appears to be affected by host factors other than the host immune response (Coadwell and Ward, 1977). This was concluded from observations of infections in worm-free, and thus non-immune, sheep in which the worm growth and arrested development varied throughout the year. This host effect was hypothesized to be due to factors present within the host which vary through the year and which are sensed by developing worms, and that this is the basis on which worms make the developmental decision between arrested and continued development (Coadwell and Ward, 1977). The presence of adult worms of *Ostertagia ostertagi* in a host appears to favour the arrest of developing larvae. This was shown from experiments in which the removal of adult *O. ostertagi* from calves by drug treatment stimulated the development of extant arrested larvae into adults (Michel, 1971). The environmental temperature experienced by the L3 in the external environment also affects the probability of larvae arresting their development within a host, with lower temperatures favouring arrested development within the host (Gibbs, 1986). However, this effect may be limited to parasite strains obtained from temperate regions (Gibbs, 1986). The rabbit stomach worm, *Obeliscoides cuniculi*, was selected for propensity to arrest development within a host following cold treatment of larvae prior to infection. This selection regime was effective but appeared to require a continued selection pressure to maintain high levels of arrested development (Watkins and Fernando, 1984).

Examination of propensity for arrested development in isolates from different locations revealed further controls of arrested development. A comparison of different isolates of *O. ostertagi* for their arrest phenotype compared isolates from cattle managed for beef production and cattle managed for dairy production at the same location. This found that greater arrest occurred in the beef compared with the dairy cattle (Smeal and Donald, 1982). This was thought to be due to the different timing and persistence of beef and dairy cattle on pasture, with the effect that there was a selective advantage for arrested development on beef pasture, but no selective advantage for arrested development on dairy pasture (Smeal and Donald, 1982). Comparison of isolates of *O. ostertagi* from the northern (Ohio) and southern (Louisiana) USA showed that the parasites varied in the timing and extent of their arrested development when tested in either location. This indicated that these differences in arrested development were genetic (Frank *et al.*, 1986). In these two cases the variation is likely to be the result of differing local selection pressures for arrested-development phenotypes.

Arrested development has also been observed with human hookworm infections in West Bengal, India (Schad *et al.*, 1973). In this epidemiological setting, infections acquired by a host in one rainy season delay development until the following rainy season. This ensures that infective stages are liberated into the environment at the onset of the monsoon season, which gives optimal conditions for the development and survival of infective larvae before the end of that monsoon season (Hominick *et al.*, 1987). The cues controlling this are not known, but are likely to be similar to those operating in non-human species.

The overall picture that therefore emerges is that species which undergo facultative arrested development use a number of cues from their environment both within and outside their hosts to make this developmental decision. The nature of the cues probably varies among different species of nematode and much work remains to be done on specifying them and defining their effects. It is clear that the host immune response is an important cue, but, as for *S. ratti*, the detailed nature of the cues or their sensation is not clear. Furthermore, it is necessary to understand how different cues interact in their effects. For example, and analogously to *S. ratti*, how do developing larvae use both the host immune response acting against larvae within the host and the temperature experienced by

L3s in the environment to make the developmental decision between arresting and continuing their development?

The Sensory Biology of Parasitic Nematodes

It is clear that many nematode species, including *Strongyloides* spp., sense aspects of their environment both within and outside the host and that this information is used in making developmental decisions. However, it is also clear that we know relatively little about the details of the cues used by these species. In *C. elegans*, some of the neuronal and genetic bases of temperature and chemosensation are beginning to be understood (Bargmann and Mori, 1997) and this information is likely to readily apply to parasitic nematodes. However, our lack of knowledge is especially apparent when considering how parasitic nematodes sense an immune response that is acting against them. It often seems to be assumed that nematodes use the generally 'poor conditions' of their within-host environment as their means of sensing an immune response. This may be so. However, with the perspective of studies on the sensory biology of *C. elegans*, it seems more probable that there are very specific cues used by nematodes in sensing all aspects of their environment *per se* and therefore, by extension, that parasitic nematodes use very specific cues to assess the host immune response acting against them, or at least its consequences. Indeed, given the importance of the immune response in a parasitic nematode's environment, it seems unreasonable that they would do otherwise. Furthermore, it seems likely that in natural infections in wild hosts there will potentially be significant variation between hosts in the quality and quantity of the immune response that a host will mount against a parasitic nematode; i.e. there will be significant variation in the immune environment in which a parasitic nematode may find itself. For this reason, it seems more reasonable still that parasitic nematodes will sense or otherwise assess the host immune response acting against them, and use this information in developmental decisions.

Summary

S. ratti responds in a complex and sophisticated way to its environment within and outside its host. Of particular note is its developmental response to host anti-*S. ratti* immune responses. Other nematodes also have specific responses to host immunity. In trying to understand these developmental responses, it is clear that we need to determine the specific cues used by nematodes and to determine how parasites sense these cues. Here I have suggested some possibilities for how nematodes may do this, largely by analogy with environmental sensation by other nematodes, such as *C. elegans*. It is clear that understanding this area of parasitic nematode biology is a research challenge for the future.

References

Ashton, F.T. and Schad, G.A. (1996) Amphids in *Strongyloides stercoralis* and other parasitic nematodes. *Parasitology Today* 12, 187–194.

Ashton, F.T., Bhopale, V.M., Holt, D., Smith, G. and Schad, G.A. (1998) Developmental switching in the parasitic nematode *Strongyloides stercoralis* is controlled by the ASF and ASI amphidial neurons. *Journal of Parasitology* 84, 691–695.

Bargmann, C.I. and Mori, I. (1997) Chemotaxis and thermotaxis. In: Riddle, D.L., Blumenthal, T., Meyer, B.J. and Priess, J.R. (eds) *C. elegans II*. Cold Spring Harbor Laboratory Press, Cold Spring Harbor, New York, pp. 717–737.

Barton, N.H. and Charlesworth, B. (1998) Why sex and recombination? *Science* 281, 1986–1990.

Behnke, J.M., Hannah, J. and Pritchard, D. (1983) *Nematospiroides dubius* in the mouse: evidence that adult worms depress the expression of homologous immunity. *Parasite Immunology* 5, 397–408.

Blackman, R.L. (1974) *Aphids*. Ginn, London, 175 pp.

Blackman, R.L. (1980) Chromosomes and parthenogenesis in aphids. In: Blackman, R.L., Hewitt, G.M. and Ashburner, M. (eds) *Insect Cytogenetics, 10th Symposium, Royal Entomological Society*. Blackwell, Oxford, pp. 133–148.

Coadwell, W.J. and Ward, P.F.V. (1977) Annual variation in the growth of *Haemonchus contortus* in experimental infections of sheep and its relation to arrested development. *Parasitology* 74, 121–132.

Frank, G.R., Herd, R.P., Marbury, K.S. and Williams, J.C. (1986) Effects of transfer of *Ostertagia ostertagi* between northern and southern USA on the pattern and frequency of hypobiosis. *International Journal for Parasitology* 16, 391–398.

Gibbs, H.C. (1986) Hypobiosis in parasitic nematodes – an update. *Advances in Parasitology* 25, 129–174.

Golden, J.W. and Riddle, D.L. (1984) A pheromone-induced developmental switch in *Caenorhabditis elegans*: temperature-sensitive mutants reveal a wild-type temperature dependent process. *Proceedings of the National Academy of Sciences USA* 81, 819–823.

Goldstein, P. and Curtis, M. (1987) Age-related changes in the meiotic chromosomes of the nematode *Caenorhabditis elegans*. *Mechanisms of Ageing and Development* 40, 115–130.

Grove, D.I., Heenan, P.J. and Northern, C. (1983) Persistent and disseminated infections of *Strongyloides stercoralis* in immunosuppressed dogs. *International Journal for Parasitology* 13, 483–490.

Harvey, S.C. and Viney, M.E. (2001) Sex determination in the parasitic nematode *Strongyloides ratti*. *Genetics* 158, 1527–1533.

Harvey, S.C., Gemmill, A.W., Read, A.F. and Viney, M.E. (2000) The control of morph development in the parasitic nematode *Strongyloides ratti*. *Proceedings of the Royal Society of London Series B* 267, 2057–2063.

Herbert, P.D.N. (1987) Genotypic characteristics of cyclic parthenogens and their obligatory asexual derivatives. In: Stearns, S.C. (ed.) *The Evolution of Sex and its Consequences*. Birkhauser Verlag, Basle, pp. 175–195.

Hodgkin, J. (1988) Sexual dimorphism and sex determination. In: Wood, W.B. (ed.) *The Nematode Caenorhabditis elegans*. Cold Spring Harbor Laboratory, Cold Spring Harbor, New York, pp. 243–279.

Hodgkin, J., Horvitz, R.H. and Brenner, S. (1979) Nondisjunction mutants of the nematode *Caenorhabditis elegans*. *Genetics* 91, 67–94.

Hominick, W.M., Dean, C.G. and Schad, G.A. (1987) Population biology of hookworms in West Bengal: analysis of numbers of infective larvae recovered from damp pads applied to the soil surface at defaecation sites. *Transactions of the Royal Society of Tropical Medicine and Hygiene* 81, 978–986.

Hughes, R.N. (1989) *A Functional Biology of Clonal Animals*. Chapman & Hall, London, 331 pp.

Kimura E., Shintoku, Y., Kadosaka, T., Fujiwara, M., Kondo, S. and Itoh, M. (1999) A second peak of egg excretion in *Strongyloides ratti*-infected rats: its origins and biological meaning. *Parasitology* 119, 221–226.

LaMontagne, J.M. and McCauley, E. (2001) Maternal effect in *Daphnia*: what mothers are telling their offspring and do they listen? *Ecology Letters* 4, 64–71.

Maizels R.M., Bundy, D.A.P., Selkirk, M.E., Smith, D.F. and Anderson, R.M. (1993) Immunological modulation and evasion by helminth parasites in human populations. *Nature* 365, 797–805.

Michel, J.F. (1971) Adult worms as a factor in the inhibition of development of *Ostertagia ostertagi* in the host. *International Journal for Parasitology* 1, 31–36.

Michel, J.F., Lancaster, M.B. and Hong, C. (1979) The effect of age, acquired resistance, pregnancy and lactation on some reactions of cattle to infection with *Ostertagia ostertagi*. *Parasitology* 79, 157–168.

Minematsu, T., Mimori, T., Tanaka, M. and Tada, I. (1989) The effect of fatty acids on the developmental direction of *Strongyloides ratti* first-stage larvae. *Journal of Helminthology* 63, 102–106.

Moqbel, R. and McLaren, D.J. (1980) *Strongyloides ratti*: structural and functional characteristics of normal and immune-damaged worms. *Experimental Parasitology* 49, 139–152.

Nigon, V. and Roman, E. (1952) Le déterminisme du sexe et le développement cyclique de *Strongyloides ratti*. *Bulletin Biologique de la France et de la Belgique* 86, 404–448.

Riddle, D.L. and Albert, P.S. (1997) Genetic and environmental regulation of dauer larva development. In: Riddle, D.L., Blumenthal, T., Meyer, B.J. and Priess, J.R. (eds) *C. elegans II*. Cold Spring Harbor Laboratory Press, Cold Spring Harbor, New York, pp. 739–768.

Schad, G.A. (1989) Morphology and life history of *Strongyloides stercoralis*. In: Grove, D.I. (ed.) *Strongyloidiasis: a Major Roundworm Infection of Man*. Taylor and Francis, London, pp. 85–104.

Schad, G.A., Chowdhury, A.B., Dean, C.G., Kochar, V.K., Nawalinski, T.A., Thomas, J. and Tonascia, J.A. (1973) Arrested development in human hookworm infections: an adaptation to a seasonally unfavourable external environment. *Science* 180, 502–504.

Schad G.A., Hellman, M.E. and Muncet, D.W. (1984) *Strongyloides stercoralis*: hyperinfection in immunosuppressed dogs. *Experimental Parasitology* 57, 287–296.

Schad G.A., Thompson, F., Talham, G., Holt, D., Nolan, T.J., Ashton, F.T., Lange, A.M. and Bhopale, V.M. (1997) Barren female *Strongyloides stercoralis* from occult chronic infections are rejuvenated by transfer to parasite-naïve recipient hosts and give rise to an autoinfective burst. *Journal of Parasitology* 83, 785–791.

Selkirk, M.E., Henson, S.M., Russell, W.S. and Hussein, A.S. (2001) Acetyl-cholinesterase secretion by nematodes. In: Kennedy, M.W. and Harnett, W. (eds) *Parasitic Nematodes: Molecular Biology, Biochemistry and Immunology*. CAB International, Wallingford, UK, pp. 211–228.

Skorping, A., Read, A.F. and Keymer, A.E. (1991) Life history covariation in intestinal nematodes of mammals. *Oikos* 60, 365–372.

Smeal, M.G. and Donald, A.D. (1982) Inhibited development of *Ostertagia ostertagi* in relation to production systems for cattle. *Parasitology* 85, 21–25.

Sulston, J. and Hodgkin, J. (1988) Methods. In: Wood, W.B. (ed.) *The Nematode Caenorhabditis elegans*. Cold Spring Harbor Laboratory Press, Cold Spring Harbor, New York, pp. 587–606.

Taylor, A.P. and Weinstein, P.P. (1990) The effect of oxygen and carbon dioxide on the development of the free-living stages of *Strongyloides ratti* in axenic culture. *Journal of Parasitology* 76, 545–551.

Tindall, N.R. and Wilson, P.A.G. (1988) Criteria for a proof of migration routes of immature parasites inside hosts exemplified by studies of *Strongyloides ratti* in the rat. *Parasitology* 96, 551–563.

Viney, M.E. (1994) A genetic analysis of reproduction in *Strongyloides ratti*. *Parasitology* 109, 511–515.

Viney, M.E. (1996) Developmental switching in the parasitic nematode *Strongyloides ratti*. *Proceedings of the Royal Society of London Series B* 263, 201–208.

Viney, M.E. (1999) Exploiting the life-cycle of *Strongyloides ratti*. *Parasitology Today* 15, 231–235.

Viney, M.E., Matthews, B.E. and Walliker, D. (1992) On the biological and biochemical nature of cloned populations of *Strongyloides ratti*. *Journal of Helminthology* 66, 45–52.

Viney, M.E., Matthews, B.E. and Walliker, D. (1993) Mating in the parasitic nematode *Strongyloides ratti*: proof of genetic exchange. *Proceedings of the Royal Society of London Series B* 254, 213–219.

Watkins, A.R.J. and Fernando, M.A. (1984) Arrested development of the rabbit stomach worm *Obeliscoides cuniculi*: manipulation of the ability to arrest through processes of selection. *International Journal for Parasitology* 14, 559–570.

West, S.A., Gemmill, A.W., Graham, A., Viney, M.E. and Read, A.F. (2001) Immune stress and facultative sex in a parasitic nematode. *Journal of Evolutionary Biology* 14, 333–337.

Yamada, M., Matsuda, S., Nakazawa, M. and Arizono, N. (1991) Species-specific differences in heterogonic development of serially transferred free-living generations of *Strongyloides planiceps* and *Strongyloides stercoralis*. *Journal of Parasitology* 77, 592–594.

The Interactions between 7
Larval Stage Parasitoids and Their Hosts

Michael R. Strand

Department of Entomology, 420 Biological Sciences Building, University of Georgia, Athens, GA 30602–2603, USA

Introduction

Parasitoids are free-living insects as adults, but are parasites of other arthropods during their immature stages. Hosts are usually located by the adult female, who lays her eggs in or on the host's body. The parasitoid larva then completes its development by feeding on host tissues. A few parasitoids oviposit in only the general proximity of where hosts occur. Hosts are then parasitized by either ingesting a parasitoid egg or are located by mobile parasitoid larvae. Most parasitoids have limited host ranges and parasitize only a specific life stage (egg, larval, pupal or adult). Most parasitoids also complete their development by feeding on only one host, and hosts are almost always killed as a consequence of being parasitized. Parasitoids in which a single offspring develops per host are referred to as solitary species, while those that produce two or more offspring per host are called gregarious. Parasitoids are also distinguished by whether larvae feed on the outside (ectoparasitoids) or inside (endoparasitoids) of hosts, and whether hosts cease to develop after parasitism (idiobionts) or remain mobile and continue to grow after parasitism (koinobionts). Idiobionts include ectoparasitoids that permanently paralyse their hosts and endoparasitoids that attack sessile host stages, such as eggs or pupae. Most koinobionts are endoparasitoids that parasitize insect larvae. Taxonomically, approximately 80% of all parasitoids belong to the order Hymenoptera (ants, bees and wasps), 15% to the order Diptera (flies), and the remaining species are restricted to a few genera of Coleoptera (beetles), Neuroptera (lacewings, ant-lions) and Lepidoptera (moths and butterflies) (Quicke, 1997). There are approximately 70,000 described species of parasitoids, which constitute about 9% of all insects. However, some estimates suggest that up to 20% of all insects are parasitoids and that more than 800,000 species of parasitoids exist worldwide (LaSalle and Gauld, 1991; Godfray, 1994; Quicke, 1997; Whitfield, 1998).

This suite of life-history traits has made parasitoids favoured organisms for many studies in behavioural and evolutionary ecology. Since parasitoid larvae derive all resources for development from one host, a direct relationship exists between host quality and the fitness of parasitoid offspring. The fitness of adult parasitoids is likewise directly affected by the number and quality of hosts a female parasitizes. There is no certainty that a foraging parasitoid will locate any hosts or that after parasitism a parasitoid larva will successfully develop. Offspring die as a result of immunological defence responses by the host or because the host provides inadequate resources for successful development. Hosts are also sometimes parasitized more than once by the same species (super-parasitism) or a different species (multiparasitism) of parasitoid, which results in intense competition among offspring for host resources. The larvae of solitary parasitoids usually fight to the death for possession of the host, while larvae of gregarious species compete for the limited resources available in a single host. Many parasitoids are also at risk of being parasitized by other species of parasitoids called hyperparasitoids.

While few studies focus specifically on the behaviour of parasitoid larvae, a large literature exists on the interactions between parasitoid larvae and their hosts. The goal of this chapter is to review these inter-actions and their impact on parasitoid fitness. I shall first review the morphological adaptations of parasitoid larvae for survival in different host environments and how host quality influences parasitoid fitness. Next, I shall examine how parasitoid larvae overcome host defence responses, and some of the behavioural and physiological strategies used by parasitoid larvae to exploit host resources. The last section will focus on hyperparasitism and intra- and interspecific competition among parasitoid larvae.

Larval Adaptations to the Parasitoid Lifestyle

Parasitoid larvae vary greatly in size and morphology. This variation is especially high during the first instar, where Clausen (1940) described 14 different larval forms. The variation in morphology of first instars reflects in large measure adaptations for development in different types of environments. For example, endoparasitoids possess several adapta-tions for development in the nutrient-rich, aquatic environment of the arthropod haemocoel. These include a very thin cuticle, which provides no protection from desiccation but which is well suited for respiration by passive diffusion and uptake of nutrients through structures such as the anal vesicle (see below). In addition, most solitary endoparasitoids produce first instars with heavily sclerotized heads, enlarged mandibles and caudal structures, which are used in combat against competing parasitoid larvae. These adaptations are typified by the teleaform first instars of scelionid egg parasitoids and the mandibulate first instars produced by larval endoparasitoids in the family Braconidae (Fig. 7.1a,b). In contrast, gregarious parasitoids, such as trichogrammatids have sacciform first instars that lack any adaptations for combat (Fig. 7.1c). Most hymenopteran ectoparasitoids paralyse their host at oviposition, which prevents the host from removing parasitoid larvae.

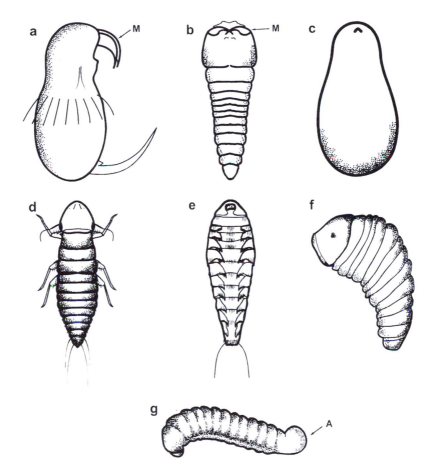

Fig. 7.1. Various forms of parasitoid larvae. (a) Teleaform first instar with large sickle-shaped mandibles (M) typical of parasitoids in the family Scelionidae. (b) Mandibulate first instar of endoparasitic braconids in the subfamily Microgastrinae. (c) Sacciform larva of gregarious parasitoids in the family Trichogrammatidae. (d) Triungulin larva of parasitic Coleoptera in the family Ripiphoridae. (e) Planidial larva of parasitoids in the family Eucharitidae. (f) Hymenopteriform larva typically produced during later instars by parasitic Hymenoptera. (g) Vesiculate larva with anal vesicle (A) produced after the first instar by endoparasitic braconids in the subfamily Microgastrinae. In (a)–(f), the head is orientated towards the top of the page. In (g), the head is orientated to the left. Drawn by J.A. Johnson after the classification scheme of Clausen (1940).

Correspondingly, ectoparasitoid larvae usually lack specialized adaptations for attaching to hosts or fighting. In contrast, ectoparasitoids that do not paralyse their hosts often produce larvae that have setae or suctorial discs to facilitate attachment to the host's cuticle. Ectoparasitoid larvae also have exoskeletons, which provide some protection from desiccation, and often possess a tracheal system for respiration.

As the majority of adult parasitoids oviposit into hosts directly, parasitoid larvae usually lack adaptations for moving or surviving outside the host. Yet, as noted above, a few parasitoids oviposit away from the host and rely on a mobile first instar for host location. In the Hymenoptera, the families Perilampidae and Eucharitidae are comprised exclusively of species that produce mobile larvae. Many parasitic Diptera in the family Tachinidae and all parasitoids in the Coleoptera, Neuroptera and Lepidoptera also produce mobile first instars. Two main types of searching larvae are distinguished (Askew, 1971). Triungulin larvae produced by parasitoid beetles, moths and neuropterans move using conventional legs, while planidium-type larvae produced by the aforementioned taxa of Hymenoptera and Diptera move via elongated setae (Fig. 7.1d, e). Many tachinids lay eggs that hatch into planidial larvae or directly lay planidial larvae, which hatch from eggs brooded in the female's reproductive tract (i.e. they larviposit). The flies ovi- or larviposit in only the general proximity of hosts, which are then located by the planidium (Clausen, 1940). Other parasitoids that produce planidial larvae do not locate hosts directly but instead are transported phoretically by the adult host. This is especially common among species that parasitize larvae of social Hymenoptera. For example, the Eurcharitidae exclusively parasitize ant larvae. Adult females deposit their eggs in locations that are frequented by foraging ants. After hatching, the planidium remains motionless until jumping on to a passing ant. The ant then carries the parasitoid back to the colony, where it parasitizes an ant larva (Heraty and Barber, 1990).

Unlike the variation in morphology and behaviour of first instars, subsequent larval stages are much more uniform among parasitoids. The vast majority of parasitic Hymenoptera are sacciform or hymenopteriform during the second and later instars (Fig. 7.1c, f). These larvae have reduced mandibles and very limited mobility and are adapted primarily for consumption of host haemolymph or tissues. The middle and late instars of endoparasitic braconids and some banchine ichneumonids are similar to this but are classified as vesiculate because of the presence of a structure called the anal vesicle (Fig. 7.1g). The anal vesicle is in essence an eversion of the hindgut, which during the larval stage is unconnected to the midgut. Everted, the anal vesicle is involved in both respiration and uptake of nutrients (Edson and Vinson, 1976). The parasitoid larva then retracts the anal vesicle after moulting to its final instar in preparation for metamorphosis.

The Impact of Host Quality on Fitness

Host quality influences three main components of parasitoid fitness: survival to the adult stage, size and development time (Waage and Godfray, 1985; Godfray, 1994). Size and age are usually considered the most important measures of host quality, although other factors, such as the diet a host feeds upon, are also important. Studies with a diversity of parasitoids indicate that host size strongly influences offspring survival and adult size, and that parasitoid size is positively correlated with other measures of lifetime reproductive success, such as fecundity, mating success and longevity. The relationship between host and parasitoid size is most direct for idiobionts, such as egg or pupal parasitoids, whose hosts are closed resources that do not change in size after parasitism. For solitary species, offspring fitness will be determined primarily by host size alone, while, for gregarious species, offspring fitness will be affected by both host size and the total number of other progeny in the host. Idiobionts are predicted to grow at a constant rate and to maximize adult size per unit of host resource consumed (Mackauer and Sequeira, 1993). Wasps will take longer to develop on larger hosts than on smaller hosts, because it takes longer to consume the resources available. However, adult size and presumably fitness will be correspondingly greater. Several empirical studies with idiobiont parasitoids demonstrate that adult females accurately assess host size before ovipositing, and that gregarious species accurately adjust clutch sizes in relation to available host resources (Wylie, 1967; Strand and Vinson, 1983; Schmidt and Smith, 1985; Takagi, 1986; Hardy *et al.*, 1992; Mayhew, 1998).

The relationship between host and parasitoid size is less direct for koinobionts such as larval endoparasitoids, because hosts continue to grow after parasitism. Koinobionts can either consume hosts rapidly or delay their development, allowing the host to increase in mass before consumption, to yield a larger parasitoid. If size is the primary target of selection, koinobiont offspring that are oviposited into a low-quality (i.e. small) host should exhibit a lag phase in development to allow the host to increase in size before consumption. In contrast, a koinobiont in a high-quality (i.e. large) host should grow at a constant rate. Parasitoid development time in turn should vary with host quality at oviposition, and selection should act on larvae to complete development coincident with when the host attains an optimal size. Several studies lend support for these predictions (Mackauer and Sequeira, 1993; Strand, 2000). For example, Harvey *et al.* (2000a) found that development time of the solitary braconid *Apanteles carpatus* is much longer in small hosts than in large hosts, because first-instar larvae delay their first moult until the host moults to its final instar. Because of this delay, however, the average size of wasp progeny produced from small hosts does not significantly differ from those produced from large hosts.

Some conditions can diminish the importance of size for fitness. Among the most important of these are high predation risks, which could favour accelerated development time at the cost of reduced size (Cole, 1954; Curio, 1989). In insects, evidence for mortality increasing with development time derives primarily from studies on insect herbivores that develop in exposed locations (Price *et al.*, 1980; Grossmueller and Lederhouse, 1985; Leather, 1985; Benrey and Denno, 1997). In contrast, prolonged development times appear to have lower costs for herbivores that feed in concealed sites, because risks of predation and parasitism are usually lower (Clancy and Price, 1987; Craig *et al.*, 1990; Johnson and Gould, 1992). Studies of parasitoid species loads lend support for the importance of feeding niche in host mortality and intraguild competition among natural enemies. In their surveys of the literature, Hawkins and Lawton (1987) and Hawkins (1988) determined that external foliar feeders, leaf-rollers and leaf-miners had much higher parasitoid species loads than concealed species, such as gallers, borers and root feeders. Given that insect herbivores are the main hosts for parasitoids, these patterns suggest that rapid development time at the expense of size will also be favoured in parasitoid larvae whose hosts confront high risks of predation. The trade-off between development time and size should also be most apparent among koinobionts, who, as noted above, have the option to consume hosts rapidly or to delay their development until their larval host attains a larger size. Comparing development times and offspring sizes in large versus small hosts, Harvey and Strand (2002) found that ichneumonoid koinobionts of exposed foliar-feeding larvae usually favoured short development time over size. In contrast, species that parasitized concealed hosts almost always favoured size over development time.

Host Defences and Parasitoid Counterstrategies

The most significant challenge confronting parasitoid larvae is survival in the host. The first line of defence available to hosts is to avoid being parasitized at all by either hiding or fending off oviposition by the parasitoid adult. Once oviposition occurs, however, the host's immune system serves as the main defence against the parasitoid egg and larva.

Encapsulation

Insect blood contains cells called haemocytes, which have many different functions in immunity. The main immune response to large, multicellular invaders, such as parasitoids, is encapsulation. Although some Diptera form melanotic capsules without the apparent participation of haemocytes (Carton and Nappi, 1997), haemocytes are responsible for encapsulation in most other insects (Strand and Pech, 1995). The insect immune system does not recognize specific entities, but it is able to recognize classes of foreign objects, using pattern-recognition receptors (PRRs) (Medzhitov and Janeway, 1997). For example, certain PRRs recognize lipopolysaccharides that are ubiquitous on the surface of bacteria but absent from the surface of most eukaryotic cells. Insects are also able to recognize parasitoids and other pathogenic eukaryotes, such as nematodes. Genetic analysis suggests that host resistance to parasitoids is determined by one or only a few genes and that resistance is restricted to certain types of parasitoids as opposed to all parasitoids (Benassi *et al.*, 1998; Hita *et al.*, 1999; Fellowes *et al.*, 1999). However, the identity of the molecules on parasitoids that allow hosts to recognize them as foreign and the PRRs involved in recognition are currently unknown (Schmidt *et al.*, 2001). What is known is that the receptors involved in recognition of parasitoids are present both in haemolymph and on the surface of specific types of haemocytes (Lavine and Strand, 2001). In the lepidopteran *Pseudoplusia includens*, a particular class of haemocytes, called granular cells, are specifically responsible for recognition of many foreign targets. Once granular cells have attached to the foreign invader, they release cytokines, which induce a second type of haemocyte, called plasmatocytes, to attach to the target and form the capsule (Pech and Strand, 1996; Clark *et al.*, 1997).

The ability to encapsulate parasitoids is affected by other conditions, including host age, temperature and stress (Salt, 1970). For example, parasitoids in the genus *Metaphycus* are readily encapsulated by older instars of soft brown scale but are not encapsulated by younger instars (Blumberg and DeBach, 1981). Encapsulation responses to the same parasitoid species can also vary among strains of the same host species, as well as among closely related host species (Kraaijeveld and Godfray, 1997; Fellowes *et al.*, 1998). Why parasitoid larvae die in capsules is not well understood, although asphyxiation and secretion of cytotoxic molecules have both been implicated as killing mechanisms (Salt, 1968; Strand and Pech, 1995; Nappi and Ottaviani, 2000). In addition, while encapsulation is the main defence against parasitoid larvae, non-cellular immune responses can also kill parasitoids (Vinson, 1990; Carton and Nappi, 1997). For example, parasitoid eggs and larvae fail to develop in some lepidopteran and aphid hosts, even though they are never encapsulated or melanized (Henter and Via, 1995; Trudeau and Strand, 1998). Whether this is due to a humoral molecule, nutritional incompatibility or some other factor is unknown.

Evasion of host defence responses

Many parasitoids avoid encapsulation passively by developing either on hosts that lack a functional immune system or in a location that is inaccessible to host immune cells or effector molecules. The best example of the former are parasitoids that attack hosts when they are still in the egg stage, while examples of the latter include ectoparasitoids. The feeding activity of ectoparasitoid larvae probably stimulates the host's immune system, given that simple abrasion of the insect cuticle activates both haemocytes and humoral defence molecules, such as antimicrobial peptides (Brey *et al.*, 1993). However, ectoparasitoid larvae obviously cannot be encapsulated and also tend to rapidly kill their host by secreting proteases, which preorally digest internal tissues. Another strategy for passively avoiding encapsulation is for parasitoids to develop in specific host organs, so that larvae are protected from immune recognition. As previously noted, some Diptera in the family Tachinidae develop from eggs that are ingested by the host (Salt, 1968). After hatching, first-instar larvae move through the midgut and enter ganglia, salivary glands or flight muscle. Platygasterids (Hymenoptera) likewise spend most of their immature life in the gut or ganglia of their dipteran hosts (Leiby and Hill, 1923). A final means by which parasitoids passively avoid elimination is by having surface features that are either not recognized as foreign or that haemocytes do not adhere to (Strand and Pech, 1995). For example, host haemocytes do not attach to proteins that coat the eggs and larvae of parasitoids in the genus *Cotesia* (Hayakawa and Yazaki, 1997; Asgari *et al.*, 1998). At the other extreme, the eggs of parasitoids in the genus *Asobara* strongly adhere to the fat body of *Drosophila* larvae, with the strength of adhesion being directly proportional to the egg's ability to avoid encapsulation (Kraaijeveld and van Alphen, 1994). While some tachinids develop in host organs, others develop directly in the host's haemocoel. Larvae of these species have hooks on their posterior spiracles, which they use to puncture the host's body or trachea to form a respiratory funnel as a conduit for oxygen. The host then encapsulates the fly larva but, possibly because of the respiratory funnel, the capsule is unable to kill the fly larva, which continues to develop normally (Clausen, 1940; Salt, 1968).

Active suppression of the host immune system is usually mediated by maternal factors injected into the host at oviposition, rather than by the parasitoid larva. The most studied of these factors are polydnaviruses (PDVs) which are divided into the ichno- and bracoviruses, on the basis of their association with endoparasitoid wasps in the families Ichneumonidae and Braconidae, respectively (Webb, 1998). PDVs are the only viruses with segmented DNA genomes, and all appear to share a similar life cycle (Fig. 7.2). PDVs persist as stably integrated proviruses in the genome of the wasp and replicate asymptotically in the ovaries of females. When the wasp lays an egg into its host, it injects a quantity of virus, which then infects host immune cells and other tissues. PDVs do not replicate in the wasp's host, but specific viral gene products, whose primary function is to suppress the host's immune system, allow the wasp's progeny to successfully develop. Thus, a true mutualistic relationship exists, as viral transmission depends on wasp survival and wasp survival depends on PDV-mediated immunosuppression of the host. PDVs disrupt encapsulation by inducing specific alterations in

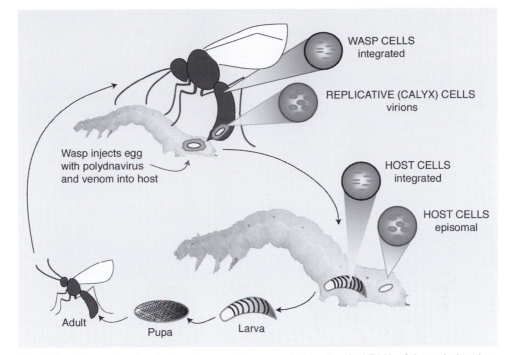

Fig. 7.2. Life cycle of polydnavirus-carrying parasitoids. Proviral DNA of the polydnavirus is integrated into the genome of all cells in the adult wasp including eggs. Polydnaviruses replicate in a specialized type of cell in the ovaries of female wasps called calyx cells. Circularized polydnavirus DNAs excise out of the genome of calyx cells, circularize and are packaged into infectious particles (virions). Virus particles are then stored in the lateral oviducts. The female wasp injects a quantity of virus particles along with an egg, venom and other proteins when it oviposits into a host. Hosts are usually larval-stage Lepidoptera. Virus particles enter and migrate to the nuclei of host cells, such as haemocytes. Once in the nucleus, the viral DNA molecules usually remain episomal (i.e. do not integrate into the host's genome). However, certain viral genes are transcribed in host cells, and the resulting viral proteins disrupt encapsulation and other physiological processes in the host (see text). The wasp egg is already infected with polydnavirus, as its nucleus contains integrated provirus. Thus, when the egg hatches all cells of the parasitoid larva also contain polydnavirus in a proviral form. The parasitoid larva completes its development by feeding on the host. At maturity, the larva emerges from the host and pupates and, during the pupal phase, the virus then begins to replicate again in calyx cells. The parasitoid later ecloses as an adult wasp and mates, and female wasps then seek hosts to repeat the life cycle.

haemocytes, which include cytoskeletal rearrangements, inhibition of adhesion, alterations in the number of circulating haemocytes and induction of apoptosis (reviewed by Webb, 1998; Schmidt *et al.*, 2001). In some PDV–parasitoid systems, disruption of haemocyte function requires that cells be directly infected by PDV, whereas, in others, virus infection of possibly other tissues results in production of secreted proteins, which alter haemocyte function. Viral transcripts are expressed in some host–parasite systems at near steady-state levels beginning a few hours after wasp oviposition and continuing until the wasp's progeny complete development (Theilmann and Summers, 1986; Strand *et al.*, 1992). In other systems, viral expression is transient, lasting less than 36 h after wasp oviposition (Harwood *et al.*, 1994; Asgari *et al.*, 1996; Strand *et al.*, 1997). Other viral and microbial symbionts besides polydnaviruses have also been reported in other groups of parasitoids, but these remain poorly characterized (Quicke, 1997).

Other factors potentially involved in overcoming host defences are teratocytes and the parasitoid larva itself. Teratocytes are derived from the serosal membrane that surrounds the embryos of hymenopteran parasitoids in the families Braconidae and Scelionidae. From 30 to more than 800 teratocytes per parasitoid are released into the host's haemocoel, where they continue to grow but not divide. Several studies implicate teratocytes in altering both host development and encapsulation (reviewed by Vinson and Iwantsch, 1980; Strand, 1986; Dahlman, 1990). The molecular bases for these effects are unclear, although recent studies suggest that teratocytes from the braconid *Microplitis croceipes* secrete a protein that disrupts host protein synthesis (B.A. Webb, personal communication). Endoparasitoid larvae also secrete factors that have been implicated in disrupting encapsulation and host development (Fuhrer and Willers, 1986; Brown and Reed-Larson, 1991).

Parasitoid-induced Changes in Host Development

As previously noted, the hosts of idiobionts cease to develop after parasitism, while the hosts of koinobionts continue to grow. Hosts parasitized by ectoparasitoids usually cease to develop, because adults inject paralysing venoms (Piek, 1986). After hatching, ectoparasitoid larvae and egg parasitoids secrete enzymes while feeding that extraorally digest host tissues and, in the process, accelerate host death (Strand, 1986). The developmental interactions between koinobiont larvae and their hosts, in contrast, are more complex. As a general rule, hosts parasitized by solitary endoparasitoids usually grow more slowly and attain smaller sizes than non-parasitized individuals, whereas hosts parasitized by gregarious endoparasitoids attain sizes only modestly smaller or sometimes actually larger than non-parasitized hosts. Hosts parasitized by larval endoparasitoids are also usually unable to initiate metamorphosis.

These changes in host development are due in part to alterations in endocrine physiology. The growth and metamorphosis of insects are controlled primarily by the interplay between ecdysteroids released by the prothoracic glands, which induce an insect to moult, and juvenile hormone (JH) from the corpora allata, which controls the type of moult an insect undergoes (Riddiford, 1985; Nijhout, 1994). A high JH titre during ecdysteroid release results in moulting to another larval or nymphal stage, whereas a low JH titre during ecdysteroid release results in developmental reprogramming and metamorphosis. As discussed in the section on 'The Impact of Host Quality on Fitness', many larval endoparasitoids oviposit into early-instar hosts and progeny remain in the first instar and delay their development until the host attains a larger size. Studies with several parasitoid–host systems indicate that the parasitoid larva recognizes when the host attains this larger size by moulting to a feeding second instar in response to the endocrine signals in the host that control initiation of metamorphosis (Lawrence and Lanzrein, 1993; Beckage, 1997).

Parasitoids also induce changes in host endocrine state by altering rates of hormone biosynthesis and catabolism or by inactivating hormones (Beckage, 1997). For example, endoparasitic braconids, such as *Cotesia congregata*, *Glyptapanteles liaridis* and *Microplitis demolitor*, arrest the development of their lepidopteran hosts during their final instar and inhibit metamorphosis. This occurs because JH titres remain elevated in parasitized hosts, and is of benefit to the parasitoid because larvae are unable to emerge from the host if it pupates (Beckage and Riddiford, 1982; Balgopal *et al.*, 1996; Schopf *et al.*, 1996). Studies with the braconid *Diachasmimorpha longicaudatus* suggest that elevation of host JH titres is due to secretion of JH by the parasitoid larva (Lawrence *et al.*, 1990), while other studies indicate that the parasitoid larva and factors like polydnaviruses and teratocytes interact to arrest host development. Development of the lepidopteran *Manduca sexta*, for example, is partially arrested when injected with polydnavirus from *C. congregata*, but PDV-injected larvae always develop further than naturally parasitized hosts. In contrast, co-injection of PDV and parasitoid eggs results in arrested host development, which is very similar to that of parasitized hosts (Alleyne and Beckage, 1997). Developmental arrest and/or inhibition of pupation can also occur as a result of alterations in ecdysteroids. As examples, polydnavirus from the ichneumonid *Campoletis sonorensis* causes degeneration of prothoracic glands in last-instar host larvae (Dover *et al.*, 1987), while teratocytes from the braconid *Toxoneuron* (= *Cardiochiles*) *nigriceps* inactivate 20-hydroxyecdysone released from the host's prothoracic glands (Pennachio *et al.*, 1994).

Parasitoid-induced Changes in Host Behaviour

Many arthropods exhibit changes in behaviour in response to the risk of parasitism or as a direct consequence of being parasitized (Moore, 1995). Among the best examples of parasitoids as a selective force in host populations is in field crickets, *Teleogryllus oceanicus*, attacked by tachinids that prefer to forage at dusk (Zuk *et al.*, 1993). In turn, cricket populations where tachinids occur sing primarily during scotophase, whereas cricket populations in areas without tachinids sing during both scotophase and dusk. Crickets also exhibit alterations in male aggression and female egg-laying behaviour after being parasitized (Adamo *et al.*, 1995). However, it is unclear whether these changes are induced by the parasitoid or are an indirect effect from the trauma of parasitism. Changes in the movement or feeding of hosts have also been described for a number of other parasitoids, but it is similarly unclear whether these alterations have any adaptive significance (Godfray, 1994; Moore, 1995).

In contrast, there are a few studies that support the idea that some parasitoid-induced changes in host behaviour have measurable consequences for fitness. Observations of the aphid parasitoid *Aphidius nigripes* revealed that hosts containing diapausing parasitoids move out of the aphid colony to concealed locations, where the parasitoid larva then consumes the aphid and pupates. Aphids containing non-diapausing parasitoid larvae also tend to leave colonies, but generally move to the underside of leaves rather than to concealed sites. Experiments with both diapausing and non-diapausing parasitoids indicate that these alterations in host behaviour result in reduced levels of hyperparasitism (Brodeur and McNeil, 1989, 1992). Another example occurs among conopid flies that parasitize adult bumblebees while flying in midair. The fly larva takes many days to develop in the bumblebee's abdomen, during which time the host continues to forage (Schmid-Hempel and Schmid-Hempel, 1991). However, parasitized bees tend to visit different flowers and collect more nectar than normal bees, which benefits the development of the parasitoid.

An even more sophisticated change in host behaviour occurs in the spider *Plesiometa argyra* after parasitism by the ichneumonid ectoparasitoid *Hymenoepimecis* sp. (Eberhard, 2000). During the first few days after parasitism, *P. argyra* spins webs to capture prey that are identical to non-parasitized spiders. On the day that the parasitoid larva kills the spider, however, the spider builds a completely different type of web, called a cocoon web. After the spider finishes spinning the cocoon web, the parasitoid consumes the spider and then spins a pupal cocoon, which hangs by a thread from the cocoon web. Manipulative experiments reveal that changes in web-spinning behaviour are induced by chemical factors secreted by the parasitoid larva. Cocoon webs support the parasitoid cocoon more durably and prevent it from being removed by heavy rains.

Competitive Interactions between Larval Parasitoids

Beyond contending with the potentially hostile environment of the host, parasitoid larvae also face threats from other parasitoids. These include hyperparasitoids that oviposit directly into the parasitoid larva, and intra- (superparasitism) or interspecific (multiparasitism) competitors that oviposit into the same host and compete for available host resources.

Defensive behaviour against hyperparasitoids

Hyperparasitoids occur in 21 families of Hymenoptera and parasitize a wide range of other parasitoids, which are also mostly hymenopterans (Brodeur, 2000). Most hyperparasitoids attack parasitoids whose hosts are other arthropods, but a few species parasitize conspecifics or other species of hyperparasitoids. Thus, typical parasitoids that attack other arthropods are sometimes called primary parasitoids, hyperparasitoids that attack primary parasitoids are called secondary parasitoids and hyperparasitoids that parasitize hyperparasitoids are called tertiary parasitoids. There are even a few documented cases of quaternary parasitism (Gauld and Bolton, 1988). While hyperparasitoids are common to many parasitoid guilds, surprisingly little is known about their biology, with the possible exception of those that attack parasitoids of aphids. As noted above, some aphid parasitoids reduce their risk of being located by hyperparasitoids by altering the behaviour of their hosts. In contrast, mature larvae of the hyperparasitoid *Dendrocerus carpenteri* (Megaspilidae) defend themselves from tertiary parasitism by possessing spine-like projections on their body, which function as armour. These larvae also thrash violently when approached by another hyperparasitoid adult (Bennett and Sullivan, 1978; Carew and Sullivan, 1993).

Competition among solitary and gregarious parasitoids

Parasitoids are solitary, with only one offspring surviving to adulthood per host, or gregarious, with multiple offspring produced per host. Solitary parasitoids sometimes lay more than one egg per host (Rosenheim and Hongkham, 1996) and hosts attacked by solitary parasitoids can also be superparasitized by a conspecific or multiparasitized by another species. Oviposition of multiple egg clutches by some solitary parasitoids can be favoured if they increase the probability that one offspring will survive in the host (Rosenheim, 1993). For example, studies with parasitoids of *Drosophila* and scale insects have found that laying of multiple eggs reduces the host's encapsulation response, which increases the chance that at least one offspring will survive (Blumberg and Luck, 1990; van Alphen and Visser, 1990). Although adult parasitoids are usually able to distinguish parasitized from non-parasitized hosts, super- or multiparasitism can be favoured if offspring have some chance of surviving competition (van Alphen and Visser, 1990). The extent of the disadvantage in attacking an already parasitized host depends on the amount of time between ovipositions, with recently parasitized hosts generally affording the greatest probability for survival (Strand and Godfray, 1989; Visser, 1993).

For solitary species, hosts usually provide only enough host resources for the development of one individual, which results in intense competition when more than one parasitoid is present. Solitary parasitoids eliminate competitors by either combat or physiological suppression (Salt, 1968; Fisher, 1971; Vinson and Iwantsch, 1980). As discussed in the section on 'Larval Adaptations to the Parasitoid Lifestyle', most solitary endoparasitic Hymenoptera and some Diptera produce first instars with enlarged mandibles (Salt, 1961; Vinson and Iwantsch, 1980). First instars are able to move through the host haemocoel and locate competitors, possibly by perception of as yet unidentified chemical cues (S.B. Vinson, personal communication). Once located, larvae fight with their mandibles until one of the opponents is killed. Although ectoparasitoids lack enlarged mandibles, some species are able to move across the surface of hosts and kill competitors by eating them (Clausen, 1940). Ectoparasitoids are also usually competitively superior to endoparasitoids, because endo-parasitoid larvae cannot attack ectoparasitoids and ectoparasitoids tend to consume hosts much more rapidly. Since parasitoid larvae lack fighting mandibles after the first instar, older larvae use other strategies to eliminate competitors. Some species outcompete younger larvae simply by consuming available host resources before the younger parasitoid can attain a large enough size to pupate. Others physiologically suppress the development of younger larvae. Studies with the ichneumonid *Venturia canescens* indicate that older larvae inhibit development of younger com-petitors by reducing oxygen levels in the host (Fisher, 1963). Experiments with other species, however, suggest that physiological suppression is due to factors injected into the host by the ovipositing female (e.g. viruses) or secretion of cytotoxic factors by the older larvae or teratocytes (Strand, 1986).

Gregarious parasitoids do not fight, but the presence of additional larvae diminishes the amount of host resources available to any one offspring. Competition is resolved by the race to consume host resources, with older larvae usually having a clear advantage over younger larvae. This is well illustrated by Strand and Godfray (1989), who used eye-colour mutants of the ectoparasitoid *Bracon hebetor* to show that survival of larvae in a second clutch depends on the size of the first clutch and the amount of time between ovipositions (Fig. 7.3).

Larval specialization in polyembryonic parasitoids

One group of gregarious parasitoids that engage in larval fighting is polyembryonic parasitoids from the family Encyrtidae (Hymenoptera) (Strand and Grbic, 1997). Polyembryony is a form of clonal development in which a single egg produces two or more genetically identical offspring. Polyembryonic encyrtids lay their eggs into the egg stage of Lepidoptera and offspring complete their development at the end of the host's larval stage. The wasp egg first develops into a single embryo, which then proliferates into an assemblage of embryos, called a polygerm or polymorula. The majority of embryos in the polymorula develop into larvae when the host moults to its final instar. These so-called

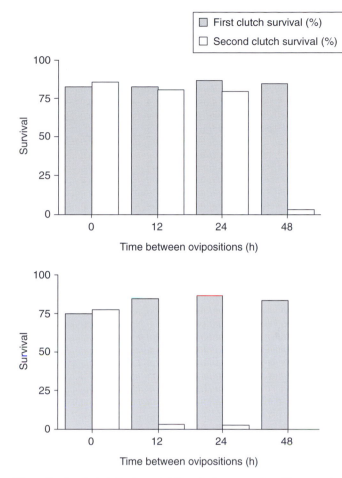

Fig. 7.3.   The effects of clutch size and time between ovipositions on larval competition in the gregarious ectoparasitoid *Bracon hebetor*. In these experiments, host size was held constant. The first wasp was allowed to lay a clutch (four or ten eggs) and then a second wasp was allowed to lay a clutch of equal size at the same time (0 h) or 12, 24 or 48 h later. When the first wasp laid four eggs, the amount of host resources available exceeded the capacity of the larvae to consume them. As a result, most offspring in the second clutch survived, even if laid 24 h later than the first clutch. Survival of a second clutch was low with a 48 h penalty, because the host desiccates from the feeding damage caused by the first clutch. When the first wasp laid ten eggs, offspring of the first clutch were able to rapidly consume all available host resources, resulting in low survival of a second clutch with any temporal penalty. For example, when two females simultaneously oviposited ten eggs each, 78% of larvae from both clutches survived to adulthood. However, when oviposition by females was separated by 12 h, 84% of the first clutch survived, but only 4% of the second clutch survived. (Adapted from Strand and Godfray, 1989.)

reproductive larvae consume the host, pupate and emerge as adult wasps, which seek mates (males) or new hosts (females) (Fig. 7.4). However, some embryos in the polymorula develop into what are called precocious

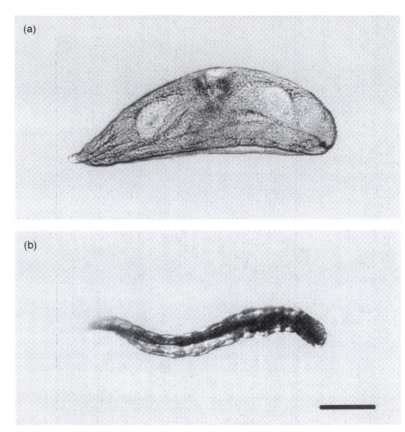

Fig. 7.4. Light micrograph of a reproductive larva (a) and precocious larva (b) from the polyembryonic parasitoid *Copidosoma floridanum*. The head of each larva is orientated to the right. Scale bar, 100 µm. (Adapted from Harvey *et al.*, 2000b.)

larvae, which never moult and always die when the reproductive larvae finish consuming the host. Precocious larvae are also morphologically distinguished from reproductive larvae by having larger mandibles and a distinctly serpentine body, which facilitates movement in the host (Fig. 7.4). Cruz (1981) determined that precocious larvae function as a sterile soldier caste, which kill interspecific competitors by piercing them with their mandibles. This altruistic behaviour is advantageous, because precocious larvae increase their own fitness by assuring the survival of their reproductive siblings. Since selection acts at the level of both the individual and the brood, the ratio of investment in reproductive and precocious larvae would also be predicted to vary depending on how large or small the threat from competitors might be. Such adaptive phenotypic plasticity occurs in the polyembryonic wasp *Copidosoma floridanum*, where the proportion of embryos developing into soldiers changes from 4% in hosts that are not attacked by another parasitoid to 24% in hosts that are attacked by a competitor (Harvey *et al.*, 2000b).

A second, more complex function is that precocious larvae in some species also kill male siblings, which significantly distorts the sex ratio of adult wasps that emerge from the host (Grbic' *et al.*, 1992). Like most Hymenoptera, polyembryonic wasps are haplodiploid, with male offspring developing from unfertilized eggs and female offspring developing from fertilized eggs. Many species of polyembryonic wasps lay only a single egg per host and, as a result, produce broods comprised exclusively of one sex or the other. In these species, precocious larvae presumably develop from both male and female eggs and function exclusively as soldiers that defend their reproductive siblings from competitors. Other species, including *C. floridanum,* usually lay two eggs per host (one male and one female) and produce broods comprised of both sexes. In *C. floridanum*, precocious larvae develop almost exclusively from female eggs, such that, in most broods, soldiers are fully related to their sisters but not to their brothers. Female precocious larvae attack their brothers while still embryos, but not their sisters, resulting in most of the adult wasps (> 95%) that emerge from the host being females. These females then mate with the surviving male wasps before dispersing to find new hosts. This interaction may have arisen as a consequence of genetic conflict between siblings. Female-biased sex ratios are predicted in a situation called local mate competition, where female wasps mate with their brothers at the site of emergence and mating opportunities for males are insignificant away from the brood. Under such conditions no genetic conflict exists between parent, sons and daughters, and sex ratios are predicted to be strongly female-biased. However, if males can obtain a significant number of matings away from the natal brood, the male clone would favour a more male-biased sex ratio to capitalize on opportunities to mate with other females, but the female clone would favour a more female-biased sex ratio to reduce competition for the limited food resources of the host. Any difference in evolutionary optima would be increased by the genetic asymmetries between males and females due to haplodiploidy. Field data suggest that *C. floridanum* males do obtain additional matings away from the natal host and that the resulting conflict between siblings has been resolved in favour of sisters by precocious larvae.

Summary

Adult parasitoids are free-living and mobile, while immature stages are largely immobile, obligate parasites of other organisms. This combination of traits provides tremendous opportunity for dispersal to novel habitats by winged adults, while simultaneously favouring larval specialization to exploit particular types of hosts. These traits are also probably responsible for the high species diversity of parasitoids and the evolution of diverse taxa that parasitize almost all groups of terrestrial arthropods and molluscs. As emphasized in this chapter, the behaviour of larval parasitoids cannot be separated from that of adult parasitoids or the hosts they attack. In most instances, the adult parasitoid locates hosts, assesses their quality and decides how many and what sex of eggs to lay. The importance of the adult parasitoid to offspring also extends beyond oviposition, as maternal factors, such as polydnaviruses, are essential for successful development. However, parasitoid larvae are by no means passive entities totally dependent on their parent. They too influence the development and behaviour of hosts and are also able to effectively compete for resources.

Among the most important areas for future study of larval parasitoids are: (i) understanding the molecular mechanisms that underlie the developmental and behavioural changes in hosts; and (ii) determining whether these changes have any functional significance for the parasitoid. In the area of endocrine physiology, for example, it is relatively easy to measure changes in hormone titres, but it is very difficult to unambiguously determine whether a given alteration is due to a specific parasitoid factor or is an indirect consequence of parasitoid feeding or tissue damage. Likewise, while many studies report changes in the behaviour of hosts after parasitism, far fewer studies have experimentally tested whether these alterations have any adaptive relevance to the reproductive success of the parasitoid. Identification of additional virulence genes and development of *in vivo* and *in vitro* methods for conducting manipulative experiments in parasitized hosts are strongly needed to enhance our understanding of physiological interactions. Greater attention to conducting manipulative field experiments is also needed to address the importance of behavioural changes in hosts for parasitoid fitness. Another area of importance is understanding the molecular basis for immunological resistance by hosts and how parasitoid virulence factors overcome host resistance mechanisms. The coevolution of resistance and virulence, their costs to hosts and parasitoids and how variability in these traits are maintained in natural populations are all issues of broad significance to the field of parasitology that parasitoids are well suited to address.

Acknowledgements

I would like to thank Jeff Harvey, Laura Corley, Mark Lavine and Markus Beck for discussions about parasitoid–host interactions, and Jena Johnson for creating Figs 7.1 and 7.2.

References

Adamo, S.A., Robert, D. and Hoy, R.R. (1995) The effect of a tachinid parasitoid, *Ormia ochacea*, on the behaviour and reproduction of its host, the field cricket. *Journal of Insect Physiology* 41, 269–277.

Alleyne, M. and Beckage, N.E. (1997) Parasite density-dependent effects on host growth and metabolic efficiency in tobacco hornworm larvae parasitized by *Cotesia congregata. Journal of Insect Physiology* 43, 407–424.

Asgari, S., Hellers, M. and Schmidt, O. (1996) Host haemocyte inactivation by an insect parasitoid: transient expression of a polydnavirus gene. *Journal of General Virology* 77, 2653–2662.

Asgari, S., Theopold, U., Wellby, C. and Schmidt, O. (1998) A protein with protective properties against the cellular defence reactions in insects. *Proceedings of the National Academy of Sciences USA* 95, 3690–3695.

Askew R.R. (1971) *Parasitic Insects.* Elsevier, New York, 316 pp.

Balgopal, M.M., Dover, B.A., Goodman, W.G. and Strand, M.R. (1996) Parasitism by *Microplitis demolitor* induces alterations in the juvenile hormone titer of its host, *Pseudoplusia includens. Journal of Insect Physiology* 42, 337–345.

Beckage, N.E. (1997) New insights: how parasites and pathogens alter the endocrine physiology and development of insect hosts. In: Beckage, N.E. (ed.) *Parasites and Pathogens: Effects on Host Hormones and Behavior.* Chapman & Hall, New York, pp. 3–36.

Beckage, N.E. and Riddiford, L.M. (1982) Effects of parasitism by *Apanteles congregatus* on the endocrine physiology of the tobacco hornworm *Manduca sexta. General and Comparative Endocrinology* 47, 308–322.

Benassi, V., Frey, F. and Carton, Y. (1998) A new specific gene for wasp cellular immune resistance in *Drosophila. Heredity* 80, 347–352.

Bennett, A.W. and Sullivan, D.J (1978) Defensive behaviour against tertiary parasitism by the larva of *Dendrocerus carpenteri*, an aphid hyperparasitoid. *Journal of the New York Entomological Society* 86, 153–160.

Benrey, B. and Denno, R.F. (1997) The slow growth–high mortality hypothesis: a test using the cabbage butterfly. *Ecology* 78, 987–999.

Blumberg, D. and DeBach, P. (1981) Effects of temperature and host age upon the encapsulation of *Metaphycus stanleyi* and *Metaphycus helvolus* eggs by brown soft scale *Coccus hesperidum* biological control. *Journal of Invertebrate Pathology* 37, 73–79.

Blumberg, D. and Luck, R.F. (1990) Differences in the rates of superparasitism between two strains of *Comperiella bifasciata* (Howard) (Hymenoptera: Encyrtidae) parasitizing California red scale (Homoptera: Diaspididae): an adaptation to circumvent encapsulation? *Annals of the Entomological Society of America* 83, 591–597.

Brey, P.T., Lee, W.J., Yamakawa, M., Koizumi, Y., Perrot, S., Francois, M. and Ashida, M. (1993) Role of the integument in insect immunity: epicuticular abrasion and induction of cecropin synthesis in cuticular epithelial cells. *Proceedings of the National Academy of Sciences USA* 90, 6275–6279.

Brodeur, J. (2000) Host specificity and trophic relationships of hyperparasitoids. In: Hochberg, M.E. and Ives, A.R. (eds) *Parasitoid Population Biology.* Princeton University Press, Princeton, New Jersey, pp. 163–183.

Brodeur, J. and McNeil, J.N. (1989) Seasonal microhabitat selection by an endoparasitoid through adaptive modification of host behaviour. *Science* 244, 226–228.

Brodeur, J. and McNeil, J.N. (1992) Host behaviour modified by the endoparasitoid *Aphidius nigripes*: a strategy to reduce hyperparasitism. *Ecological Entomology* 17, 97–104.

Brown, J.J. and Reed-Larson, D. (1991) Ecdysteroids and insect host/parasitoid interactions. *Biological Control* 1, 136–143.

Carew, W.P. and Sullivan, D.J. (1993) Interspecific parasitism between two aphid hyperparasitoids, *Dendrocerus carpenteri* (Hymenoptera; Megaspilide) and *Asaphes lucens* (Hymenoptera: Pteromalidae). *Annals of the Entomological Society of America* 86, 794–798.

Carton, Y. and Nappi, A.J. (1997) *Drosophila* cellular immunity against parasitoids. *Parasitology Today* 6, 218–227.

Clancy, K.M. and Price, P.W. (1987) Rapid herbivore growth enhances enemy attack: sublethal plant defences remain a paradox. *Ecology* 68, 736–738.

Clark, K.D., Pech, L.L. and Strand, M.R. (1997) Isolation and identification of a plasmatocyte-spreading peptide from the hemolymph of the lepidopteran insect *Pseudoplusia includens*. *Journal of Biological Chemistry* 272, 23440–23447.

Clausen, C.P. (1940) *Entomophagous Insects*. McGraw-Hill, New York, 688 pp.

Cole, L.C. (1954) The population consequences of life history phenomena. *Quarterly Review of Biology* 29, 103–137.

Craig, T.P., Itami, J.K. and Price, P.W. (1990) The window of vulnerability of a shoot galling sawfly to attack by a parasitoid. *Ecology* 71, 1471–1482.

Cruz, Y.P. (1981) A sterile defender morph in a polyembryonic hymenopterous parasite. *Nature* 294, 446–447.

Curio, E. (1989) Is avian mortality preprogrammed? *Trends in Ecology and Evolution* 4, 81–82.

Dahlman, D.L. (1990) Evaluation of teratocyte functions: an overview. *Archives of Insect Biochemistry and Physiology* 13, 159–166.

Dover, B.A., Davies, D.H., Strand, M.R., Gray, R.S., Keeley, L.L. and Vinson S.B (1987) Ecdysteroid titer reduction and developmental arrest of last-instar *Heliothis virescens* larvae by calyx fluid from the parasitoid *Campoletis sonorensis*. *Journal of Insect Physiology* 33, 333–338.

Eberhard, W.G. (2000) Spider manipulation by a wasp larva. *Nature* 406, 255–256.

Edson, K.M. and Vinson, S.B. (1976) The function of the anal vesicle in respiration and excretion in the braconid wasp, *Microplitis croceipes*. *Journal of Insect Physiology* 22, 1037–1043.

Fellowes, M.D.E., Kraaijeveld, A.R. and Godfray, H.C.J. (1998) Trade-off associated with selection for increased ability to resist parasitoid attack in *Drosophila melanogaster*. *Proceedings of the Royal Society of London – Biological Sciences* 265, 1553–1558.

Fellowes, M.D.E., Kraaijeveld, A.R. and Godfray, H.C.J. (1999) Cross-resistance following artificial selection for increased defence against parasitoids in *Drosophila melanogaster*. *Evolution* 53, 966–972.

Fisher, R.C. (1963) Oxygen requirements and the physiological suppression of supernumerary insect parasitoids. *Journal of Experimental Biology* 40, 531–540.

Fisher, R.C. (1971) Aspects of the physiology of endoparasitic Hymenoptera. *Biological Reviews* 46, 243–278.

Fuhrer, E. and Willers, D. (1986) The anal secretion of the endoparasitic larva *Pimpla turionellae*: sites of production and effects. *Journal of Insect Physiology* 32, 361–367.

Gauld, I.D. and Bolton, B. (1988) *The Hymenoptera*. Oxford University Press, Oxford, UK, 332 pp.

Godfray, H.C.J. (1994) *Parasitoids: Behavioral and Evolutionary Ecology*. Princeton University Press, Princeton, New Jersey, 473 pp.

Grbic', M., Ode, P.J. and Strand, M.R. (1992) Sibling rivalry and brood sex ratios in polyembryonic wasps. *Nature* 360, 254–256.

Grossmueller, D.W. and Lederhouse, R.C. (1985) Oviposition site selection: an aid to rapid growth and development in the tiger swallowtail butterfly, *Papilio glaucus*. *Oecologia* 66, 68–73.

Hardy, I.C.W., Griffiths, N.T and Godfray, H.C.J. (1992) Clutch size in a parasitoid wasp: a manipulation experiment. *Journal of Animal Ecology* 61, 121–129.

Harvey, J.A. and Strand, M.R. (2002) The developmental strategies of endo-parasitoid wasps vary with host feeding ecology. *Ecology* (in press).

Harvey, J.A., Kadash, K. and Strand, M.R. (2000a) Differences in larval feeding behaviour correlate with altered developmental strategies in two parasitic wasps: implications for the size–fitness hypothesis. *Oikos* 88, 621–629.

Harvey, J.A., Corley, L.S. and Strand, M.R. (2000b) Competition induces adaptive shifts in caste ratios of a polyembryonic wasp. *Nature* 406, 183–186.

Harwood, S.H., Grosovsky, A.J., Cowles, E.A., Davis, J.W. and Beckage, N.E. (1994) An abundantly expressed hemolymph glycoprotein isolated from newly parasitized *Manduca sexta* larvae is a polydnavirus gene product. *Virology* 205, 381–392.

Hawkins, B.A. (1988) Do galls protect endophytic herbivores from parasitoids? A comparison of galling and non-galling Diptera. *Ecological Entomology* 13, 473–477.

Hawkins, B.A. and Lawton, J.H. (1987) Species richness for parasitoids of British phytophagous insects. *Nature* 326, 788–790.

Hayakawa, Y. and Yazaki, K. (1997) Envelope protein of parasitic wasp symbiont virus, polydnavirus, protects the wasp eggs from cellular immune reactions by the host insect. *European Journal of Biochemistry* 246, 820–826.

Henter, H. and Via, S. (1995) The potential for coevolution in a host–parasitoid system. 1. Genetic variation within an aphid population in susceptibility to a parasitic wasp. *Evolution* 49, 427–438.

Heraty, J.M. and Barber, K.N. (1990) Biology of *Obeza floridna* and *Pseudochalcura gibbosa* (Hymenoptera: Eucharitidae). *Proceedings of the Entomological Society of Washington* 92, 248–258.

Hita, M.T., Poirie, M., Leblanc, N., Lemeunier, F., Lutcher, F., Frey, F., Periquet, G. and Carton, Y. (1999) Genetic localization of a *Drosophila melanogaster* resistance gene to a parasitoid wasp and physical mapping of the region. *Genome Research* 9, 471–481.

Johnson, T.M. and Gould, F. (1992) Interaction of genetically engineered host plant resistance and natural enemies of *Heliothis virescens* (Lepidoptera: Noctuidae) in tobacco. *Environmental Entomology* 21, 586–597.

Kraaijeveld, A.R. and Godfray, H.C.J. (1997) Trade-off between parasitoid resistance and larval competitive ability in *Drosophila melanogaster*. *Nature* 389, 278–280.

Kraaijeveld, A.R. and van Alphen, J.J.M. (1994) Geographic variation in resistance of the parasitoid *Asobara tabida* against encapsulation by *Drosophila melanogaster* larvae: the mechanism explored. *Physiological Entomology* 19, 9–14.

LaSalle, J. and Gauld, I.D. (1991) Parasitic Hymenoptera and the biodiversity crisis. *Redia* 74, 315–334.

Lavine, M.D. and Strand, M.R. (2001) Surface characteristics of foreign targets that elicit an encapsulation response by the moth *Pseudoplusia includens*. *Journal of Insect Physiology* 47, 965–974.

Lawrence, P.O. and Lanzrein, B. (1993) Hormonal interactions between insect endoparasites and their host insects. In: Beckage, N.E., Thompson, S.N. and Federici, B.A. (eds) *Parasites and Pathogens of Insects*, Vol. 1. Academic Press, San Diego, California, pp. 59–86.

Lawrence, P.O., Baker, F.C., Tsai, L.W., Miller, C.A., Schooley, D.A. and Geddes, L.G. (1990) JH III levels in larvae and pharate pupae of *Anastrepha suspensa* (Diptera: Tephritidae) and in larvae of the parasitic wasp *Biosteres longicaudatus* (Hymenoptera: Braconidae). *Archives of Insect Biochemistry and Physiology* 13, 53–62.

Leather, S.R. (1985) Oviposition preferences in relation to larval growth rates and survival in the pine beauty moth, *Pannolis flammea*. *Ecological Entomology* 10, 213–217.

Leiby, R.W. and Hill, C.C. (1923) The twinning and monoembryonic development of *Platygaster hiemalis*, a parasite of the Hessian fly. *Journal of Agricultural Research* 25, 237–249.

Mackauer, M. and Sequeira, R. (1993) Patterns of development in insect parasites. In: Beckage, N.E., Thompson, S.N. and Federici, B.A. (eds) *Parasites and Pathogens of Insects*, Vol. 1. Academic Press, San Diego, California, pp. 1–24.

Mayhew, P.J. (1998) The evolution of gregariousness in parasitoid wasps. *Proceedings of the Royal Society of London B, Biological Sciences* 265, 1–7.

Medzhitov, R. and Janeway, C.A. (1997) Innate immunity: impact on the adaptive immune respsonse. *Current Opinion in Immunology* 9, 4–9.

Moore, J. (1995) The behaviour of parasitized animals. *Bioscience* 45, 89–96.

Nappi, A.J. and Ottaviani, E. (2000) Cytotoxicity and cytotoxic molecules in invertebrates. *BioEssays* 22, 469–480.

Nijhout, H. F. (1994) *Insect Hormones*. Princeton University Press, Princeton, New Jersey, 310 pp.

Pech, L.L. and Strand, M.R. (1996) Granular cells are required for encapsulation of foreign targets by insect haemocytes. *Journal of Cell Science* 109, 2053–2060.

Pennachio, F., Vinson, S.B., Tremblay, E. and Ostuni, A. (1994) Regulation of ecdysone metabolism in *Heliothis virescens* (F.) (Lepidoptera: Noctuidae) larvae induced by *Cardiochiles nigriceps* Viereck (Hymenoptera: Braconidae) teratocytes. *Insect Biochemistry and Molecular Biology* 24, 383–394.

Piek, T. (1986) *Venoms of the Hymenoptera: Biochemical, Pharmacological and Behavioural Aspects*. Academic Press, London, 570 pp.

Price, P.W., Bouton, C.E., Gross, P., McPheron, B.A., Thompson, J.N. and Weis, A.E. (1980) Interactions among three trophic levels: influence of the plant on interactions between insect herbivores and natural enemies. *Annual Review of Ecology and Systematics* 11, 41–65.

Quicke, D.L.J. (1997) *Parasitic Wasps*. Chapman & Hall, Cambridge, UK, 469 pp.

Riddiford, L.M. (1985) Hormone action at the cellular level. In: Kerkut, G.A. and Gilbert, L.I. (eds) *Comprehensive Insect Physiology, Biochemistry, and Pharmacology,* Vol. 8. Pergamon Press, Oxford, pp. 37–84.

Rosenheim, J.A. (1993) Single-sex broods and the evolution of nonsiblicidal parasitoid wasps. *American Naturalist* 141, 90–104.

Rosenheim, J.A. and Hongkham, H. (1996) Clutch size in an obligately siblicidal parasitoid wasp. *Animal Behaviour* 51, 841–852.

Salt, G. (1961) Competition among insect parasitoids: mechanisms in biological competition. *Symposium of the Society for Experimental Biology* 15, 96–119.

Salt, G. (1968) The resistance of insect parasitoids to the defence reactions of their hosts. *Biological Reviews* 43, 200–232.

Salt, G. (1970) *The Cellular Defence Reactions of Insects*. Cambridge University Press, Cambridge, 240 pp.

Schmid-Hempel, P. and Schmid-Hempel, R. (1991) Endoparasitic flies, pollen-collection by bumblebees and potential parasite–host conflict. *Oecologia* 87, 227–232.

Schmidt, J.M. and Smith, J.J.B. (1985) Host volume measurement by the parasitoid wasp *Trichogramma minutum*: the roles of curvature and surface area. *Entomologia Experimentalis et Applicata*, 39, 213–221.

Schmidt, O., Theopold, U. and Strand, M.R. (2001) Innate immunity and its evasion and suppression by hymenopteran endoparasitoids. *BioEssays* 23, 344–351.

Schopf, A., Nussbaumer, C., Rembold, H. and Hammock, B.D. (1996) Influence of the braconid *Glyptapanteles liparidis* on the juvenile hormone titer of its larval host, the gypsy moth, *Lymantria dispar*. *Archives of Insect Biochemistry and Physiology* 31, 337–351.

Strand, M.R. (1986) The physiological interactions of parasitoids with their hosts and their influence on reproductive strategies. In: Waage, J. and Greathead, D. (eds) *Insect Parasitoids*. Academic Press, London, pp. 97–136.

Strand, M.R. (2000) Developmental traits and life-history evolution in parasitoids. In: Hochberg, M.E. and Ives, A.R. (eds) *Parasitoid Population Biology*. Princeton University Press, Princeton, New Jersey, pp. 139–162.

Strand, M.R. and Godfray, H.C.J. (1989) Superparasitism and ovicide in parasitic Hymenopera: theory and a case study of the ectoparasitoid *Bracon hebetor*. *Behavioral Ecology and Sociobiology* 24, 421–432.

Strand, M.R. and Grbic, M. (1997) The development and evolution of polyembryonic insects. *Current Topics in Developmental Biology* 35, 121–158.

Strand, M.R. and Pech, L.L. (1995) Immunological compatibility in parasitoid–host relationships. *Annual Review of Entomology* 40, 31–56.

Strand, M.R. and Vinson, S.B. (1983) Host acceptance behaviour of *Telenomus heliothidis* toward the eggs of *Heliothis virescens*. *Annals of the Entomological Society of America* 76, 781–785.

Strand, M.R., McKenzie, D.I., Grassl, V., Dover, B.A. and Aiken, J.M. (1992) Persistence and expression of *Microplitis demolitor* polydnavirus in *Pseudoplusia includens*. *Journal of General Virology* 73, 1627–1635.

Strand, M.R., Witherell, S.A. and Trudeau, D. (1997) Two related *Microplitis demolitor* polydnavirus mRNAs expressed in hemocytes of *Pseudoplusia includens* contain a common cysteine-rich domain. *Journal of Virology* 71, 2146–2156.

Takagi, M. (1986) The reproductive strategy of the gregarious parasitoid, *Pteromalus puparum* (Hymenoptera: Pteromalidae): host size discrimination and regulation of the number and sex ratio of progeny in a single host. *Oecologia* 70, 321–325.

Theilmann, D.A. and Summers, M.D. (1986) Molecular analysis of *Campoletis sonorensis* virus DNA in the lepidopteran host *Heliothis virescens*. *Journal of General Virology* 67, 1061–1069.

Trudeau, D. and Strand, M.R. (1998) The role of *Microplitis demolitor* polydnavirus in parasitism by its associated wasp, *Microplitis demolitor* (Hymenoptera: Braconida). *Journal of Insect Physiology* 44, 795–805.

van Alphen, J.J.M. and Visser, M.E. (1990) Superparasitism as an adaptive strategy for insect parasitoids. *Annual Review of Entomology* 35, 59–79.

Vinson, S.B. (1990) How parasitoids deal with the immune system of their host: an overview. *Archives of Insect Biochemistry and Physiology* 13, 3–27.

Vinson, S.B. and Iwantsch, G. (1980) Host regulation by insect parasitoids. *Quarterly Review of Biology* 55, 145–165.

Visser, M.E. (1993) Adaptive self- and conspecific superparasitism in the solitary parasitoid *Leptopilina heterotoma*. *Behavioural Ecology* 4, 22–28.

Waage, J.K. and Godfray, H.C.J. (1985) Reproductive strategies and population ecology of insect parasitoids. In: Sibly, R.M. and Smith, R.H. (eds) *Behavioural Ecology*. British Ecological Society Symposium 25, Blackwell Scientific Publications, Oxford, pp. 449–470.

Webb, B.A. (1998) Polydnavirus biology, genome structure and evolution. In: Miller, L.K. and Ball, A. (eds) *The Insect Viruses*. Plenum Press, New York, pp. 105–139.

Whitfield, J.B. (1998) Phylogeny and evolution of host–parasitoid interactions in Hymenoptera. *Annual Review of Entomology* 43, 129–151.

Wylie, H.G. (1967) Some effects of host size on *Nasonia vitripennis* and *Muscidifurax raptor* (Hymenoptera: Pteromalidae). *Canadian Entomologist* 99, 742–748.

Zuk, M., Simmons, L.W. and Cupp, L. (1993) Calling characteristics of parasitized and unparasitized populations of the field cricket *Teleogryllus oceanicus*. *Behavioral Ecology and Sociobiology* 33, 339–343.

Interspecific Interactions in Trematode Communities

8

Kevin D. Lafferty

USGS, Western Ecological Research Center, Marine Science Institute, University of California, Santa Barbara, CA 93106–6150, USA

Introduction

This chapter reviews the behavioural ecology of trematodes, focusing on intramolluscan stages. In some cases, the strategies exhibited by larval trematodes (e.g. excretion of a protein) may not be strictly behavioural. However, their consequences are analogous to behavioural strategies in other organisms and their inclusion in a book on parasite behaviour seems justifiable. In the complex life cycle of a generalized trematode, a free-swimming cercaria leaves the first intermediate host snail and then finds and encysts in a second intermediate host as a metacercaria; if a final host eats an infected second intermediate host, a worm will emerge from the cyst and parasitize the alimentary tract of the final host, where the adult worm will feed and mate; eggs shed from the final host hatch into free-swimming miracidia, which seek out and penetrate an appropriate mollusc host. This life cycle will act as my outline, allowing me to touch briefly on cercarial, adult and miracidial behaviours (some of these are reviewed in more detail by Claude Combes *et al.*, Chapter 1, this volume) and end with a more detailed focus on intramolluscan stages. This latter focus will be in the context of the limited resources that the molluscan host provides because much of the behaviour of larval trematodes relates to competition with other trematodes for nutrition and habitat. I shall then discuss interspecific competitive interactions, their frequency, dominance hierarchies and resolutions. I shall also review strategies employed by larval trematode species to adapt to a hostile, competitive environment, how these adaptations shape larval trematode communities and how altered trematode communities can affect ecosystems and human health.

Cercariae

A discussion of life cycles needs to start somewhere and, for this chapter, it makes the most sense to start with the free-swimming cercarial stage. Nearly all trematode species shed cercariae from the infected molluscs, which act as first intermediate hosts. Infectivity starts to decline a few hours after cercariae leave the mollusc and nearly all cercariae are dead after 1 day (Olivier, 1966). Many employ strategies to conserve energy, to avoid predation (Haas, 1994) and to time shedding (Théron, 1984) so as to overlap with the activity patterns of the next host. Most cercariae are sophisticated swimmers with muscular tails and, in many cases, eye-spots to aid in their orientation to light. Using gravity, light and temperature as cues, cercariae attempt to disperse from the mollusc habitat to areas frequented by the target host (Haas, 1994). At close range, they can respond to specific stimuli, such as shadows, water turbulence and chemical compounds (Haas, 1994). Once they contact something, temperature and chemical signals help the cercariae to distinguish if the object is a potential host (Feiler and Haas, 1988). Some species are able to distinguish among infected and uninfected hosts of the same species (Nolf and Court, 1933; Campbell, 1997), while others will penetrate a wide variety of inappropriate host species (e.g. swimmer's itch in humans). Some cercariae mimic prey items and achieve transmission as they are approached and eaten by their second intermediate hosts (e.g. Martorelli, 1994). After attaching to a potential host, a cercaria can use its suckers to creep and find a suitable place to penetrate and drop its tail (Haas, 1988). It then moves through the tissues, often to a particular site, and encysts as a metacercaria (schistosome cercariae are unusual in that they transform into adult worms in the circulatory system of their vertebrate final hosts).

Metacercariae

Once encysted, the metacercaria is traditionally thought to be relatively passive. However, because metacercariae are trophically transmitted to the definitive host, they may benefit from actions that lead to parasite-increased trophic transmission (PITT) (Lafferty, 1999), a subject reviewed by Robert Poulin (Chapter 12, this volume). The site of infection (muscles, central nervous system, eyes, skin) may allow some metacercariae to influence the behaviour or coloration of their host and predispose them to predation (Lafferty, 1999). Metacercariae on fish brains (Lafferty and Morris, 1996), on amphipod brains (Helluy, 1983), in coral polyps (Aeby, 1991), in crab claws (Lafferty, 1999) and in cockles (Thomas and Poulin, 1998) can manipulate host behaviour. How they do this is not known, but some metacercariae seem capable of secreting substances that affect host behaviour (Lafferty, 1999). In some cases, metacercariae may share their intermediate host with other trematodes (or other parasites, for that matter). For example, several trematodes use the California killifish as an intermediate host. *Euhaplorchis californiensis* alters killifish behaviour, while others, such as *Renicola buchanani*, seem not to (Lafferty and Morris, 1996). Yet all species presumably can use the same definitive host (a bird). In multiple infections, *R. buchanani* clearly benefits from increased transmission resulting from the PITT of *E. californiensis*. Intensities of the two species associate positively with each other, consistent with the possibility that *R. buchanani* may actively seek out fish infected with *E. californiensis* to obtain increased transmission (though other simpler hypotheses for this and other such associations exist). Such 'hitch-hiking' (as per Thomas *et al.*, 1997) has been investigated in more detail with the trematode *Microphallus subdolum*, which infects amphipods as second intermediate hosts. *M. subdolum* does not alter the amphipod host's behaviour but tends to coinfect with *Microphallus papillorobustus* (Thomas *et al.*, 1998), which infects the amphipod's brain, making it swim closer to the water's surface and increasing its susceptibility to predation by birds (Helluy, 1983). The positive association between the manipulator and the hitch-hiker seems more than coincidental, because the cercariae of the hitch-hiker, *M. subdolum*, swim close to the water's surface, where they seem more likely to penetrate amphipods already modified by *M. papillorobustus* (Thomas *et al.*, 1997).

Adults

If an appropriate vertebrate consumes a second intermediate host infected with one to several thousand metacercariae, the digestive process releases immature worms from the cysts. Excysted worms then migrate through the host to a specific site, which varies widely among species, but is often in the intestinal tract. To find an appropriate site within the host, helminths must respond to a number of cues in the host environment (Sukhdeo and Bansemir, 1996). Most ingested metacercariae do not become established, probably due to several levels of host defence, starting with mastication and including both general and specific immune responses. If the host is uninfected, the establishment of new infections can increase along with the number of metacercariae given in laboratory exposures, presumably because large numbers can catch the relaxed immune system off guard (Christensen *et al.*, 1988). Some schistosomes may present trematode-derived antigens that mimic host antigens (Damian, 1967) or may coat themselves with host-derived materials to escape detection (Smithers *et al.*, 1977). In most cases, trematodes enter a host already infected with other parasites. If these parasites suppress the host's immune system, they may gain an increased chance of survivorship and longevity; in most cases, however, trematodes may be thwarted by a hostile, sensitized host immune system, especially if there is cross-reactivity between their antigens and antigens of established parasites (Christensen *et al.*, 1987). Trematodes mature and mate with their hermaphroditic conspecifics (although schistosomes have separate sexes) and pass eggs (often with the host's faeces) out of the host. If resources become limited in the host, due to competition with other parasites, growth and egg production can suffer, as can be seen in experimentally induced, high-intensity infections with echinostomes (Mohandas and Nadakal, 1978).

Miracidia

If the trematode egg is fortunate, it ends up in an aquatic environment with a suitable host mollusc population (usually the host is a snail). Inside the egg is a free-swimming, ciliated larval stage, called a miracidium (pl. miracidia). Miracidia have two strategies, depending on the family of trematode; they can wait for a molluscan host to ingest them (where, if the host is the appropriate species, consumption will release a miracidium from its egg) or, once fully developed and under appropriate environmental conditions, they can hatch out of the egg and use a variety of sense receptors to find a new host (Wright, 1971).

Trematodes are highly specific for their mollusc host (van der Knaap and Loker, 1990). The first behaviour of the swimming miracidium is to find the habitat of its specific mollusc host; this is done primarily by seeking appropriate depths where the molluscs live (a target that varies among mollusc species) (Wright and Ross, 1966). Miracidia are also known to be sensitive to light, temperature and gravity (Nollen, 1994). This allows them to adjust their depth through taxis or by following gradients. During this focused phase, they may even be incapable of attacking a mollusc (Isseroff and Cable, 1968). Once in a suitable mollusc habitat, the miracidium begins a random search by moving straight ahead with an occasional turn, searching for a chemical signal that indicates the presence of a mollusc (in some cases, even non-host snails will produce a sufficiently tempting odour (Kawashima *et al.*, 1961)). Each miracidium has only a few hours before it stops functioning (Wright, 1971). However, if it crosses a mollusc's odour plume, the miracidium's behaviour changes suddenly, as it begins a klinokinetic 'devil dance', which allows it to sample the odour concentration gradient and move towards the mollusc (Wright and Ross, 1966). Trematodes that rely on having a mollusc consume their eggs do not seek out their hosts, but may shape their eggs so as to mimic or become entangled in food (Wright, 1971).

On contacting a prospective host, the miracidium uses suction to attach and begins to secrete substances to lyse the mollusc's epithelium. In some trematodes, the miracidium sheds its ciliated epithelium as it squeezes through the hole it has made in the snail's foot, where it then metamorphoses into what is termed the primary (or mother) sporocyst (a sporocyst is a worm-shaped sac that lacks a pharynx or gut and absorbs nutrients through the integument). At this point, the success of the infection requires compatibility between mollusc and trematode genotypes, which is thought to be determined by the ability of the mollusc to aggressively resist infection (van der Knaap and Loker, 1990). Recent research suggests that it is the toxicity of the mollusc's plasma (Sapp and Loker, 2000a), not the activity of the haemocytes (Sapp and Loker, 2000b), that helps determine host specificity.

Sporocysts and Rediae

The primary sporocyst usually stays near the point of penetration (or in the digestive gland if the species infects molluscs by having its egg ingested). This is the mollusc's first opportunity to defend itself against the parasite. If the primary sporocyst survives, it asexually produces more larvae, which, depending on the family of trematode, can be classified as either rediae (larvae with a pharynx and gut that actively feed on mollusc tissue) or secondary sporocysts (also called daughter sporocysts). These larval stages typically migrate towards the gonad and/or digestive gland (in some species, they reside in other parts of the mollusc, such as the mantle). They may also alter the host's allocation of resources by manipulating the mollusc's hormones (de Jong-Brink *et al.*, 1988). For example, the first rediae of *Ribeiroia marini guadeloupensis* migrate to the snail's brain and make the snail stop producing eggs, even before the trematode has had the opportunity to consume much host tissue (Nassi *et al.*, 1979). Rediae and sporocysts produce either additional secondary stages or, when the gonad of the mollusc is completely filled with secondary stages, they produce cercariae (Dönges, 1971). The consequence of asexual reproduction for the mollusc is castration. The consequence for the trematode is that nutrients in the form of the host's reproductive allocation eventually become the limiting resource needed for producing cercariae.

Before I can discuss trematode behaviours inside the molluscan host, it is necessary to describe, in detail, how snail defences pose a challenge for trematodes. Bayne (1983) argues that molluscs do not have an immune system *per se*, and he and others have repeatedly chosen to use the term internal defence system (IDS) to describe a mollusc's ability to distinguish self and non-self and remove bacteria. Haemocytes that move, encapsulate, release cytotoxic superoxides and phagocytose are the main parts of the IDS that attack trematodes (Bayne *et al.*, 1980), and the more haemocytes the mollusc has at its disposal, the more effective the response is. This is why adult snails are generally less susceptible to infection than juvenile snails (Loker *et al.*, 1987); a larger blood volume translates into more haemocytes that the IDS can deploy. It is possible to infect normally resistant adult snails by exposing them to many miracidia simultaneously, presumably because this reduces the number of haemocytes per parasite (Loker *et al.*, 1987). Snail size is not the only factor affecting haemocyte counts; the snail must balance the energetic costs of defence against allocations to other life-history strategies, such as growth and avoidance of predators (Rigby and Jokela, 2000). This allocation to defence is plastic; snails can respond to infection attempts by increasing their investment in the IDS for at least 2 months (van der Knaap and Loker, 1990). The mollusc's IDS is not the only obstacle. Trematodes are subject to their own parasites, particularly microsporans of the genus *Nosema*, which infect larval trematodes and prevent cercariae from developing (Cort *et al.*, 1960).

Not surprisingly, trematodes have evolved defensive and offensive strategies to evade the mollusc's IDS. Some species seem to be able to prevent haemocytes from recognizing the carbohydrates on their surface that would otherwise denote non-self to the IDS. For example, the IDS will not detect a trematode if the trematode does not produce molecules that the IDS can recognize as non-self (by presenting either unrecognizable or unique epitopes (van der Knaap and Loker, 1990)). Alternatively, some trematodes can masquerade as host tissue, thereby escaping attack by the IDS. One way to do this is by expressing hostlike molecules on the tegument of the trematode's surface (Damian, 1987). Another approach is to use host substances found in the blood as a coating to mask the trematode's identity (van der Knaap and Loker, 1990). A final strategy is to interfere with the mollusc's IDS, primarily by impairing the spreading ability of haemocytes (Loker, 1994).

Most of what we know about trematode strategies to deal with the snail IDS is from studies of schistosomes and echinostomes (see reviews by van der Knaap and Loker, 1990; Loker *et al.*, 1992; Loker, 1994). Strategies employed by schistosomes may occur in other trematodes that have only sporocyst stages, whereas echinostomes may provide insight into the strategies of other trematodes with rediae (Lim and Heyneman, 1972).

It is increasingly clear that echinostomes possess the ability to interfere with the snail's IDS. For example, echinostomes appear to repel haemocytes from rediae (Adema *et al.*, 1994) by synthesizing and releasing > 100 kDa secretory/excretory products (SEPs), which are heat- and trypsin-labile (Loker *et al.*, 1992). In addition, haemocytes near rediae lose their adherence (Noda and Loker, 1989; Adema *et al.*, 1994), phagocytotic ability (Loker *et al.*, 1989) and normal spreading behaviour (Loker *et al.*, 1992). DeGaffe and Loker (1997) observed that the susceptibility of *Biomphalaria glabrata* to *Echinostoma paraensei* increases with the extent that SEPs interfere with the spreading behaviour of haemocytes, an effect that can vary with snail strain and be diluted in large snails. Although a snail's haemocyte production increases in response to infection, a successful repellent effect by SEPs can limit haemocytes to irrelevant areas of the snail (Lie *et al.*, 1977b). The localized nature of the effect on haemocytes is important, because it means that the mollusc maintains some ability to fight off bacterial or other infections that might shorten the mollusc's and the trematode's life (Loker, 1994). For example, it is probably important for *E. paraensei* that its host is still able to phagocytose foreign objects such as larval nematodes (*Angiostrongylus*), and to repair wounds (Lie *et al.*, 1981).

In comparison with echinostomes, schistosomes do little to the host's immune system (Lie *et al.*, 1976, 1977a, 1979). Perhaps as a result, schistosomes do not evoke an IDS response to future invading miracidia (Yoshino and Boswell, 1986). In some cases, snail haemocytes bind to the sporocysts readily but with no apparent effect (Loker *et al.*, 1989). Encapsulation by haemocytes does not necessarily kill sporocysts, probably due to phagocytotic or cytotoxic inhibitory substances that sporocysts release or indirectly mediate through the mollusc's central nervous system (de Jong-Brink, 1980; Riley and Chappell, 1992).

The host's IDS is not the only hostile component of the host environment. There are somewhere around 5000–10,000 trematode species (van der Knaap and Loker, 1990) and several trematode species can infect the same mollusc species in a particular location. Because a single miracidium has the capacity to fully convert the mollusc's reproductive output into cercarial production, if two trematode species successfully infect the same mollusc and coexist, each trematode's cercarial production will decline (DeCoursey and Vernberg, 1974; Robson and Williams, 1970; Walker, 1979). For this reason, one might expect: (i) miracidia to avoid infected molluscs; (ii) established trematodes to prevent infection by miracidia; and (iii) competitive strategies to displace a prior resident or prevent displacement.

Early on, Sewell (1922), faced with the observation that rarely did more than one species of trematode infect an individual snail, proposed that snails infected by trematodes might somehow lose their chemical attractiveness to miracidia. At present, there is little evidence that miracidia avoid already infected snails, even though this could be to their advantage if the target snail has been infected with a superior competitor (Lie, 1966; Sousa, 1992, 1993). However, there is some evidence that established rediae or sporocysts can reduce the ability of subsequent miracidia to establish. After infection, echinostomes temporarily induce the snail to produce a substance that immobilizes the miracidia of other echinostomes (it does not work against schistosome miracidia), such that, after infection with an echinostome, snails acquire resistance to other echinostomes (Lie *et al.*, 1979). However, most studies indicate that a miracidium is more likely to successfully attack an infected snail with a compromised IDS, a phenomenon termed 'acquired susceptibility' (Loker, 1994). A general pattern is that most opportunists (those that seem to do better in an already infected host) have sporocysts (Loker, 1994). Snails normally resistant to *Schistosoma mansoni* can become susceptible if infected with *E. paraensei*, even though the former does not eventually displace the latter under normal circumstances (Lie *et al.*, 1977a,b). Similarly, the rediae of *Calicophoron microbothrium* make the snail *Bulinis tropicus* more susceptible to infection by *Schistosoma bovis* (Southgate *et al.*, 1989). In some cases, sporocyst species, even with their limited impacts on the IDS, can increase susceptibility to secondary invasion (Lie and Heyneman, 1976; Lie *et al.*, 1973b). For example, *E. paraensei* preferentially invades *S. mansoni*-infected snails and then displaces the schistosome (Heyneman *et al.*, 1972).

Some trematode species may fully depend on a compromised mollusc IDS. For example, the estuarine schistosome *Austrobilharzia terrigalensis* is an obligate secondary invader of infected snails (Walker, 1979). Such species may be obligate secondary invaders because they require a mollusc host with a compromised IDS and have found ways to coexist (though not necessarily peacefully) with other species. They may even alter the host environment in such a way as to allow other species to coexist that might not normally be able to, as evidenced by the unusually high proportion of snails in which *Austrobilharzia* sp. occurs in triple-species infections (Martin, 1955). Once *Austrobilharzia* sp. invades, it reduces the growth and development of the preceding species that facilitated its invasion (Appleton, 1983). Because suppression of the mollusc's IDS is local, the safest place for an opportunist may be next to a suppressor, so long as the suppressor is not aggressive (Loker, 1994).

How do trematodes interact? Early investigators (Wesenberg-Lund, 1934) found that the most dramatic interactions involved species with rediae. The redia, with its muscular pharynx and ability to consume large chunks of host tissue, is capable of ingesting other trematodes (Heyneman and Umathevy, 1968), an interaction most appropriately termed intra-guild predation (Polis *et al.*, 1989). Such a strategy is particularly important if a trematode suppresses the IDS and therefore faces more frequent challenges from invading species (Loker, 1994). Trematodes with rediae tend to dominate trematodes with sporocysts. If an echinostome infects a schistosome-infected snail, the echinostome rediae will move about the periphery of a schistosome sporocyst mass, occasionally heading inside the mass to devour sporocysts, and will eventually displace the schistosome (Heyneman and Umathevy, 1968). After infection by an echinostome, the resident schistosome produces cercariae only as long as the echinostome daughter rediae are still young; both echinostome and schistosome cercariae will shed simultaneously for a brief period of time, indicating that the echinostome daughter rediae mature before displacing the schistosome (Heyneman and Umathevy, 1968). In addition to preying on sporocysts, rediae can actively seek out and ingest other rediae and cercariae. In some cases, they swallow their prey whole, while, in others, they bite a hole in the integument and suck the contents out. These abilities may be why redial species can afford to compromise the IDS of their host (Loker, 1994).

It may be possible for an established trematode to prevent a challenge from another species by preying on new invaders that may be at a dis-advantage in terms of their size or stage. Prior residence (Lie, 1966; Anteson, 1970) and the extent to which the trematode normally infects the host snail species (Heyneman and Umathevy, 1968) can influence dominance. For instance, among several similar-sized echinostomes, the dominant species is the one that first infects the snail (Lie and

Umathevy, 1965a,b; Lie *et al.*, 1965). However, large rediae tend to be dominant over small rediae (Kuris, 1990), a pattern that holds in observations of replacements of small species by large ones in the field (Sousa, 1993). For example, a large echinostome displaces a medium-sized heterophyid, which displaces a small cyathocotylid (Yoshino, 1975). There is even some evidence that redial size is a plastic trait that can change according to the need to compete; rediae of a particular species may be larger in double infections than they are in single infections (Lie, 1973; Lie *et al.*, 1973a). Recent observations have found that the *E. paraensei* mother sporocyst first produces a precocious mother redia (PMR) that remains adjacent to the mother sporocyst for at least a month (Sapp *et al.*, 1998). The PMR is long with a large pharynx (Sapp *et al.*, 1998), morphology consistent with dominance (Kuris, 1990). Although the PMR is not seen in all echinostomes, it does occur in other systems (T. Huspeni, personal observation). The size–dominance relationship allows the construction of potential dominance hierarchies based on redial measurements (Kuris, 1990). Exceptions can occur, however. An example of a non-linear dominance hierarchy is that a notocotylid is dominant to a small cyathocotylid, but not to a small microphallid, while the small cyathocotylid is dominant over the small microphallid (Kuris, 1990).

A few trematodes employ alternative competitive strategies. Most commonly, some species, particularly those with sporocysts, may secrete products that are toxic to other trematodes (Basch *et al.*, 1969; Lim and Heyneman, 1972; Lie, 1982). A particularly unusual interaction is hyperparasitism of other larval trematodes, as exhibited by *Cotylurus flabelliformis* (Cort *et al.*, 1941).

The ability of dominant trematodes to competitively exclude other species seems the best explanation for why infections with more than one species of trematode are much less frequent than one would expect if miracidia infected molluscs at random (Kuris and Lafferty, 1994). In fact, for some species pairs, it seems likely that the only times they occur in double infections are during transition from invasion to replacement. Although most researchers agree that intraguild predation makes it difficult for two species to coexist in the same individual mollusc, several authors have suggested that such opportunities are probably rare, asserting that the lack of double infections in molluscs is, in large part, because spatial and temporal heterogeneity in the distribution of miracidia isolates species from one another (Cort *et al.*, 1937; Curtis and Hubbard, 1990; Fernandez and Esch, 1991; Sousa, 1993). New methods of simultaneously analysing heterogeneity and competition (Lafferty *et al.*, 1994) have found that heterogeneity, while significant in many systems, usually has the unexpected effect of concentrating trematodes into a geographical subset of the snail population, thus intensifying competitive interactions among trematode species (Kuris and Lafferty, 1994; Stevens, 1996; Smith, 1999; Huspeni, 2000). Therefore, intraguild predation can significantly structure trematode communities in areas where prevalence is high, an effect that is easy to underestimate unless one uses appropriate sampling and analytical techniques.

Because molluscs are repeatable sampling units, larval trematode communities make tractable model systems for studying how inter-specific competition for limited resources can alter community structure, shedding light on some of the current competing paradigms in community ecology (Kuris, 1990). A consequence of interactions among trematodes is that competition with dominants can greatly alter the prevalence of sub-ordinate species, an effect that can be quantified (Kuris, 1990; Lafferty *et al.*, 1994). If subordinate trematode species play an important role in the ecosystem, such as affecting the behaviour or morbidity of intermediate or final hosts, the reduction in their abundance by dominant trematodes may alter whole ecosystems. Take the case of *E. californiensis*, the heterophyid that makes killifish easy for birds to catch (Lafferty and Morris, 1996). This trematode's abundance in a salt-marsh may increase foraging efficiency for piscivorous birds and decrease the killifish population. However, *E. californiensis* is subordinate to five other species (Kuris, 1990) and Sousa (1993) estimated it to be the most commonly excluded trematode species among the guild he studied. Lafferty *et al.* (1994) estimated that dominants reduced the prevalence of *E. californiensis* by 21% (for three other subordinate species, the reduction in prevalence due to competition was greater than 50%). Such a decrease in prevalence seems sufficient to significantly alter the ecology of the salt-marsh in that the loss of *E. californiensis* to dominant competitors could result in lower levels of predation on fish by birds.

A particularly relevant situation occurs when subordinate species affect human health. Schistosomiasis is one of the major impediments to public health and socio-economic progress in the developing world (WHO, 1993). Though some countries have been successful in reducing or eradicating the disease, in the poorest nations human modification of the environment (dams, irrigation canals, rice-fields and aquaculture), which leads to increased contact between humans and host snails, increases transmission (Lafferty and Kuris, 1999). Continued failures in developing effective, low-cost vaccines for schistosomes in humans (Gryseels, 2000) suggest that other effective, self-sustaining and ecologically benign alternatives should be explored. An approach missing from textbooks and strategic planning is to use trematodes that do not generally cause human disease (such as some echinostomes) against schistosomes. Firstly, adding such trematodes to an environment should have a negative effect on the snail population, providing fewer hosts for disease-causing schistosomes (Lafferty, 1993). In addition, many researchers have noted that, because schistosomes are subordinate to echinostomes, there should be negative associations between the two types of trematodes. Pilot studies intro-ducing redial species into populations of snails harbouring schistosomes have resulted in dramatic declines in, or even extirpation of, the schisto-some population, suggesting that redial species could actually be effective biological control agents against schistosomes (see review by Combes,

1982). As an example, I plot data from a field experiment that added schistosome eggs and then echinostome eggs to a pond of uninfected snails (Fig. 8.1 (from Table 1 in Lie and Ow-Yang, 1973)). The schistosomes initially infect half the snails, after which the echinostomes completely replace them and, several weeks later, drive the snail population to extinction through the effects of parasitic castration and increased mortality resulting either from multiple penetration by miracidia and cercariae or a build-up of metacercariae (Lie and Ow-Yang, 1973; Kuris and Warren, 1980). Despite this evidence, such an approach has not been adopted, largely due to theoretical models which find that the low global prevalence of schistosomes in snails should make efforts at reducing infected snails ineffective (Anderson and May, 1979). However, because spatial heterogeneity concentrates schistosomes into focal areas of high prevalence and high transmission, such locations could be targeted for treatment with echinostomes, with outcomes considerably more beneficial than those expected by general theory (Kuris and Lafferty, 1994).

Summary

As they move through their complex life cycles, trematodes have evolved behavioural adaptations to help them find the host habitat and to locate hosts within that habitat. They have found means to hide from or thwart

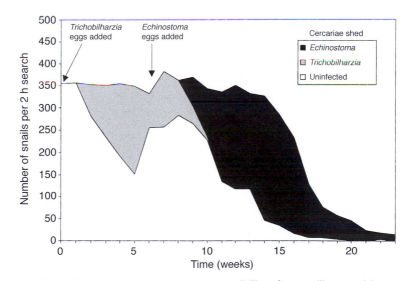

Fig. 8.1. A field experiment to assess the possibility of controlling a schistosome with an echinostome, demonstrating a complete displacement of the subordinate species through intraguild predation, followed by extirpation of the snail population (my plotting of published data (Lie and Ow-Yang, 1973)).

the defences of their hosts and deal with competitors for limited resources within the molluscan host, actions that structure trematode communities in accordance with dominance hierarchies based on size and aggressive (redial) morphology.

Acknowledgements

My appreciation to Todd Huspeni, Nick Kalodimos and Karen Miller for assistance in preparing the manuscript. Mike Sukhdeo and Ed Lewis provided valuable comments on an earlier draft of the manuscript. Finally, thanks to Armand Kuris for getting me interested in trematodes and Sam Loker and his colleagues for doing all the hard work on the IDS that helped me to make sense of everything.

References

Adema, C.M., Arguello, D.F.I., Stricker, S.A. and Loker, E.S. (1994) A time-lapse study of interactions between *Echinostoma paraensei* intramolluscan larval stages and adherent hemocytes from *Biomphalaria glabrata* and *Helix aspersa*. *Journal of Parasitology* 80, 719–727.

Aeby, G.S. (1991) Behavioral and ecological relationships of a parasite and its hosts within a coral reef system. *Pacific Science* 45, 263–269.

Anderson, R.M. and May, R.M. (1979) Prevalence of schistosome infection within molluscan populations: observed patterns and theoretical predictions. *Parasitology* 79, 63–94.

Anteson, R.K. (1970) On the resistance of the snail, *Lymnaea catascopium pallida* (Adams) to concurrent infection with sporocysts of the strigeid trematodes, *Cotylurus flabelliformis* (Faust) and *Diplostomum flexicaudum* (Court and Brooks). *Tropical Medicine and Parasitology* 64, 101–107.

Appleton, C.C. (1983) Studies on *Austrobilharzia terrigalensis* (Trematoda, Schistosomatidae) in the Swan estuary, Western-Australia – observations on the biology of the cercaria. *International Journal for Parasitology* 13, 239–247.

Basch, P.F., Lie, K.J. and Heyneman, D. (1969) Antagonistic interaction between strigeid and schistosome sporocysts within a snail host. *Journal of Parasitology* 55, 753–758.

Bayne, C.J. (1983) Molluscan immunobiology. In: Saleuddin, A.S.M. and Wilbur, K.M. (eds) *The Mollusca*. Academic Press, New York, pp. 407–486.

Bayne, C.J., Buckley, P.M. and Dewan, P.C. (1980) Macrophage-like hemocytes of resistant *Biomphalaria glabrata* are cyto-toxic for sporocysts of *Schistosoma mansoni* in vitro. *Journal of Parasitology* 66, 413–419.

Campbell, R.A. (1997) Host-finding behaviour of *Cotylurus flabelliformis* (Trematoda: Strigeidae) cercariae for snail hosts. *Folia Parasitologica* 44, 199–204.

Christensen, N.Ø., Nansen, P., Fagbemi, B.O. and Monrad, J. (1987) Heterologous antagonistic and synergistic interactions between helminths and between helminths and protozoans in concurrent experimental infection of mammalian hosts. *Parasitology Research* 73, 387–410.

Christensen, N.Ø., Odaibo, A.B. and Simonsen, P.E. (1988) *Echinostoma* population regulation in experimental rodent definitive hosts. *Parasitology Research* 75, 83–87.

Combes, C. (1982) Trematodes: antagonism between species and sterilizing effects on snails in biological control. *Parasitology* 84, 151–175.

Cort, W.W., McMullen, D.B. and Brackett, S. (1937) Ecological studies on the cercariae in *Stagnicola emarginata angulata* (Sowerby) in the Douglas Lake region. *Journal of Parasitology* 23, 504–552.

Cort, W.W., Oliver, L. and Brackett, S. (1941) The relation of physid and planorbid snails to the life cycle of the strigeid trematode *Cotylurus flabelliformis*. *Journal of Parasitology* 27, 437–448.

Cort, W.W., Hussey, K.L. and Ameel, D.J. (1960) Studies on a microsporidian hyperparasite of strigeoid trematodes. I. Prevalence and effect on the parasitized larval trematodes. *Journal of Parasitology* 46, 317–325.

Curtis, L. and Hubbard, K. (1990) Trematode interactions in a gastropod host misrepresented by observing shed cercariae. *Journal of Experimental Marine Biology and Ecology* 143, 131–137.

Damian, R.T. (1967) Common antigens between adult *Schistosoma mansoni* and the laboratory mouse. *Journal of Parasitology* 53, 60–64.

Damian, R.T. (1987) Molecular mimicry revisited. *Parasitology Today* 3, 263–266.

DeCoursey, J. and Vernberg, W.B. (1974) Double infections of larval trematodes: competitive interactions. In: Vernberg, W.B. (ed.) *Symbiosis in the Sea*. University of South Carolina Press, Columbia, pp. 93–109.

DeGaffe, G. and Loker, E.S. (1997) Susceptibility of *Biomphalaria glabrata* to infection with *Echinostoma paraensei*: correlation with the effect on parasite secretory–excretory products on host hemocyte spreading. *Journal of Invertebrate Pathology* 71, 64–72.

de Jong-Brink, M. (1980) How trematode parasites interfere with reproduction of their intermediate hosts, freshwater snails. *Journal of Medical and Applied Malacology* 2, 101–133.

de Jong-Brink, M., Elasaadany, M.M. and Boer, H.H. (1988) *Trichobilharzia ocellata*: interference with the endocrine control of female reproduction of its host *Lymnaea stagnalis*. *Experimental Parasitology* 68, 93–98.

Dönges, J. (1971) The potential number of redial generations in echinostomatids (trematoda). *International Journal for Parasitology* 1, 51–59.

Feiler, W. and Haas, W. (1988) Host-finding in *Trichobilharzia ocellata* cercariae: swimming and attachment to the host. *Parasitology* 96, 493–506.

Fernandez, J. and Esch, G.W. (1991) The component community structure of larval trematodes in the pulmonate snail *Helisoma anceps*. *Journal of Parasitology* 77, 540–550.

Gryseels, B. (2000) Schistosomiasis vaccines: the devil's advocate's final plea. *Parasitology Today* 16, 357–358.

Haas, W. (1988) Host-finding – a physiological effect. In: Mehlhorn, H. (ed.) *Parasitology in Focus*. Springer Verlag, New York, pp. 454–464.

Haas, W. (1994) Physiological analyses of host-finding behaviour in trematode cercariae: adaptations for transmission success. *Parasitology* 109(S), S15-S29.

Helluy, S. (1983) Un mode de favorisation de la transmission parasitaire: la manipulation du comportement de l'hôte intermédiare. *Revue d'Écologie, la Terre et la Vie* 38, 211–223.

Heyneman, D. and Umathevy, T. (1968) Interaction of trematodes by predation within natural double infections in the host snail *Indoplanorbis exustus*. *Nature* 217, 283–285.

Heyneman, D., Lim, H.K. and Jeyarasa, U. (1972) Antagonism of *Echinostoma liei* (Trematoda–Echinostomatidae) against trematodes *Paryphostomum segregatum* and *Schistosoma mansoni*. *Parasitology* 65, 223–233.

Huspeni, T. (2000) A molecular genetic analysis of host specificity, continental geography, and recruitment dynamics of a larval trematode in a salt marsh snail. PhD thesis, University of California, Santa Barbara.

Isseroff, H. and Cable, R.M. (1968) Fine structure of photoreceptors in larval trematodes. *Zeitschrift für Zellforschung und Mikroskopische Anatomie* 86, 511–534.

Kawashima, K., Tada, I. and Miyazaki, I. (1961) Host preference of miracidia of *Paragonimus ohirai* Miyazaki, 1939 among three species of snails of the genus *Assiminea*. *Kyushu Journal of Medical Science* 12, 99–106.

Kuris, A.K. (1990) Guild structure of larval trematodes in molluscan hosts: prevalence, dominance and significance of competition. In: Esch, G.W., Bush, A.O. and Aho, J.M. (eds) *Parasite Communities: Patterns and Processes*. Chapman & Hall, New York, pp. 69–100.

Kuris, A.M. and Lafferty, K.D. (1994) Community structure: larval trematodes in snail hosts. *Annual Review of Ecology and Systematics* 25, 189–217.

Kuris, A.M. and Warren, J. (1980) Echinostome cercarial penetration and metacercarial encystment as mortality factors for a second intermediate host, *Biomphalaria glabrata*. *Journal of Parasitology* 66, 630–635.

Lafferty, K.D. (1993) Effects of parasitic castration on growth, reproduction and population dynamics of the marine snail *Cerithidea californica*. *Marine Ecology Progress Series* 96, 229–237.

Lafferty, K.D. (1999) The evolution of trophic transmission. *Parasitology Today* 15, 111–115.

Lafferty, K.D. and Kuris, A.M. (1999) How environmental stress affects the impacts of parasites. *Limnology and Oceanography* 44, 925–931.

Lafferty, K.D. and Morris, A.K. (1996) Altered behavior of parasitized killifish increases susceptibility to predation by bird final hosts. *Ecology* 77, 1390–1397.

Lafferty, K.D., Sammond, D.T. and Kuris, A.M. (1994) Analysis of larval trematode communities. *Ecology* 75, 2275–2285.

Lie, K.J. (1966) Antagonistic interaction between *Schistosoma mansoni* sporocysts and echinostome rediae in the snail *Australorbis glabratus*. *Nature* 211, 1213–1215.

Lie, K.J. (1973) Larval trematode antagonism: principles and possible application as a control method. *Experimental Parasitology* 33, 343–349.

Lie, K.J. (1982) Survival of *Schistosoma mansoni* and other trematode larvae in the snail *Biomphalaria glabrata* – a discussion of the interference theory. *Tropical Geography Medicine* 34, 111–122.

Lie, K.J. and Heyneman, D. (1976) Studies on resistance in snails. 6. Escape of *Echinostoma lindoense* sporocysts from encapsulation in snail heart and subsequent loss of host's ability to resist infection by same parasite. *Journal of Parasitology* 62, 298–302.

Lie, K.J. and Ow-Yang, C.K. (1973) A field trial to control *Trichobilharzia brevis* by dispersing eggs of *Echinostoma audyi*. *Southeast Asian Journal of Tropical Medicine and Public Health* 4, 208–217.

Lie, K.J. and Umathevy, T. (1965a) Studies on Echinostomatidae (Trematoda) in Malaya. VIII. The life history of *Echinostoma audyi*, sp. n. *Journal of Parasitology* 51, 781–788.

Lie, K.J. and Umathevy, T. (1965b) Studies on Echinostomatidae (Trematoda) in Malaya. X. The life history of *Echinoparyphium dunni* sp. n. *Journal of Parasitology* 51, 793–799.

Lie, K.J., Basch, P.F. and Umathevy, T. (1965) Antagonism between two species of larval trematodes in the same snail. *Nature (London)* 206, 422–423.

Lie, K.J., Lim, H.K. and Ow-Yang, C.K. (1973a) Antagonism between *Echinostoma audyi* and *Echinostoma hystricosum* in the snail *Lymnaea rubiginosa* with a discussion on patterns of trematode interaction. *Southeast Asian Journal of Tropical Medicine and Public Health* 4, 504–508.

Lie, K.J., Lim, H.K. and Ow-Yang, C.K. (1973b) Synergism and antagonism between two trematode species in the snail *Lymnaea rubiginosa*. *International Journal for Parasitology* 3, 719–733.

Lie, K.J., Heyneman, D., and Jeong, K.H. (1976) Studies on resistance in snails. 7. Evidence of interference with defense reaction in *Biomphalaria glabrata* by trematode larvae. *Journal of Parasitology* 62, 608–615.

Lie, K.J., Heyneman, D. and Richards, C.S. (1977a) *Schistosoma mansoni* – temporary reduction of natural resistance in *Biomphalaria glabrata* induced by irradiated miracidia of *Echinostoma paraensei*. *Experimental Parasitology* 43, 54–62.

Lie, K.J., Heyneman, D. and Richards, C.S. (1977b) Studies on resistance in snails – interference by nonirradiated echinostome larvae with natural resistance to *Schistosoma mansoni* in *Biomphalaria glabrata*. *Journal of Invertebrate Pathology* 29, 118–125.

Lie, K.J., Jeong, K.H. and Heyneman, D. (1979) Inducement of miracidia-immobilizing substance in the hemolymph of *Biomphalaria glabrata*. *International Journal for Parasitology* 10, 183–188.

Lie, K.J., Jeong, K.H. and Heyneman, D. (1981) Selective interference with granulocyte function induced by *Echinostoma paraensei* (Trematoda) larvae in *Biomphalaria glabrata* (Mollusca). *Journal of Parasitology* 67, 790–796.

Lim, H.K. and Heyneman, D. (1972) Intramolluscan inter-trematode antagonism: a review of factors influencing the host–parasite system and its possible role in biological control. *Advances in Parasitology* 10, 191–268.

Loker, E.S. (1994) On being a parasite in an invertebrate host: a short survival course. *Journal of Parasitology* 80, 728–747.

Loker, E.S., Cimino, D.F., Stryker, G.A. and Hertell, L.A. (1987) The effect of size of M line *Biomphalaria glabrata* on the course of development of *Echinostoma paraensei*. *Journal of Parasitology* 73, 1090–1098.

Loker, E.S., Boston, M.E. and Bayne, C.J. (1989) Differential adherence of M line *Biomphalaria glabrata* hemocytes to *Schistosoma mansoni* and *Echinostoma paraensei* larvae, and experimental manipulation of hemocyte binding. *Journal of Invertebrate Pathology* 54, 260–268.

Loker, E.S., Cimino, D.F. and Hertell, L.A. (1992) Excretory–secretory products of *Echinostoma paraensei* sporocysts mediated interference with *Biomphalaria glabrata* hemocyte functions. *Journal of Parasitology* 78, 104–115.

Martin, W.E. (1955) Seasonal infections of the snail, *Cerithidea californica* Haldman, with larval trematodes. In: *Essays in Natural Science in Honor of Captain Alan Hancock on the Occasion of his Birthday*, pp. 203–210.

Martorelli, S.R. (1994) A new cystophorous cercaria (Digenea, Hemiuriformes) in *Potamolithus agapetus* (Mollusca, Hydrobiidae): behaviour of host attraction. *Iheringia Serie Zoologia* 0(77), 15–19.

Mohandas, A. and Nadakal, A.M. (1978) *In vivo* development of *Echinostoma malayanum* Leiper 1911 with notes on effects of population density, chemical composition and pathogenicity and *in vitro* excystment of the metacercaria (Trematoda: Echinostomatidae). *Zeitschrift für Parasitenkunde* 55, 139–151.

Nassi, H., Pointier, J.P. and Golvan, Y.J. (1979) Bilan d'un essai de contrôle de *Biomphalaria glabrata* en Guadalupe à l'aide d'un trématode stérilisant. *Annales de Parasitologie* 52, 277–323.

Noda, S. and Loker, E.S. (1989) Effects of infection with *Echinostoma paraensei* on the circulating haemocyte population of the host snail *Biomphalaria glabrata*. *Parasitology* 98, 35–41.

Nolf, L. and Court, W.W. (1933) On immunity reactions of snails to the penetration of the cercariae of the strigeid trematode, *Cotylurus flabelliformis*. *Journal of Parasitology* 20, 38–48.

Nollen, P.M. (1994) The hatching behavior of *Echinostoma trivolvis* miracidia and their responses to gravity, light and chemicals. *International Journal for Parasitology* 24, 637–642.

Olivier, L.J. (1966) Infectivity of *Schistosoma mansoni* cercariae. *American Journal of Tropical Medicine and Hygiene* 15, 882–885.

Polis, G.A., Myers, C.A. and Holt, R.D. (1989) The ecology and evolution of intraguild predation: potential competitors that eat each other. *Annual Review of Ecology and Systematics* 20, 292–330.

Rigby, M.C. and Jokela, J. (2000) Costs and trade-offs between predator avoidance and immune defense in snails. *Proceedings of the Royal Academy of London B* 267, 171–176.

Riley, E.M. and Chappell, L.H. (1992) Effect of infection with *Diplostomum spathaceum* on the internal defense system of *Lymnaea stagnalis*. *Journal of Invertebrate Pathology* 59, 190–196.

Robson, E.M. and Williams, I.C. (1970) Relationships of some species of Digenea with the marine prosobranch *Littorina littorea* (L.) I. The occurrence of larval Digenea in *L. littorea* on the North Yorkshire Coast. *Journal of Helminthology* 44, 153–168.

Sapp, K.K. and Loker, E.S. (2000a) Mechanisms underlying digenean-snail specificity: role of miracidial attachment and host plasma factors. *Journal of Parasitology* 86, 1012–1019.

Sapp, K.K. and Loker, E.S. (2000b) A comparative study of mechanisms underlying digenean-snail specificity: *in vitro* interactions between hemocytes and digenean larvae. *Journal of Parasitology* 86, 1020–1029.

Sapp, K.K., Meyer, K.A. and Loker, E.S. (1998) Intramolluscan development of the digenean *Echinostoma paraensei*: rapid production of a unique mother redia that adversely affects development of conspecific parasites. *Invertebrate Biology* 117, 20–28.

Sewell, S. (1922) Cercariae indicae. *Indian Journal of Medical Research* 10, 1–327.

Smith, N.F. (1999) Life history dynamics of the mangrove snail, *Cerithidea scalariformis*, and the recruitment patterns of its trematode parasites. PhD thesis, University of California, Santa Barbara.

Smithers, R.S., McLaren, D.J. and Ramalho-Pinto, F.J. (1977) Immunity to schistosomes: the target. *American Journal of Tropical Medicine and Hygiene* 26(S), 11–19.

Sousa, W.P. (1992) Interspecific interactions among larval trematode parasites of fresh-water and marine snails. *American Zoologist* 32, 583–592.

Sousa, W.P. (1993) Interspecific antagonism and species coexistence in a diverse guild of larval trematode parasites. *Ecological Monographs* 63, 103–128.

Southgate, V.R., Brown, D.S., Warlow, A., Knowles, R.J. and Jones, A. (1989) The influence of *Calicophoron microbothrium* on the susceptibility of *Bulinus tropicus* to *Schistosoma bovis*. *Parasitology Research* 75, 381–391.

Stevens, T. (1996) The importance of spatial heterogeneity and recruitment in organisms with complex life cycles: an analysis of digenetic trematodes in a salt marsh community. PhD thesis, University of California, Santa Barbara.

Sukhdeo, M.V.K. and Bansemir, A.D. (1996) Critical resources that influence habitat selection decisions by gastrointestinal helminth parasites. *International Journal for Parasitology* 26, 483–498.

Théron, A. (1984) Early and late shedding patterns of *Schistosoma mansoni* cercariae: ecological significance in transmission to human and murine hosts. *Journal of Parasitology* 4, 652–655.

Thomas, F. and Poulin, R. (1998) Manipulation of a mollusc by a trophically transmitted parasite: convergent evolution or phylogenetic inheritance? *Parasitology* 116, 431–436.

Thomas, F., Mete, K., Helluy, S., Santalla, F., Verneau, O., De Meeüs, T., Cézilly, F. and Renaud, F. (1997) Hitch-hiker parasites or how to benefit from the strategy of another parasite. *Evolution* 51, 1316–1318.

Thomas, F., Poulin, R. and Renaud, F. (1998) Nonmanipulative parasites in manipulated hosts: 'hitch-hikers' or simply 'lucky passengers'? *Journal of Parasitology* 84, 1059–1061.

van der Knaap, W.P.W. and Loker, E.S. (1990) Immune mechanisms in trematode–snail interactions. *Parasitology Today* 6, 175–182.

Walker, J.C. (1979) *Austrobilharzia terrigalensis*: a schistosome dominant in interspecific interactions in the mollusc host. *International Journal for Parasitology* 9, 137–140.

Wesenberg-Lund, C. (1934) Contributions to the development of the Trematoda Digena. Part II: the biology of the freshwater cercariae in Danish fresh waters. *Mémoires de l'Académie Royale des Sciences et des Lettres de Danemark Section des Sciences 9* 5, 1–223.

WHO (1993) *The Control of Schistosomiasis. Second Report of the WHO Expert Committee*, Vol. 830. WHO, Geneva.

Wright, C.A. (1971) *Bulinus* on Aldabra and the sub-family Bulininae in the Indian Ocean area. *Philosophical Transactions of the Royal Society of London B* 260, 299–313.

Wright, C.A. and Ross, G.C. (1966) Electrophoretic studies on planorbid egg-proteins: the *Bulinus africanus* and *B. forskali* species group. *Bulletin of the World Health Organization* 35, 727–731.

Yoshino, T.P. (1975) A seasonal and historical study of larval Digenea infecting *Cerithidea californica* (Gastropoda: Prosobranchia) from Goleta Slough, Santa Barbara County, California. *Veliger* 18, 156–161.

Yoshino, T.P. and Boswell, C.A. (1986) Antigen sharing between larval trematodes and their snail hosts: how real a phenomenon in immune evasion? In: Lackie, A. (ed.) *Immune Mechanisms in Invertebrate Vectors*. Clarendon Press, New York, pp. 221–238.

Niche Restriction and Mate Finding in Vertebrate Hosts

9

Klaus Rohde

School of Biological Sciences, University of New England,
Armidale, NSW 2351, Australia

Introduction

Niche restriction is universal among parasites, and it has been suggested that it is at least partly the result of selection which favours mate finding and thus cross-fertilization, leading to greater genetic diversity ('mating hypothesis'). In this chapter, I review evidence for niche restriction at the level of hosts (host range and host specificity), populations (aggregation) and host individuals (microhabitat specificity). I also discuss mating behaviour and particularly mate finding, the morphological basis of mating, using some recent light- and electron-microscopic studies, evidence for the role of niche restriction in facilitating mating, and the relative importance of niche segregation and differences in copulatory organs for reinforcing reproductive barriers. No attempt is made to discuss all these aspects for each of the parasite groups in detail. Instead, I select examples from those groups that have been examined best, in order to clarify patterns and to test the mating hypothesis of niche restriction and segregation. It is emphasized that only a holistic approach can clarify the significance of mating in niche restriction, i.e. an approach that considers all evidence jointly.

Niche Restriction

Restriction to host species (host specificity and host range)

There is no 'universal' parasite, i.e. a parasite that infects all species, although the number of hosts used varies among parasite species. It is useful to distinguish host range and host specificity. Host range is defined as the number of host species used by a parasite, irrespective of how heavily and how frequently particular host species are infected. Host specificity takes intensity and prevalence (frequency) of infection into consideration. Rohde (1980) has developed specificity indices that measure host specificity. He defined the host specificity index based on intensity of infection as:

$$S_i \text{ (intensity)} = \Sigma(x_{ij}/n_j h_{ij})/\Sigma(x_{ij}/n_j)$$

where S_i = host specificity in the ith parasite species, x_{ij} = number of parasite individuals of the ith species in the jth host species, n_j = number of host individuals of the jth species examined, and h_{ij} = rank of host species j based on intensity of infection. If all parasite species of a community are considered, the specificity index of the parasite community can be defined as $S_i \text{ (intensity)} = \dfrac{\Sigma S_i}{n_p}$, where n_p = number of parasite species in the community.

The index can be modified to take frequencies (prevalences) of infection instead of intensities into consideration. Parameters in the formula are now changed as follows: x_{ij} = number of host individuals of jth species infected with parasite species i, n_j = number of host individuals of jth species examined, and h_{ij} = rank of host species based on frequency of infection. One can also make use of intensities as well as frequencies of infection, for instance by using a combined index S_i (intensity) + S_i (frequency).

Numerical values for the indices vary between 0 and 1: the closer to 1, the higher the degree of host specificity, the closer to 0, the lower. For example, many Digenea from the White Sea have a wide host range, i.e. they infect many host species, but few of them heavily (Shulman and Shulman-Albova, 1953): *Lecithaster gibbosus* was found in 12 of 31 fish species examined, but the vast majority of all parasites of this species were found in a single host species, and its S_i (intensity) is consequently 0.99. An example of a parasite with very wide host range and low host specificity is *Toxoplasma gondii*, which infects many species of mammals; examples of parasites with very high host specificity, infecting a single host species, are many monogeneans: *Polystomoides malayi* has been found only in the freshwater turtle *Cuora amboinensis*, and some other species of the same genus are restricted to single host species as well. Rohde (1978) has shown that host ranges of marine monogeneans are small at all latitudes, but that host ranges of marine digenean trematodes show a gradient of decreasing host ranges from cold to warm seas. In contrast, host specificity is similarly great at all latitudes in both groups.

Restriction to host individuals (aggregation within host populations)

Almost as universal as restriction to certain host species is aggregation within host populations, i.e. some host individuals are much more heavily infected than expected on a random basis, and some (or most) are little infected or not infected at all. Very often, aggregation can best be described by a negative binomial distribution. For example, MacKenzie and Liversidge (1975) found that the distribution of metacercariae of *Stephanostomum baccatum* from a particular locality could almost perfectly be described by a negative binomial distribution. Studies of latitudinal gradients in aggregation have not been made.

Restriction to microhabitats within or on host individuals

All parasites prefer certain microhabitats to others, although the degree of microhabitat restriction varies. Thus, larvae of the nematode *Trichinella spiralis* infect the striated muscles throughout the body of many mammal species, whereas adults of the same species are restricted to the small intestine of the same hosts. Different species of Mallophaga and ticks live on different parts of the body of particular bird species (e.g. Dubinin, 1948, 1951; review in Dogiel, 1962; Fig. 9.1). Reasons for the differential distribution of ticks, according to Dubinin (1951), are the spatial position and function of various groups of feathers, the macro- and microscopic structure of the feathers of each group, and the marked differences in the microstructure of feathers of different groups responsible for differences in aerodynamic properties. Marshall (1981) has given many other examples of site specificity in ectoparasitic insects. Thus, Mallophaga and Anoplura of mammals show considerable site specificity, although apparently less so than bird lice. The most important factors controlling distribution appear to be hair type and temperature. In cool weather the cattle anopluran *Haematopinus eurysternus* occurs largely on the neck, but in hot weather it is found around the ears, horns and tails. The crab louse of humans, *Phthirus pubis*, is found largely in the pubic and perianal regions of the body, but it has also been recovered from other areas of coarse hair, e.g. the armpits, beard and eyelashes. Certain trematode species live in the stomach, others in certain sections of the small intestine and others again in the rectum of teleost fishes and other hosts. The aspidogastrean *Rugogaster hydrolagi* infects the rectal glands and the aspidogastrean *Multicalyx elegans* the gall-bladder and bile-ducts of chimaerid fishes, whereas the nematodes *Dirofilaria immitis* and *Dirofilaria repens* infect the heart and associated blood-vessels, or the skin of some mammals, respectively. Site restriction has been particularly well investigated in monogeneans of fishes. For example, five species of monogeneans are found on the gills of the mackerel *Scomber australasicus*. One species is found only on the pseudobranchs, a second species at the base of the filaments of the main gills, a third species in the middle of the filaments, a fourth only on the most posterior (and probably most anterior) filaments and a fifth species is scattered over all gill filaments along their whole length (Fig. 9.2). This site restriction is observed even in the absence of other species, i.e. the parasites are

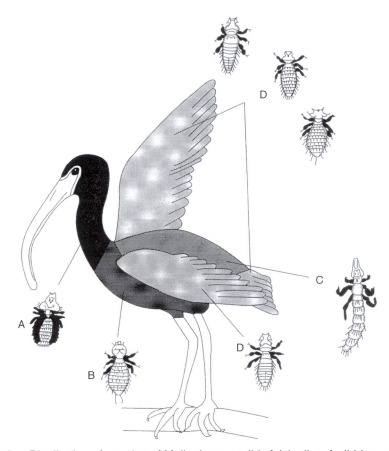

Fig. 9.1.   Distribution of species of Mallophaga on *Ibis falcinellus*. A. *Ibidoecus bisignatus*, feathers of head and neck. B. *Menopon plagadis*, feathers of chest, sides and abdomen. C. *Esthiopterum rhaphidium*, feathers of the dorsum.
D. Species of *Colpocephalum* and *Ferribia*, feathers of wings and tail. Redrawn and strongly modified after V.B. Dubinin (in Dogiel, 1962).

genetically programmed to select certain microhabitats, using particular (but unknown) cues (for a recent study of niche preferences in gill Monogenea, see also Gutiérrez and Martorelli, 1999).

Mating

The role of mating in hermaphroditic parasites

The common occurrence of sexual reproduction in most animal (and plant) species clearly indicates that it has significant selective advantages, but there is still debate on what these advantages are. Escape from parasites and the accelerated evolution of beneficial traits have been

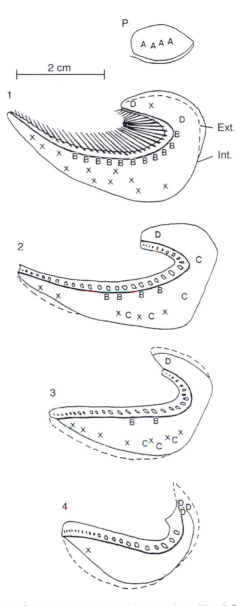

Fig. 9.2. Distribution of monogenean parasites on the gills of *Scomber australasicus* on the coast of south-eastern Australia. P, pseudobranch; 1–4, gills 1–4; Ext., external gill filaments; Int., internal gill filaments; A, B, C, three species of *Kuhnia*; D, *Grubea australis*; x, *Pseudokuhnia minor*. Note: all species of *Kuhnia* and *Grubea* share an identical male copulatory organ, while *Pseudokuhnia* has a different copulatory organ (see Fig. 9.6). (From Rohde, 2001a.)

suggested (e.g. Ochoa and Jaffe, 1999), although an increase in diversity is probably the most commonly accepted explanation. However, among angiosperms, dioecious clades have fewer species than hermaphroditic ones, i.e. dioecy does not promote radiation, at least in these plants (Heilbuth, 2000). The importance of cross-fertilization is made particularly clear by the fact that even many hermaphroditic animals mate. Most parasitic Platyhelminthes, i.e. the Monogenea, Trematoda and Cestoda, are hermaphroditic. Kearn (1970) described mating in the skin-parasitic monogenean *Entobdella soleae*; he found no evidence for self-fertilization (see also Kearn *et al.*, 1993). In the viviparous *Gyrodactylus alexanderi*, the first daughter animal is produced without cross-fertilization, although cross-fertilization may occur in the production of the second embryo (Lester and Adams, 1974). Lo (1999) concluded that mating is the 'main focus of monogenean life', based on an extensive study of a monogenean infecting small tropical reef fish (details below). Finally, many, if not most, monogeneans have very complex copulatory organs, and this alone suggests that mating is of extreme importance.

Didymozoid trematodes and the Chinese liver fluke, parasitic in the tissues of vertebrates, usually encapsulate in pairs, apparently to ensure cross-fertilization. Several authors have observed that echinostome trematodes self-fertilize and produce viable eggs (e.g. Beaver, 1937). Nollen (1968, 1978, 1990, 1993, 1994a,b, 1996a,b, 1997a,b, 1999, 2000) made thorough experimental studies of the mating behaviour of trematodes, using unlabelled flukes and flukes labelled with ^{3}H-tyrosine and transplanted into mice or hamsters. He found that three species of eye flukes (*Philophthalmus*) never self-inseminated when occurring in groups ('restricted mating'), but that several species of echinostomes, as well as *Zygocotyle lunata* and *Megalodiscus temperatus*, either self- or cross-inseminated when occurring in groups ('unrestricted mating'). In one study, Nollen (1996a), using unlabelled worms and worms labelled with ^{3}H-tyrosine transplanted into mice and recovered after 5 days, has shown that, of nine labelled *Echinostoma paraensei,* four (44%) had self-inseminated. Of the 39 unlabelled worms found with the labelled ones, only eight (21%) had cross-inseminated, and six of the labelled worms (75%) had also self-inseminated. Trouve *et al.* (1999b) used double infections with *Echinostoma caproni* in which the two individuals did or did not belong to the same isolate from a certain geographical area, and triple infections in which two of the three individuals originated from the same isolate and the third one from a different isolate. They found no differences between intra- and interisolate selfing rates in the first experiment, and they found preferential outcrossing between individuals from the same isolate, apparently in order to avoid hybrid breakdown, in the second experiment. Saito (1984) noted that most *Echinostoma hortense* had a cirrus inserted into their own metraterm. However, such observations are not evidence for successful self-insemination. Rohde (1973) observed that the eggs of three specimens of *Lobatostoma manteri* (Aspidogastrea) in the intestine of naturally infected *Trachinotus blochi,* occurring singly, appeared normal, but egg cells had the haploid number of chromosomes and did not divide at all or development ceased at a very early stage. Nevertheless, specimens of this species with the cirrus inserted into their own metraterm were seen (Fig. 4 in Rohde, 1973).

Older references (in Rohde, 1993) indicate that self-fertilization appears to be common in the hermaphroditic cestodes, but cross-fertilization may occur. Schärer and Wedekind (1999) found that the facultatively self-fertilizing cestode *Schizocephalus solidus* produced fewer eggs when paired, but that egg production lasted longer.

Thus, in summary, both self- and cross-fertilization occurs in the parasitic Platyhelminthes, but some species at least do not self-fertilize, and it is not known whether or not species for which self-fertilization has been demonstrated need cross-fertilization over longer periods and many generations to maintain the viability of the species.

In this context, the detailed studies of the free-living hermaphroditic nematode *Caenorhabditis elegans* are relevant. Lamunyon and Ward (1995) found that the male's sperm outcompete the hermaphrodite's own sperm in mating between male and hermaphroditic worms, suggesting that this trait maximizes outcrossing and increases genetic diversity and heterozygosity of offspring whose parents may be highly homozygous due to self-fertilization. However, mating occurs at a cost. Mating with males reduces the lifespan of hermaphroditic worms for reasons not related to egg production or receipt of sperm, whereas males do not seem to be affected (Gems and Riddle, 1996).

It has been suggested that the nematode *Strongyloides ratti* reproduces by pseudogamy, but Viney *et al.* (1993) have shown by minisatellite DNA fingerprinting that there is genetic exchange apparently due to sexual reproduction.

Mate Finding

Lo (1999) examined 365 humbugs, *Dascyllus aruanus*, a small (few cm long) coral-reef fish, infected with a single species of gill monogenean, *Haliotrema* sp., and recorded the number of 'couples', i.e. monogeneans showing microhabitat overlap, on fish with low intensities of infection (fewer than five monogeneans per gill). A very high percentage of monogeneans occurred in couples, suggesting (possibly chemical) attraction by worms, because intraspecies microhabitat overlap should be very scarce if coupling occurred on a random basis. Attraction seems to be effective over relatively great distances, for example between gill arches. Haseeb and Fried (1988) also suggested that adult monogeneans, *Diclidophora merlangi*, migrate towards each other. Lo (1999) concluded that 'mating is a main focus of monogenean life', supporting Rohde's mating hypothesis. Kearn *et al.* (1993) made extensive studies of the monogenean *E. soleae*, a parasite of the sole, *Solea solea*, in England. They concluded that the parasite is unable to self-fertilize in the wild. Mathematical modelling showed that random locomotion with searching movements of the body would lead to mate finding on small fish within the reproductive lifetime of the worm. However, it is unlikely that mating will occur soon after maturation. Attraction by pheromones could ensure early mating and the thin layer of water, stagnant between the lower surface of the fish and the sea bottom, would be an ideal environment for the action of such pheromones. The authors suggest that this may explain the preferred habitat of the worm: the lower surface of the fish. However, pheromones have not yet been demonstrated.

Nollen (1997b) has given a review of mating and chemical attraction in trematodes, Basch (1991) reviewed the reproductive biology in the dioecious schistosomes and Haas (2000) included a brief section on mating in his review of the behavioural ecology of echinostomes. *In vitro* experiments by Fried and collaborators showed chemoattraction between immature and mature adults of echinostomes. The free sterol fraction of lipophilic extracts of the worms was found to be responsible (Fried *et al.*, 1980; Fried and Diaz, 1987; Fried and Haseeb, 1990). Interspecific attraction between adult *E. caproni* and *Echinostoma trivolvis* also occurred, but was less significant than that between individuals of *E. trivolvis*. Interestingly, *E. trivolvis*, which occurs along the whole small intestine, showed greater intraspecific chemoattraction than *E. paraensei*, which is very site-specific – to the duodenum. Trouve and Coustau (1998) demonstrated differences in the excretory–secretory products of three strains of *E. caproni*, which may be the basis of selective mating. In this species, populations are highly subdivided along their microhabitat, the small intestine, promoting local mate competition (Trouve *et al.*, 1999b). An alternative mechanism of reproductive isolation, not dependent on mate attraction, is suggested by the findings of Trouve and Coustau (1999). These authors examined mating between two geographical isolates of *E. caproni* and another species of the same genus. They recorded similar attraction for intraisolate, interisolate and interspecific combinations and suggested that reproductive isolation may be due to sperm selection. Some studies have demonstrated chemoattraction by neutral lipids in mating of schistosomes (review in Nollen, 1997b). *In vitro*, excretion of such lipids is increased in the presence of other worms, and males and females that were separated and allowed to reunite paired more frequently with the original partner than with new ones. Males of schistosomes may have more than one female in their gynaecophoric groove, or they may share females or also hold males if there is an undersupply of females. Also, female reproductive organs do not fully develop in unpaired worms, and females of some species (review in Basch, 1991) need stimulation by males to continue growth and production of eggs. Female reproductive organs even regress in the absence of a male, but production of eggs begins again after re-pairing. Females obtain glucose from males, and male extracts stimulate development of female vitellaria. Some schistosomes, at least, may also reproduce by parthenogenesis (references in Nollen, 1997b). In the two species *Schistosoma haematobium* (primarily parasitic in humans) and *Schistosoma mattheei* (parasitic in livestock and wild ungulates), hybridization, leading to the production of viable offspring, occurs in experimentally infected mice (Southgate *et al.*, 1995), indicating that mate attraction is not limited to males and females of the same species. However, hybridization in nature is highly unlikely, considering the different hosts. Also, specific mate choice systems do not appear to exist in some other combinations of different *Schistosoma* species, whereas they do in others (Southgate *et al.*, 1998). Nevertheless, overall, most schistosomes infecting the same final host maintain their genetic identity (Southgate *et al.*, 1998).

Among the acanthocephalans, endoparasites of vertebrates, aggregation for mating is known to occur in some species. In a study on *Acanthocephaloides propinquus* infecting the gobiid fish *Gobius bucchichii*, Sasal *et al.* (2000) determined infection intensities, sex ratios and testicular size and found that, when the percentage of males increased (suggesting increased competition between males), the size of the testes increased as well. The authors also concluded that competition for access to females was more important than competition for space.

Numerous studies of mating of nematodes have been made, but most of them deal with the free-living *C. elegans*. Copulation involves a series of complicated steps (e.g. Liu and Sternberg, 1995), copulatory plugs are used to assure paternity (Barker, 1994), and even mutations that affect neurons involved in certain copulatory behaviour patterns have been identified (Loer and Kenyon, 1993). In *T. spiralis*, differences in distributions of mixed, single male and single female infections suggest that males migrate in search of females (Sukhdeo and Bansemir, 1996), and experimental studies have indeed shown that males and females are attracted to each other (references in Sukhdeo and Bansemir, 1996). The same authors report that, when females and males of *Heligmosomoides polygyrus* are transplanted together into the terminal ileum of mice, males migrate faster to the duodenum (their normal habitat) than females, leaving the females behind, implying that males choose habitat over female. This is hardly an argument against the importance of mate selection, since mating occurs once both sexes are established in the duodenum.

Mating behaviour of copepods has been studied in a number of species, but most of them are free-living. Sophisticated behaviour patterns lead to mate encounters, and males show distance chemoreception, as well as recognition at close range, which may involve fluid mechanism signals of short duration and contact chemical signals. i.e. species- and sex-specific glycoproteins (Boxshall, 1998, and references therein). Precopulatory mate-guarding appears to be common not only in free-living species, but in parasitic species as well (e.g. in *Lernaeocera branchialis* (Heuch and Schram, 1996)). The mating behaviour of *Lepeophtheirus salmonis* and some other parasitic copepods was described by Anstensrud (1990a–d, 1992) and Ritchie *et al.* (1996). The latter study suggests that long-range, short-range and contact chemical stimuli play a role in pairing and mate recognition, either on their own or jointly. It must be emphasized that sea lice are highly mobile, both on and between their hosts (salmon) (Hull *et al.*, 1998), which may explain the involvement of long-range chemical stimuli. But, overall, little is known about the role of chemoreception in mating sea lice (Hull *et al.*, 1998).

The Morphological Basis of Mating and Microhabitat Finding (Copulatory Organs and Sensory Receptors)

Many Monogenea, a group for which niche restriction and segregation have been particularly well studied, have an amazing variety of complex male (and sometimes female) copulatory organs (some examples in Fig. 9.6). This alone points to the importance of mating, even in hermaphroditic species. In this group, numerous electron-microscopic studies have been made, which have shown the presence of several types of sensory receptors (Figs 9.3–9.5) possibly involved in mating, as well as perhaps in feeding and microhabitat finding. However, the functions of the receptors are not known. Comparison of Figs 9.4 and 9.5 shows certain similarities in the receptors of two monogenean species infecting the gills of different fish species and belonging to different families (Microcotylidae and

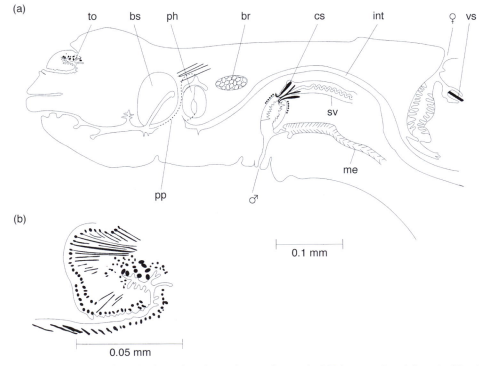

Fig. 9.3. Diagram of sagittal section through anterior end of *Pricea multae* (a) and of 'taste organ' (b). br, brain; bs, buccal sucker; cs, copulatory spicules; int, intestine; me, metraterm; ph, pharynx; pp, prepharynx; sv, seminal vesicle; to, taste organ; vs, vaginal spine. (From Rohde and Watson, 1996.)

Gastrocotylidae), but it also shows marked differences. Thus, both species possess similar multiciliate and non-ciliate receptors, but sizes and numbers of electron-dense collars vary. In the future, comparative ultrastructural and functional studies of such species may clarify the role of the various receptor types in distinguishing microhabitats and mating partners. In the aspidogastrean *L. manteri*, at least eight types of receptors, some occurring in very great numbers, have been demonstrated, and

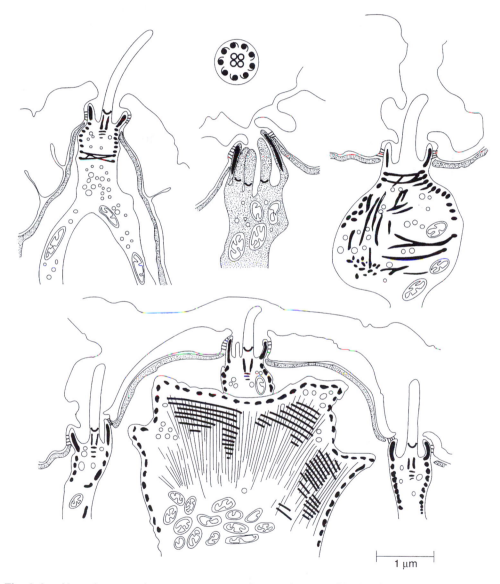

1 μm

Fig. 9.4. Non-pharyngeal sensory receptors in anterior part of body of *Pricea multae* (Monogenea, Polyopisthocotylea, Gastrocotylidae) from the gills of *Scomberomorus commerson* (from Rohde and Watson, 1996).

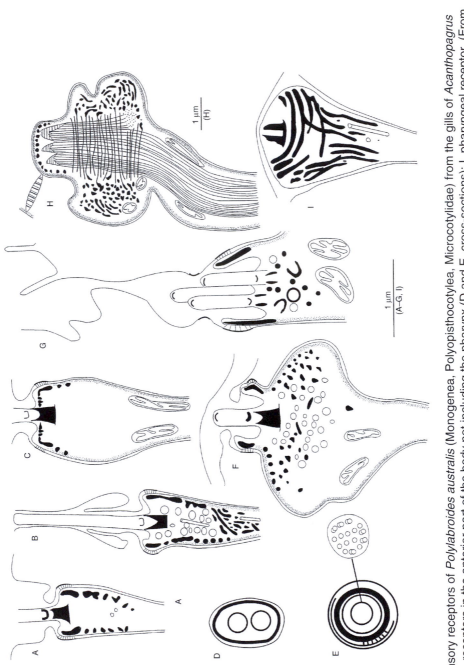

Fig. 9.5. Sensory receptors of *Polylabroides australis* (Monogenea, Polyopisthocotylea, Microcotylidae) from the gills of *Acanthopagrus australis*. A–H, receptors in the anterior part of the body not including the pharynx (D and E, cross-sections); I, pharyngeal receptor. (From Rohde and Watson, 1995.)

the total number of receptors was estimated to be about 20,000–40,000 (Rohde, 1989). All stages in the life cycle of this species are parasitic (with the exception of the egg) (Rohde, 1973); the assumption is therefore justified that the receptors are important in microhabitat finding, feeding and/or mating, but, again, functional studies of the receptors have not been made.

The free-living nematode *C. elegans* has been studied in much greater detail than parasitic ones. Mating involves a series of steps; by laser microbeam ablation of male-specific copulatory organs and their associated neurons, it was possible to identify sensory structures and neurons responsible for each step (Liu and Sternberg, 1995). Using stereo images of the amphids (the main chemoreceptory organs of nematodes) of *Strongyloides stercoralis*, three-dimensional reconstructions could be made, as previously done for *C. elegans* (Ashton and Schad, 1996; also Ashton *et al.*, 1999). A role in host finding and the control of development has been suggested, but a role in mating is also possible. Ashton *et al.* (1999) have reviewed chemo- and thermosensory neurons in several parasitic nematodes. These neurons are thought to play a role in host finding and development. The interpretation is based on comparisons with studies of *C. elegans*. Neurons involved in mating have apparently not been identified in parasitic nematodes.

In the parasitic copepod *L. salmonis*, ablation of the tip of the antennules, which carry many receptors (Gresty *et al.*, 1993), reduces mating success. However, the mechanism of action and the sensory roles of the setae on the antennules are not known. Also, even ablated males still formed pairs, although pair formation and mating were delayed (Hull *et al.*, 1998).

The Role of Niche Restriction in Facilitating Mating

Rohde (1979, also 1994, and further references therein) gave the following evidence for the mating hypothesis of niche restriction: host ranges and microhabitats of parasites on the gills (and of other parasites) are often extremely restricted, although competing species are not present; species with good locomotory ability often have larger microhabitats than less mobile species; asexual or larval stages often have wider microhabitats than adult, sexually reproducing stages; species that can establish large populations often have wider microhabitats than less populous species; and species often aggregate, i.e. reduce their microhabitat width, at the time of mating. Examples for the effect of locomotory ability are differences in microhabitat width of mobile and sessile monogeneans and copepods parasitizing fish; examples for differences in microhabitat width of asexual/larval and sexual stages are larval *Trichinella*, protozoan cysts on the gills and metacercariae vs. adult *Trichinella*, adult Monogenea and Trematoda; and examples for the effect of population size are many gill Monogenea. Examples of aggregation for mating are some acanthocephalans and monogeneans. Kamegai (1986) reported that juveniles of the monogenean *Diplozoon nipponicum* on *Carassius carassius auratus* are found on all gills until the fourth day of infection; then they gather on one gill and copulate for life (further examples in Rohde, 1994).

Some recent studies with information on the mating hypothesis are discussed in the following. Geets *et al.* (1997) tested the importance of interspecific competition and intraspecific factors in ectoparasites of the white-spotted rabbit-fish, *Siganus sutor*, on the Kenyan coast and found that intraspecific factors are much more important than interspecific factors in choosing microhabitats. They specifically tested the mating hypothesis in two species of Monogenea and found that niche restriction on the gills leads to higher intraspecific contact, consistent with the hypothesis. The finding of Kearn *et al.* (1993), discussed above, that niche restriction in a monogenean infecting European sole may play a role in allowing chemoattraction of mates also supports the hypothesis.

Echinostome trematodes use small parts of the intestine and this is thought to contribute to mate finding, although chemical communication also plays a role, as shown by the studies of Nollen (see above). In this context, the observation is important that *E. trivolvis*, which occurs along the whole small intestine, showed greater intraspecific chemoattraction than *E. paraensei*, which is very site-specific – to the duodenum. According to Sukhdeo and Mettrick (1987) and Sukhdeo and Sukhdeo (1994), non-directed activation-dependent or contact-dependent mechanisms may be more important than chemotaxis during migration within hosts; for mate finding, this would imply an even greater role of microhabitat restriction.

Nollen (1996b) found that *E. paraensei* lives in the duodenum and *E. caproni* in the ileum of mice, into which they were transplanted. Nevertheless, 25% of the worms of both species were found within 1 cm of each other. Lack of interspecies mating, demonstrated by autoradiographic studies, cannot therefore be due to microhabitat segregation alone.

Adamson and Caira (1994) have argued that the mating hypothesis implies greater site specificity in gynochoristic than in hermaphroditic and particularly selfing hermaphroditic species. However, we have seen above that, even in (selfing) hermaphrodites, cross-fertilization occurs and that it may be important in all hermaphroditic species over many generations. It could be argued that in hermaphrodites site restriction is even more important than in dioecious species, in order to force them into at least occasional cross-fertilization. Adamson and Caira (1994) further argued that greater mobility and greater densities of species with increased microhabitat might be expected even without selection for greater mating success. These arguments are less convincing if species are compared that belong to the same parasite group and live in the same habitat. For example, metacercariae almost always have wide microhabitats in their vertebrate hosts, but adult didymozoid trematodes, also parasitic in the tissues of vertebrates, often have extremely restricted microhabitats; monogeneans that are permanently sessile generally have much smaller microhabitats than vagile and populous species of the same group. There is indeed evidence that microhabitat width increases in some species when infection intensities increase, but this is very often not the case (Rohde, 1991). Nevertheless, Adamson and Caira's (1994) arguments have to be taken seriously, and only a holistic viewpoint will lead to an estimate on how widely the mating hypothesis can be applied, taking all evidence jointly into account, e.g. general saturation of niche space with species and abundance of populations, in addition to the points discussed at the beginning of this section. Also, occasional exceptions, e.g. that in the guinea-worm mating occurs prior to habitat selection, do not prove that a rule is invalid. It is also important to note that free-living insects, i.e. trypetid flies, were shown to use certain plants as a meeting place for mating (Zwölfer, 1974).

Concerning the role of aggregation of parasites in host populations, it is generally thought to reduce the net deleterious effects of parasites on host populations (e.g. Jaenike, 1996, and further references therein), but Jaenike (1996) has used mathematical modelling to show that, under certain conditions, the opposite may be true. Concerning the role in mating, May (1977) has demonstrated that, in dioecious species of parasites, mating probability is increased if both sexes share the same negative binomial distribution, but it is reduced if they have different distributions. For schistosomes, the former is indeed the case. Galvani and Gupta (1998) have examined this further and concluded that mating probability also depends on whether worms are monogamous or promiscuous. In the latter case, mating probability is increased.

The Relative Roles of Niche Segregation and Differences of Copulatory Organs in Reinforcing Reproductive Barriers

To maintain their identity, i.e. to avoid interspecific hybridization, species need isolating mechanisms, which may be differences in microhabitats, hosts, geographical area, etc. Examples for site segregation can be found in the section on microhabitat restriction (see, for example, Figs 9.1 and 9.2), and an example of geographical isolation is two species of *Grubea* on the gills of *Scomber* spp. (legend of Fig. 9.6). Further examples can be found in Rohde (1993). Intraspecific chemoattraction is another factor that may be involved. The discussion above (section on mate finding) has shown that trematodes, for example, exhibit stronger intra- than interspecific attraction, and that selection of appropriate sperm (i.e. sperm belonging to the same species) may occur. Synxenic monogeneans (i.e. monogeneans infecting the same host species) are excellent examples to demonstrate the relative importance of site segreation and morphological differences in copulatory sclerites for the maintenance of specific identity. Closely related species (usually belonging to the same genus) that possess identical copulatory organs use

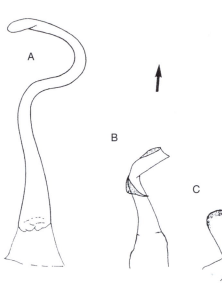

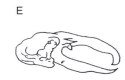

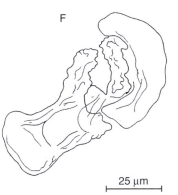

25 μm

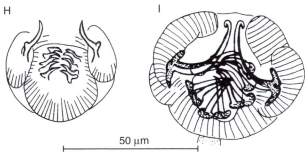

50 μm

different sites, whereas closely related species with markedly different copulatory organs may inhabit the same site (Figs 9.2 and 9.6). This suggests that site segregation has not evolved to avoid interspecific competition, but has the function of preventing interspecific hybrid-ization (Rohde, 1991). These conclusions are supported by the recent study of nine species of the monogenean *Dactylogyrus*, coexisting on the gills of the roach, *Rutilus rutilus*, by Simková *et al.* (2000, and further references therein). The authors found differences in the structure of copulatory organs between species aggregating interspecifically and no evidence for a major role of interspecific competition. A review of published work suggests that polyopisthocotylean Monogenea have fewer types of copulatory sclerites and, consequently, rely more on spatial segregation than monopisthocotylean Monogenea with their much greater variety of copulatory sclerites.

Evolutionary Maintenance of Niche Restriction

Combes and Théron (2000) discussed the mating hypothesis and two mechanisms that may maintain niche restriction after host speciation or host switching. Because of the importance of their study, it is discussed here in greater detail. The mechanisms for maintaining specificity include alloxenic speciation (i.e. speciation on different hosts) by habitat prefer-ences (in an 'encounter arms race') and alloxenic speciation by assortive survival (in a 'compatibility arms race'). The authors point out that specialization provides constraints and benefits. Among the first, the most important are the limitation of resources and the increased risk of extinction; among the second are the increased chances of outcrossing, especially when infection intensities are low, increasing genetic diversity. Limitation of interspecific competition and generally 'better adaptation' to a specialized habitat may also be important, but effects on genetic diversity may constitute the main benefit. In an encounter arms race, natural selection improves the probability of the parasite meeting a host and it improves the probability of the host avoiding the infective stages of a parasite. In a compatibility arms race, natural selection favours the probability of survival of a parasite in or on a host and it favours the probability of the host killing the parasite. At any time in evolution, an 'encounter filter' and a 'compatibility filter', representing 'crossed

Fig. 9.6. (opposite) Copulatory organs of monopisthocotylean monogeneans infecting the gills of *Lethrinus miniatus* on the Great Barrier Reef, Australia (A–G), and of polyopisthocotylean monogeneans infecting the gills of *Scomber* spp. (H, I). A–C, *Haliotrema* spp. in overlapping microhabitats. D, F, G, *Calydiscoides* spp. in overlapping microhabitats. E, *Protolamellodiscus* sp. on the pseudobranch. H, three species of *Kuhnia* and two species of *Grubea* spatially segregated in different microhabitats or in different geographical areas. I, *Pseudokuhnia minor* overlapping with four species of *Kuhnia* and *Grubea*. (A–G from Rohde *et al.*, 1994.)

phenotypes', can be used to symbolize the state of the two arms races. Behavioural adaptations may reinforce host specificity, e.g. by eliminating those host species whose behaviour or other characteristics do not correspond with those of the infective stages of the parasite. For example, each species of *Schistosoma* has a shedding pattern that is adapted to the behaviour of the host. A stable polymorphism can be established when the genetic variability of hosts and parasites permits lateral transfer (e.g. in *Schistosoma mansoni* in humans and black rats); and this may represent the beginning of reproductive isolation and alloxenic speciation. Combes and Théron (2000) also suggest that gene flow may be restricted between parasites inhabiting different microhabitats of the same host species (synxenic speciation), for example between congeneric trematodes living in different parts of the digestive tract of the same host species.

Concerning the second mechanism, habitat compatibility, Combes and Théron (2000) emphasize that parasites will find it difficult to invest in evasion mechanisms against several different host defences, which may explain adaptation to certain hosts. An example is the association between the parasitoid wasp *Leptophilina houlardi* and *Drosophila melanogaster* in different regions. In an area where the wasp attacks five instead of a single species of *Drosophila*, the defence reactions of the flies to the parasitoid are much more effective.

The question of why some parasites are much more specific than closely related species cannot be answered. Reasons may be different durations of the parasite–host association or the extreme sensitivity of the balance between benefit and constraint to slight changes in selective pressures.

Most of the arguments given by Combes and Théron (2000) are convincing. However, it must be stressed that evidence for synxenic speciation of parasites occupying different microhabitats on or in vertebrates is not available. In a study using molecular techniques, Littlewood *et al.* (1997) found that polystome monogeneans inhabiting the same microhabitat (urinary bladder/cloaca or pharynx/mouth cavity) of different host species are more closely related to each other than polystomes infecting different sites in the same host species. This suggests that speciation has been alloxenic. On the other hand, the studies by Tauber and Tauber (1977a,b) and Tauber *et al.* (1977) on plant-parasitic insects have demonstrated reproductive isolation as the result of few mutations and habitat/seasonal isolation, lending support to the view that sympatric speciation occurs.

The mating hypothesis can be further generalized by including additional niche parameters, such as seasonality, host age, sex of hosts, ecological specificity (adaptations to macroenvironmental factors of the host) and geographical isolation.

Supplementary Mechanisms for Niche Restriction: Specialization and Avoidance of Interspecific Competition

It has never been suggested that selection for more effective mating is the only mechanism leading to niche restriction. An important ultimate cause of niche restriction may be specialization not just for more effective mating, but for more effective use of resources (Price, 1980), such as food and sites for attachment. A widely accepted view is that the primary function of niche restriction is the avoidance of intra- and interspecific competition. There is indeed evidence that microhabitat width and host ranges may be affected by the presence of other species, but many studies have shown that, in parasites, competition is of minor importance (reviews in Rohde, 1991, 1994, 1999). Sukhdeo *et al.* discuss interspecific competition in Chapter 11 (this volume) and I therefore mention only some major points. Most species of marine fish harbour few parasite species, and abundance of infections is low (Fig. 9.7). Consequently, many vacant niches are available, or, in other words, resources are in oversupply and there is no need for competition to occur. Packing rules derived from spatial scaling laws, which predict unimodal distributions skewed to the left in plots of species numbers against the size of species and a decline of body-size ratios of species of adjacent sizes with increasing size of organisms, apply to few, if any, parasites (Rohde, 2001b; Fig. 9.8), because species are not densely packed and do not compete for limiting resources. Some recent studies demonstrating the availability of empty niches and the lack of competition are by Sasal *et al.* (1999): digeneans of Mediterranean fish; Buchmann (1989), Dzika and Szymanski (1989), Koskivaara *et al.* (1992), Bagge and Valtonen (1999): monogeneans of freshwater fish; and Ramasamy *et al.* (1985): monogeneans of marine fish. Sousa (1994) has reviewed the evidence for interspecific interactions in parasite communities and concluded that such effects are important in some parasite communities, but not in others. However, although interspecific effects occur, evidence for their evolutionary significance does not exist. It may well be that, generally, such effects, where they occur, are intermittent and have no lasting effect on community structure (Price, 1980).

In summary, parasite data clearly indicate that niche space is largely empty (Figs 9.7 and 9.8). This strongly suggests that avoidance of competition is not an important factor in niche restriction and segregation, much less important than the necessity to find suitable habitats for survival (by specialization) and mating partners (mating hypothesis). As stated by Rohde (1979): 'Niche diversification is self-augmenting, and in a continuously expanding niche space populations would be diluted to such a degree that mating would become impossible without the counteracting selection for niche restriction.'

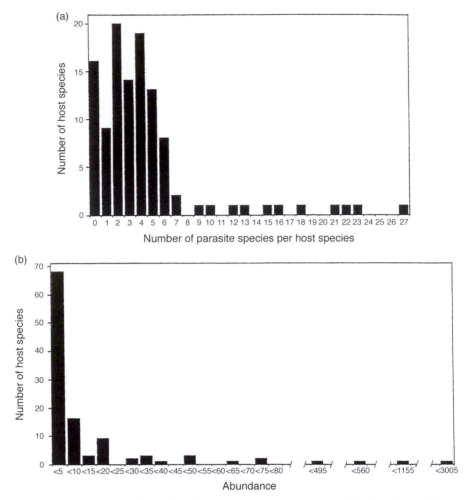

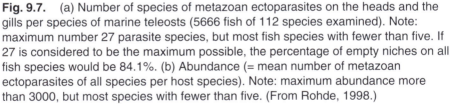

Fig. 9.7. (a) Number of species of metazoan ectoparasites on the heads and the gills per species of marine teleosts (5666 fish of 112 species examined). Note: maximum number 27 parasite species, but most fish species with fewer than five. If 27 is considered to be the maximum possible, the percentage of empty niches on all fish species would be 84.1%. (b) Abundance (= mean number of metazoan ectoparasites of all species per host species). Note: maximum abundance more than 3000, but most species with fewer than five. (From Rohde, 1998.)

Conclusions and Suggestions for Future Work

This chapter has shown that niche restriction at the levels of host species, host population and microhabitats within or on host individuals is universal among parasites. It has also been shown that, even in hermaphroditic parasites, cross-fertilization is important (and obligatory at least for some species) and that niche segregation is often not the result of interspecific competition, but of selection to reduce the chances of

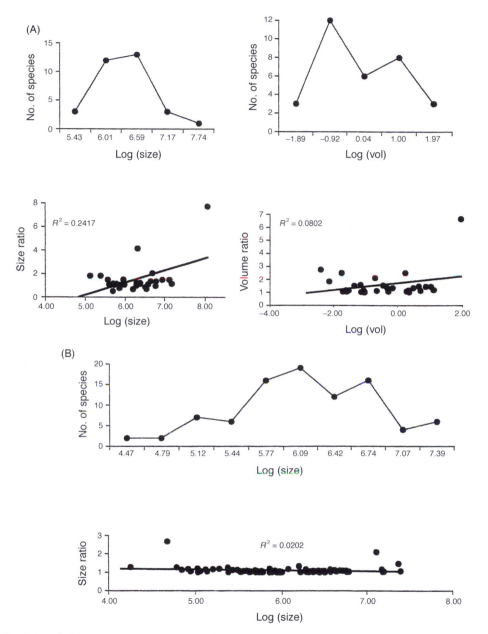

Fig. 9.8. A. Metazoan endoparasites of the marine teleost fish *Acanthopagrus australis*. Number of species of different sizes (maximum width × maximum length) and volumes (maximum width × maximum length × maximum depth) and size ratios of species with adjacent sizes. Note: 11 trematodes, ten nematodes, five cestodes, four acanthocephalans. All except two species parasites of the digestive tract. Outlier a very large tissue nematode. B. Metazoan parasites in the digestive tract of marine fishes (1808 fish of 47 species). Note that the parasites do not conform to the packing rules of Ritchie and Olff (1999). (From Rohde, 2001b.)

interspecific hybridization. Niche restriction must have the effect of facilitating contact between mating partners, although evidence is circumstantial. The morphological basis of host, site and mate finding is incompletely known. Thus, although the ultrastructure of sensory receptors of many species of platyhelminths is known, nothing is known about their function. Experimental studies are necessary to fill this gap. In particular, comparative ultrastructural and functional studies of related alloxenic and synxenic species may yield clues as to the function of receptors in finding niches and mating partners. Overall, evidence suggests that the facilitation of mating and segregation to avoid interspecific hybridization are of very great importance in structuring communities and determining niche width.

Acknowledgements

I wish to thank Sandy Hamdorf for expertly redrawing Figs 9.1 and 9.5, the editor of the *Encyclopedia of Biodiversity* and Academic Press for permission to use Fig. 9.2, the editor of *Folia Parasitologica* for permission to use Figs 9.3 and 9.4, the editor of the *International Journal for Parasitology* for permission to use Figs 9.5 and 9.6 and the editor of *Oikos* for permission to use Figs 9.7 and 9.8.

References

Adamson, M.L. and Caira, J.N. (1994) Evolutionary factors influencing the nature of parasite specificity. *Parasitology* 109, S85–S95.

Anstensrud, M. (1990a) Moulting and mating in *Lepeophtheirus pectoralis* (Copepoda, Caligidae). *Journal of the Marine Biological Association of the UK* 70, 269–281.

Anstensrud, M. (1990b) Mating strategies of two parasitic copepods (*Lernaeocera branchialis* (L.) (Pennellidae) and *Lepeophtheirus pectoralis* (Müller) (Caligidae)) on flounder: polygamy, sex-specific age at maturity and sex ratios. *Journal of Experimental Marine Biology and Ecology* 136, 141–158.

Anstensrud, M. (1990c) Effects of mating on female behaviour and allometric growth in the two parasitic copepods (*Lernaeeocera branchialis* (L. 1767) (Pennellidae) and *Lepeophtheirus pectoralis* (Müller 1776) (Caligidae)). *Crustaceana* 59, 245–258.

Anstensrud, M. (1990d) Male reproductive characteristics of two parasitic copepods, *Lernaeocera branchialis* (L.) (Pennellidae) and *Lepeophtheirus pectoralis* (Müller) (Caligidae). *Journal of Crustacean Biology* 10, 627–638.

Anstensrud, M. (1992) Mate guarding and mate choice in two copepods, *Lernaeocera branchialis* (L.) (Pennellidae) and *Lepeophtheirus pectoralis* (Müller) (Caligidae), parasitic on flounder. *Journal of Crustacean Biology* 12, 31–40.

Ashton, F.T. and Schad, G.A. (1996) Amphids in *Strongyloides stercoralis* and other parasitic nematodes. *Parasitology Today* 12, 187–194.

Ashton, F.T., Li, J. and Schad, G.A. (1999) Chemo- and thermosensory neurons: structure and function in animal parasitic nematodes. *Veterinary Parasitology* 84, 297–316.

Bagge, A.M. and Valtonen, E.T. (1999) Development of monogenean communities on the gills of roach fry (*Rutilus rutilus*). *Parasitology* 118, 479–487.

Barker, D.M. (1994) Copulatory plugs and paternity assurance in the nematode *Caenorhabditis elegans*. *Animal Behaviour* 48, 147–156.

Basch, P.F. (1991) *Schistosomes: Development, Reproduction, and Host Relations.* Oxford University Press, Oxford, 248 pp.

Beaver, P.C. (1937) Experimental studies on *Echinostoma revolutum* (Froelich), a fluke from birds and mammals. *Illinois Biological Monographs* 15, 7–96.

Boxshall, G.A. (1998) Mating biology of copepod crustaceans – preface. *Philosophical Transactions of the Royal Society of London – Series B: Biological Sciences* 353, 669–670.

Buchmann, K. (1989) Microhabitats of monogenean gill parasites on European eel (*Anguilla anguilla*). *Folia Parasitologica* 36, 321–329.

Combes, C. and Théron A. (2000) Metazoan parasites and resource heterogeneity: constraints and benefits. *International Journal for Parasitology* 30, 299–304.

Dogiel, V.A. (1962) *General Parasitology.* Leningrad University Press, Leningrad (English translation 1964, Oliver and Boyd, Edinburgh), 516 pp.

Dubinin, V.B. (1948) Studies on the adaptations of ectoparasites: ecological adaptations of the mallophagans and feather-ticks. *Parasitologitscheskij sbornik* 9, 191–222 (in Russian).

Dubinin, V.B. (1951) *Analgesid ticks. I. Introduction and the State of Knowledge.* Fauna of the USSR 6(5), Publications of the Zoological Institute, Akademii Nauk, Moscow.

Dzika, E. and Szymanski, S. (1989) Co-occurrence and distribution of Monogenea of the genus *Dactylogyrus* on gills of bream, *Abramis brama* L. *Acta Parasitologica Polonica* 34, 1–14.

Fried, B. and Diaz, V. (1987) Site-finding and pairing of *Echinostoma revolutum* (Trematoda) on the chick chorioallantois. *Journal of Parasitology* 73, 546–548.

Fried, B. and Haseeb, M.A. (1990) Intra- and interspecific chemoattraction in *Echinostoma caproni* and *E. trivolvis* adults *in vitro*. *Journal of the Helminthological Society of Washington* 57, 72–73.

Fried, B., Tancer, R.B. and Fleming, S.J. (1980) *In vitro* pairing of *Echinostoma revolutum* (Trematoda) metacercariae and adults, and characterisation of worm products involved in chemoattraction. *Journal of Parasitology* 66, 1014–1018.

Galvani, A. and Gupta, S. (1998) The effects of mating probability on the population genetics of nematodes. *Journal of Helminthology* 72, 295–300.

Geets, A., Coene, H. and Ollevier, F. (1997) Ectoparasites of the whitespotted rabbitfish, *Siganus sutor* (Valenciennes, 1835) off the Kenyan coast – distribution within the host population and site selection on the gills. *Parasitology* 115, 69–79.

Gems, D. and Riddle, D.L. (1996) Longevity in *Caenorhabditis elegans* reduced by mating, but not gamete production. *Nature* 379, 723–725.

Gresty, K.A., Boxshall, G.A. and Nagasawa, K. (1993) Antennullary sensors of the infective copepodid larvae of the salmon louse, *Lepeophtheirus salmonis* (Copepoda: Caligidae). In: Boxshall, G.A and Defaye, D. (eds) *Pathogenesis of Wild and Farmed Fish: Sea Lice.* Ellis Horwood, New York, pp. 83–98.

Gutiérrez, P.A. and Martorelli, S.R. (1999) Niche preferences and spatial distribution of Monogenea on the gills of *Pimelodus maculatus* in Río de la Plata (Argentina). *Parasitology* 119, 183–188.

Haas, W. (2000) The behavioral biology of echinostomes. In: Fried, B.J. and Graczyk, T.K. (eds) *Echinostomes as Experimental Models for Biological Research*. Kluwer, Dordrecht, the Netherlands, pp. 175–197.

Haseeb, M.A. and Fried, B. (1988) Chemical communication in helminths. *Advances in Parasitology* 27, 170–207.

Heilbuth, J.C. (2000) Lower species richness in dioecious clades. *American Naturalist* 156, 221–241.

Heuch, P.A. and Schram, T.A. (1996) Male mate choice in a natural population of the parasitic copepod *Lernaeocera branchialis* (Copepoda, Pennellidae). *Behaviour* 133, 221–239.

Hull, M.Q., Pike, A.W., Mordue, A.J. and Rae, G.H. (1998) Patterns of pair formation and mating in an ectoparasitic caligid copepod *Lepeophtheirus salmonis* (Kroyer, 1837) – implications for its sensory and mating biology. *Philosophical Transactions of the Royal Society of London – Series B: Biological Sciences* 353, 753–764.

Jaenike, J. (1996) Population-level consequences of parasite aggregation. *Oikos* 76, 155–160.

Kamegai, S. (1986) Studies on *Diplozoon nipponicum* Goto, 1891. (43) The gathering phenomenon of diporpae and the effect of cortisone acetate on the union of diporpae. In: Howell, M.J. (ed.) *Parasitology – Quo Vadis? 2. Handbook, Program and Abstracts, 6th International Congress of Parasitology*. Australian Academy of Sciences, Canberra.

Kearn, G.C. (1970) The production, transfer and assimilation of spermatophores by *Entobdella soleae*, a monogenean skin parasite of the common sole. *Parasitology* 60, 301–311.

Kearn, G.C., James, R. and Evansgowing, R. (1993) Insemination and population density in *Entobdella soleae*, a monogenean skin parasite of the common sole, *Solea solea*. *International Journal for Parasitology* 23, 891–899.

Koskivaara, M., Vltonen, E.T. and Vuori, K.M. (1992) Microhabitat distribution and co-existence of *Dactylogyrus* species (Monogenea) on the gills of roach. *Parasitology* 104, 273–281.

Lamunyon, C.W. and Ward, S. (1995) Sperm precedence in a hermaphroditic nematode (*Caenorhabditis elegans*) is due to competitive superiority of male sperm. *Experientia* 51, 817–823.

Lester, R.J.G. and Adams, J.R. (1974) *Gyrodactylus alexanderi*: reproduction, mortality, and effects on its host *Gasterosteus aculeatus*. *Canadian Journal of Zoology* 52, 827–833.

Littlewood, D.J.T., Rohde, K. and Clough, K.A. (1997) Parasite speciation within or between host species? Phylogenetic evidence from site-specific polystome monogeneans. *International Journal for Parasitology* 27, 1289–1297.

Liu, K.S. and Sternberg, P.W. (1995) Sensory regulation of male mating behavior in *Caenorhabditis elegans*. *Neuron* 14, 79–89.

Lo, C.M. (1999) Mating rendezvous in monogenean gill parasites of the humbug *Dascyllus aruanus* (Pisces: Pomacentridae). *Journal of Parasitology* 85, 1178–1180.

Loer, C.M. and Kenyon, C.J. (1993) Serotonin-deficient mutants and male mating behavior in the nematode *Caenorhabditis elegans*. *Journal of Neuroscience* 13, 5407–5417.

MacKenzie, K. and Liversidge, J.M. (1975) Some aspects of the biology of the cercaria and metacercaria of *Stepanostomum baccatum* (Nicole, 1907) Manter, 1934 (Digenea: Acanthocolpidae). *Journal of Fish Biology* 7, 247–256.

Marshall, A.G. (1981) *The Ecology of Ectoparasitic Insects*. Academic Press, London, 459 pp.

May, R.M. (1977) Togetherness among schistosomes: its effects on the dynamics of the infection. *Mathematical Biosciences* 35, 301–343.

Nollen, P.M. (1968) Autoradiographic studies on reproduction in *Philophthalmus gralli* (Cort, 1914) (Trematoda). *Journal of Parasitology* 54, 43–48.

Nollen, P.M. (1978) Studies on the reproductive system of *Philophthalmus gralli* using techniques of transplantation and autoradiography. *Journal of Parasitology* 64, 613–616.

Nollen, P.M. (1990) *Echinostoma caproni*: mating behavior and the timing of development and movement of reproductive cells. *Journal of Parasitology* 76, 784–789.

Nollen, P.M. (1993) *Echinostoma trivolvis*: mating behavior of adults raised in hamsters. *Parasitology Research* 79, 130–132.

Nollen, P.M. (1994a) The hatching behavior of *Echinostoma trivolvis* miracidia and their responses to gravity, light and chemicals. *International Journal for Parasitology* 24, 637–642.

Nollen, P.M. (1994b) The mating behaviour of *Zygocotyle lunata* adults grown in mice. *Journal of Helminthology* 68, 327–329.

Nollen, P.M. (1996a) The mating behaviour of *Echinostoma paraensei* grown in mice. *Journal of Helminthology* 70, 43–45.

Nollen, P.M. (1996b) Mating behaviour of *Echinostoma caproni* and *E. paraensei* in concurrent infections in mice. *Journal of Helminthology* 70, 133–136.

Nollen, P.M. (1997a) Mating behaviour of *Echinostoma caproni* and *E. trivolvis* in concurrent infections in hamsters. *International Journal for Parasitology* 27, 71–75.

Nollen, P.M. (1997b) Reproductive physiology and behavior of digenetic trematodes. In: Fried, B. and Graczyk, T.K. (eds) *Advances in Trematode Biology*. CRC Press, Boca Raton, Florida.

Nollen, P.M. (1999) Mating behaviour of *Echinostoma trivolvis* and *E. paraensei* in concurrent infections in hamsters. *Journal of Helminthology* 73, 329–332.

Nollen, P.M. (2000) Reproductive physiology and behavior of *Echinostomes*. In: Fried, B. and Graczyk, T.K. (eds) *Echinostomes as Experimental Models for Biological Research*. Kluwer, Dordrecht, the Netherlands, pp. 137–148.

Ochoa, G. and Jaffe, K. (1999) On sex, mate selection and the Red Queen. *Journal of Theoretical Biology* 199, 1–9.

Price, P.W. (1980) *Evolutionary Biology of Parasites*. Princeton University Press, Princeton, New Jersey, 237 pp.

Ramasamy, P., Ramalingam, K., Hanna, R.E.B. and Halton, D.W. (1985) Micro-habitats of gill parasites (Monogenea and Copepoda of teleosts (*Scomberoides* spp.)). *International Journal for Parasitology* 15, 385–397.

Ritchie, G., Mordue, A.J., Pike, A.W. and Rae, G.H. (1996) Observations on the mating and reproductive behaviour of *Lepeophtheirus salmonis* Kroyer (Copepoda, Caligidae). *Journal of Experimental Marine Biology and Ecology* 201, 283–298.

Ritchie, M.E. and Olff, H. (1999) Spatial scaling laws yield a synthetic theory of biodiversity. *Nature* 440, 557–560.

Rohde, K. (1973) Structure and development of *Lobatostoma manteri* sp. nov. (Trematoda, Aspidogastrea) from the Great Barrier Reef, Australia. *Parasitology* 66, 63–83.

Rohde, K. (1978) Latitudinal differences in host specificity of marine Monogenea and Digenea. *Marine Biology* 47, 125–134.

Rohde, K. (1979) A critical evaluation of intrinsic and extrinsic factors responsible for niche restriction in parasites. *American Naturalist* 114, 648–671.

Rohde, K. (1980) Host specificity indices of parasites and their applications. *Experientia* 36, 1369–1371.

Rohde, K. (1989) At least eight types of sense receptors in an endoparasitic flatworm: a counter-trend to sacculinization. *Naturwissenschaften* 76, 383–385.

Rohde, K. (1991) Intra- and interspecific interactions in low density populations in resource-rich habitats. *Oikos* 60, 91–104.

Rohde, K. (1993) *Ecology of Marine Parasites*, 2nd edn. CAB International, Wallingford, UK, 298 pp.

Rohde, K. (1994) Niche restriction in parasites: proximate and ultimate causes. *Parasitology* 109, S69-S84.

Rohde, K. (1998) Latitudinal gradients in species diversity: area matters, but how much? *Oikos* 82, 184–190.

Rohde, K. (1999) Latitudinal gradients in species diversity and Rapoport's rule revisited: a review of recent work and what can parasites teach us about the causes of the gradients? *Ecography* 22, 593–613.

Rohde, K. (2001a) Parasitism. In: Levin, S. (ed.) *Encyclopedia of Biodiversity*, Vol. 4. Academic Press, New York, pp. 463–484.

Rohde, K. (2001b) Spatial scaling laws do not apply to most animal species. *Oikos* 93, 499–504.

Rohde, K. and Watson, N.A. (1995) Ultrastructure of the buccal complex of *Polylabroides australis* (Monogenea, Polyopisthocotylea, Microcotylidae). *International Journal for Parasitology* 25, 307–318.

Rohde, K. and Watson, N.A. (1996) Ultrastructure of the buccal complex of *Pricea multae* (Monogenea: Polyopisthocotyles, Gastrocotylidae). *Folia Parasitologica* 43, 117–132.

Rohde, K., Hayward, C., Heap, M. and Gosper, D. (1994) A tropical assemblage of ectoparasites: gill and head parasites of *Lethrinus miniatus* (Teleostei, Lethrinidae). *International Journal for Parasitology* 24, 1031–1053.

Saito, S. (1984) Development of *Echinostoma hortense* in rats, with special reference to the genital organs. *Japanese Journal of Parasitology* 33, 51–61.

Sasal, P., Niquil, N. and Bartoli, P. (1999) Community structure of digenean parasites of sparid and labrid fishes of the Mediterranean sea: a new approach. *Parasitology* 119, 635–648.

Sasal, P., Jobet, E., Faliex, E. and Morand, S. (2000) Sexual competition in an acanthocephalan parasite of fish. *Parasitology* 120, 65–69.

Schärer, L. and Wedekind, C. (1999) Lifetime reproductive output in a hermaphrodite cestode when reproducing alone or in pairs: a time cost of pairing. *Evolutionary Ecology* 13, 381–394.

Shulman, S.S. and Shulman-Albova, R.E. (1953) [*Parasites of Fishes of the White Sea*]. Isdatelstvo Akademii Nauk SSSR, Moscow and Leningrad (in Russian).

Simková, A., Desdevises, Y., Gelnar, M. and Morand, S. (2000) Co-existence of nine gill ectoparasites (*Dactylogyrus*: Monogenea) parasitising the roach

(*Rutilus rutilus* L.): history and present ecology. *International Journal for Parasitology* 30, 1077–1088.

Sousa, W.P. (1994) Patterns and processes in communities of helminth parasites. *Trends in Ecology and Evolution* 9, 52–57.

Southgate, V.R., Tchuente, L.A.T., Vercruysse, J. and Jourdane, J. (1995) Mating behaviour in mixed infections of *Schistosoma haematobium* and *S. mattheei*. *Parasitology Research* 8, 651–656.

Southgate, V.R., Jourdane, J. and Tchuente, L.A.T. (1998) Recent studies on the reproductive biology of the schistosomes and their relevance to speciation in the Digenea. *International Journal for Parasitology* 28, 159–172.

Sukhdeo, M.V.K. and Bansemir, A.D. (1996) Critical resources that influence habitat selection decisions by gastrointestinal helminth parasites. *International Journal for Parasitology* 26, 483–498.

Sukhdeo, M.V.K. and Mettrick, D.F. (1987) Parasite behaviour: understanding platyhelminth responses. *Advances in Parasitology* 26, 73–144.

Sukhdeo, M.V.K. and Sukhdeo, S.C. (1994) Optimal habitat selection by helminths within the host environment. *Parasitology* 109, S41–S55.

Tauber, C.A. and Tauber, M.J. (1977a) Sympatric speciation based on allelic change at three loci: evidence from natural populations in two habitats. *Science* 197, 1298–1299

Tauber, C.A. and Tauber, M.J. (1977b) A genetic model for sympatric speciation through habitat diversification and seasonal isolation. *Nature* 268, 702–705.

Tauber, C.A., Tauber, M.J. and Nechols, J.R. (1977) Two genes control seasonal isolation in sibling species. *Science* 197, 592–593.

Trouve, S. and Coustau, C. (1998) Differences in adult excretory–secretory products between geographical isolates of *Echinostoma caproni*. *Journal of Parasitology* 84, 1062–1065.

Trouve, S. and Coustau, C. (1999) Chemical communication and mate attraction in echinostomes. *International Journal for Parasitology* 29, 1425–1432.

Trouve, S., Jourdane, J., Renaud, F., Durand, P. and Morand, S. (1999a) Adaptive sex allocation in a simultaneous hermaphrodite. *Evolution* 53, 1599–1604.

Trouve, S., Renaud, F., Durand, P. and Jourdane, J. (1999b) Reproductive and mate choice strategies in the hermaphroditic flatworm *Echinostoma caproni*. *Journal of Heredity* 90, 582–585.

Viney, M.E., Matthews, B.E. and Walliker, D. (1993) Mating in the nematode parasite *Strongyloides ratti* – proof of genetic exchange. *Proceedings of the Royal Society of London – Series B: Biological Sciences* 254, 213–219.

Zwölfer, H. (1974) Innerartliche Kommunikationssysteme bei Bohrfliegen. *Biologie in unserer Zeit* 4, 147–153.

Parasite Sex Determination 10

R.E.L. Paul

Biochimie et Biologie Moléculaire des Insectes, Institut Pasteur,
25 Rue du Dr Roux, 75724 Paris Cedex 15, France

Introduction

The idea that organisms evolve strategies to achieve their life cycle is purposefully teleological. The approach used in this chapter is that of modern behavioural ecology, which attempts to understand the effects of certain behaviours on the fitness of the individual: that is, can observed behaviours be considered to be those that optimize the fitness of the individual and hence be favoured by natural selection? Such an approach is certainly open to discussion, but it has proved highly insightful in the study of many evolutionary questions, most especially those pertaining to sexuality. The chapter thus focuses on the behavioural strategies underlying sex determination, rather than molecular mechanisms (for a recent review of molecular mechanisms, see Marin and Baker, 1998). The discussion will be largely based on parasites from the phyla Protozoa (responsible for several of the most serious medical and veterinary diseases), Platyhelminthes and Nematoda, and on insect parasitoids, which have been used most extensively and successfully to explore sex-determination strategies.

The environment that most organisms exploit is not homogeneous; rather, it is a mosaic of habitats of varying quality. The colonization of a habitat and the subsequent exploitation of its resources are among the most basic interactions between the environment and any organism, whether parasitic or free-living, enabling growth and replication. The extent to which an organism successfully exploits a habitat for replication represents its fitness and there will be selection for those mechanisms that maximize fitness. Habitat resource exploitation will typically lead to a deterioration in habitat quality, such that dispersal to new habitats is necessary. In no other system is this basic concept of natural selection

more evident than in the exploitation of hosts by their parasites. For parasites, the habitat, for at least some part of its life cycle, is another organism. Hosts are highly discrete habitats, which will eventually deteriorate
in quality. This presents parasites with not only the need to develop transmission stages capable of surviving in another environment outside the
host (perhaps even in another host species), but also the need to respond
to the changing and variable quality of its host by adjusting its resources
between growth and maintenance vs. the production of transmission
stages. This can be extended to free-living organisms, where conditions
becoming increasingly deleterious for growth and replication induce
dispersal-stage production. Thus, although the mechanisms used are variable, the general underlying principles are the same – that is, organisms
use strategies (complex adaptations) to optimize the exploitation of their
current habitat in order to maximize their colonizing potential (or, in
epidemiological jargon, their R_0, the reproductive rate).

In the majority of sexually reproducing parasites, transmission and
sex are coupled. Parasites mature within the host and sexual reproduction
occurs either within the host or upon exit from the host prior to dispersal.
Parasites are therefore faced with the need both to assure transmission to
new hosts and to maximize reproduction. The way in which parasites
employ behavioural strategies to optimize host exploitation for reproductive gain concerns specific behaviours, such as those involved in the
decision-making process of how many daughters vs. sons to produce
(dioecious spp.) or how much female vs. male investment to make
(hermaphrodites). This chapter will address the way in which parasites
utilize behaviours that influence the sex ratio to optimize reproductive success and transmission. Why parasites (or, indeed, free-living
species) should burden themselves with sexual reproduction is beyond
the scope of this chapter, but does warrant a brief appraisal to provide
perspective.

Why Sexual Reproduction in Parasites?

The association between sexual reproduction and dispersal to colonize
new hosts is seemingly no accident. Theories on the evolution and
maintenance of sex tend to divide into those considering the role of
sex either in the reduction in the accumulation of deleterious mutations
or for accelerating adaptation to changing environments (Williams, 1975;
Maynard-Smith, 1978; Stearns, 1987; Michod and Levin, 1988), although
both the mutational- and environmental-based models may be at work
simultaneously (West *et al.*, 1999). The most popular current environmental model is the 'red queen' hypothesis, which states that sex provides
an advantage in biotic interactions, most notably host–parasite systems
(Bell, 1982) and for which there is some evidence (Lively, 1987). The 'red
queen' hypothesis assumes that there is genetic variation in host–parasite

interactions, which results in differential fitness for both hosts and parasites (for a review, see Ladle, 1992). Sex will increase the evolutionary rate of hosts, whose reproductive (and hence evolutionary) rate is generally slower than that of their parasites (Jaenike, 1978; Hamilton, 1980). Therefore, sex accelerates the evolution of the host population, thus generating an ever-changing environment (e.g. the host immune system) for the parasite population (Seger and Hamilton, 1988; Hamilton *et al.*, 1990). In response, parasites are expected to adapt to this changing environment, resulting in a coevolutionary arms race between the parasite and the host (Hamilton, 1993). Despite the higher intrinsic evolutionary rate of the parasite, intrinsic to its rapid life cycle, sex in parasites will provide an additional increase, enabling the parasite population to track the host population. As will become clear, the parasite's intrinsic evolutionary rate advantage does appear to enable more flexibility in reproductive strategies, which are strongly influenced by short-term transmission requirements, as well as the long-term advantages of sex.

Several parasitic protozoa of medical importance, notably *Leishmania* and *Trypanosoma*, although able to undergo sexual reproduction (in laboratory crosses) (Gibson and Garside, 1991), rarely, if ever, do so in the field (Tibayrenc *et al.*, 1990). Although the persistence of clonal lineages of such parasites may seem to contradict the theories of sexual reproduction, only occasional sex is necessary to remove the disadvantages of asexual reproduction (Falconer, 1981). Disposing of sexual reproduction is, however, rather rare. Parasitic Protozoa and Platyhelminthes undergo obligate sex and tend to be hermaphrodites, which in principle allows transmission, following self-fertilization, even when the parasite finds itself alone in a host (but see next section). The majority of parasitoids and nematodes, however, are gonochorists (separate sexes), although such taxa do exhibit certain particular reproductive characteristics. Hymenopteran parasitoids have a haplodiploid genetic system, where unfertilized (haploid) eggs become males and fertilized (diploid) eggs are females; hence the offspring sex ratio can be determined by the ovipositing female. Parthenogenesis occurs not infrequently in parasitoids, where unfertilized females produce diploid daughters (thelytoky). Although it is tempting to consider that thelytoky has evolved as a mating-assurance adaptation in low-density populations (Price, 1980), there is good evidence that parthenogenesis is actually controlled by infecting microorganisms (Stouthamer *et al.*, 1990). Similarly, although the majority of parasitic nematodes are gonochorists, sex determination can be under environmental control; one well-studied species, *Strongyloides ratti*, alternates between parthenogenetic and fully sexual according to the immune state of its host (Gemmill *et al.*, 1997; Viney, 1999). Thus parasites exhibit tremendous variability in reproductive strategy, but to what extent can these strategies be interpreted as those favoured by natural selection?

Gonochorists or Hermaphrodites

For those parasites where sexual reproduction is obligatory, mating can be assured by producing both male and female organs – hermaphrodites. The adaptive significance of hermaphroditism (simultaneous rather than sequential) has been considered to be its assurance of complementarity between any two individuals mating at random in a low-density population (Ghiselin, 1969). Indeed, in the extreme, facultative selfing for reproductive assurance may help maintain hermaphroditism (Charlesworth and Jarne, 1993). The importance of hermaphroditism enabling facultative selfing is expected to be of particular relevance to parasitic organisms, which, if gonochoristic (separate sexes), would have to rely on coinfection for reproduction. The probability of coinfection is determined by the epidemiology of the parasite, which can vary substantially for even a single host–parasite species interaction. Ensuring coinfection is therefore certainly context-dependent, except when the probability of coinfection is guaranteed: (i) when the foundress is free-living and lays multiple offspring on the host (e.g. solitary parasitoids); (ii) when there is cotransmission (e.g. filarial worms); or (iii) when the duration of infection in the host is sufficiently long to ensure superinfection (e.g. non-vector-borne nematodes). Such conditions are permissive of the evolution or maintenance of gonochorism. Although the predominance of hermaphroditism as a reproductive strategy within the protozoa and platyhelminths could be explained by the uncertainty of coinfection, there exist paradoxes (e.g. trematodes are hermaphrodites but very rarely self-fertilize (Brusca and Brusca, 1990)). An alternative (or rather complementary) hypothesis to mating assurance considers the evolutionary stability of the hermaphroditic state to be the result of a resource-allocation strategy.

Resource Allocation

Sex-allocation theory unifies theories concerning sex ratio to generalize across all organisms by considering how an individual should allocate its resources in female/male progeny (gonochoristic species) or male/female organs (simultaneous hermaphrodites) or when to shift from one sex to another (i.e. state-sequential hermaphrodites) (Charnov, 1982). The principle underlying sex-allocation theory is as follows. Organisms have finite resources to partition among growth, survival and reproduction. Of those made available to reproduction, the organism must further partition its reproductive resources into male or female function (progeny, organs or state). The optimal sex-allocation strategy will be that which produces the greatest fitness returns. A foundress (mother or hermaphrodite) will invest its resources in the production of a ratio of male to female function (whether testes and ova in hermaphrodites or males and females in gonochorists) that maximizes its fitness. The optimal strategy may be fixed for a given habitat or there may be facultative adjustment by the foundress according to the precise conditions encountered. In all cases, natural selection is expected to operate on the sex-allocation strategies as long as there is heritable variation in sex allocation; chromosomal sex determination, for example, constrains the evolution of unequal sex allocation. An appreciation of Charnov's (1982) sex-allocation theory is best achieved with reference to the theories it unites and develops upon. The following sections therefore largely discuss parasite sex determination in terms of classical sex-allocation theory. As will become clear, although it is one of the best verified evolutionary theories and many of the model systems used to test and further its development involve parasites and parasitoids, the absence of a host–parasite perspective, whether epidemiological (population level) or immunological (individual level), is striking. Although this certainly reflects the bias of data collected from parasite–invertebrate host systems, parasite–vertebrate host systems do bring different parameters into play. Whilst these additions can be considered within the sex-allocation framework, most notably that of host quality, the conclusions reached can modify those based solely on resource-allocation theory as it stands to date.

Fisher's Principle

Ronald Fisher (1930) developed the first major insight in the development of the sex-ratio theory, presenting an explanation as to why equal investment in male and female progeny, hence an equal sex ratio, is commonly observed in nature. His argument essentially shows that, if the cost of producing females and males were equal, a sex ratio of 1 : 1 would maximize the number of grand-offspring (F2 descendants) attributable to an individual. The fitness of a sex-ratio genotype depends on its frequency in a population, and any deviation from an equal investment in the sexes (to either a male- or female-biased sex investment) will provide a selective advantage for a mutant genotype that invests conversely in the sexes, i.e. if a population consists of individuals that produce a male-biased offspring sex ratio, females are the more valuable sex because they will all mate, whereas the males will not. Therefore any individual that produces more females will have a selective advantage and therefore the population sex ratio will tend to move towards equal numbers of male and females. Likewise, when the sex ratio is female-biased, production of males is advantageous and once again the population will tend towards an equal sex ratio. Since then, both game theory and population-genetic models have extended Fisher's (1930) insight. An equal sex ratio is the only one that cannot be invaded by mutants and has thus been termed an unbeatable strategy (Hamilton, 1967) or an evolutionarily stable strategy (Maynard-Smith and Price, 1973). Furthermore, although an equal population sex ratio could consist of a mixture of sex-ratio phenotypes (i.e. at the extreme, some individuals producing just females and some just males), there is weak selection for all individuals to invest equally in males and females and population-genetic models have shown that, no matter how many alleles are present at a sex-determining locus, an even sex ratio is the only one that is stable to novel mutations that affect sex determination and that, if the equilibrium sex ratio is not 1 : 1, only mutations that render the equilibrium closer to 1 : 1 will spread – the evolutionary genetic stability of the even sex ratio (Eshel and Feldman, 1982).

For Fisher's (1930) principle to apply, several assumptions must be met: (i) panmixia; (ii) fitness returns from increased investment in either sex must be linear (both sexes benefit to the same degree from increased investment); (iii) the mother or foundress must be equally related to both sex offspring; and (iv) there must be heritable variation in the sex ratio, allowing evolution to an equilibrium or locally optimal strategy. The majority of work on sex ratio has considered what happens when these assumptions are not met.

Habitat Population Substructure – Local Mate Competition

Violation of the assumption of panmixia is common in organisms whose population is highly substructured, such as that resulting from low dispersal or habitat isolation. This is particularly relevant to parasites, whose populations are necessarily structured according to their hosts. When mating is not random and there is a certain degree of inbreeding, an equal sex ratio is not predicted to be the optimal investment and a female-biased sex ratio is predicted. At the extreme, when an individual finds itself alone in an isolated habitat, investing equally in male and female function is a suboptimal use of resources. This is because males can fertilize several females and, as all males from a single mother will be brothers, by producing just enough males to fertilize the females, the foundress decreases competition among brothers for mates and increases the number of daughters to be fertilized and produce the next generation and resources are not wasted on unnecessary males. This principle is known as local mate competition (LMC) and was originally developed to explain the female-biased sex ratios observed in many insect species, notably arrhenotokous parasitoids (Hamilton, 1967): when a small number of foundresses produce offspring that are to mate with each other, the optimum proportion of sons (sex ratio), r, is:

$$r = (n - 1)/2n$$

where n is the average number of foundresses contributing to the mating pool (Fig. 10.1)

Many parasitoids are haplodiploid (where the males are haploid and females diploid) and the sex of offspring is determined at conception. This enables flexibility in the offspring or brood sex ratio, and the ecology of parasitoids frequently results in competition among brothers for mates,

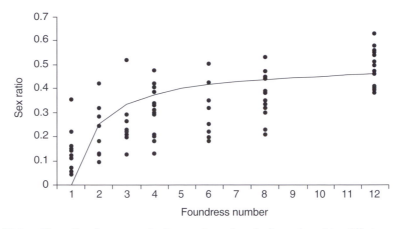

Fig. 10.1. The offspring sex ratio (proportion of males) produced by differing numbers of co-ovipositing foundresses (*Nasonia vitripennis*) (data from Werren, 1983). The solid line gives the optimal sex ratio (proportion of males) in respect of foundress number under local mate competition (Hamilton, 1979).

e.g. when solitary parasitoids parasitize gregarious hosts. The sex ratio is found to be positively correlated with the number of coinfecting foundresses, both within (Fig. 10.1; Werren, 1983) and across species (Godfray, 1994), and with the number of hosts available (Scelionidae: Waage, 1982); i.e. increasing number of foundresses decreases inbreeding and hence LMC and increasing host density will increase the probability of parasitism and thus of coinfecting with other foundresses. Similarly, in hermaphrodites, LMC predicts an increase in the investment in the male function (sperm) in comparison with the female function (ova) as the number of mates increases. This has been demonstrated experimentally in gut trematode infections of mice, where increasing mating-group size correlated with an increase in testes at the expense of ovary size, and this was independent of trematode body size (Trouvé *et al.*, 1999), another factor known to influence optimal sex allocation (cf. habitat quality). Such adaptive adjustment in resource allocation requires cues enabling assessment of the number of coinfecting trematodes. Echinostome trematodes do produce a suite of pheromones involved in interindividual attraction (for pairing and sexual stimulation) (Fried, 1986; Trouvé and Coustau, 1999), which could convey coinfection density information. It is also suggested that such facultative LMC occurs in natural cestode infections of birds (Didyk and Burt, 1998): male–female pairs of the cestode *Shipleya inermis* predominate in infections, with the occasional occurrence of hermaphrodites. The absence of unisex pairs suggests that the cestode is a protogynous hermaphrodite, where female development occurs first and the next cestode arriving would be stimulated to become male. As discussed later, there is broad support for the view that LMC also applies to parasitic protozoa.

LMC and mating assurance

Under extreme inbreeding, when there is a single foundress, LMC would predict a very female-biased sex ratio, with just enough males to fertilize their sisters. However, stochastic male mortality may select for foundresses to produce slightly more males than necessary to act as insurance (Hartl, 1971). This has been found to occur in bethylid parasitoids (Insecta: Hymenoptera) and probably malaria parasites. In 26 species of Bethylidae, the sex ratio was negatively correlated with offspring number (Griffiths and Godfray, 1988); i.e. as offspring number increases, because males can fertilize many females, the foundress produces just enough males to fertilize the females and so the optimal proportion of males (the sex ratio) decreases with increasing total number of offspring. However, the fit was approximate and they observed that, especially in large clutch sizes, more males than predicted (i.e. needed to fertilize all the females) were found; these were interpreted as insurance males.

Habitat Quality

In the 1970s, Trivers and Willard (1973), while working on mammal systems, developed the idea that the fitness of sons and daughters could vary according to the maternal condition or circumstance: that is, the Fisherian assumption that the relationship between fitness and resource allocation is identical for sons and daughters does not always hold – there are circumstances where one sex gains more from additional investment than the other. Charnov (1979) and Bull (1981) extended this idea to all situations where mothers vary in their resources available for reproduction and one sex gains more from additional investment. In solitary parasitoids, the foundress tends to lay females in large host larvae and males in small host larvae, because the female gains more from being large whereas the male suffers less from being small: that is, larval host size dictates the adult size of the developing parasitoid (Charnov, 1982; King, 1987). Size-dependent sex ratios have indeed been commonly found in parasitoid species whose hosts stop growing upon parasitism (idiobionts), but less so where the parasitized host continues to grow (koinobionts), where initial host size would not be expected to be a good indicator of resource potential (King, 1989). As well as the ability to assess the quality of the larvae, foundresses may be able to identify previously parasitized hosts. This is analogous to altering sex ratio with the number of coinfecting foundresses where individuals are expected to increase their male investment, as is the case for parasitoids and intestinal trematodes. Thus, sequential parasitism will similarly be expected to lead to the second foundress investing more in males, not only because of the predictions of LMC but also because host quality may decrease because of overexploitation. In species where sex is determined at conception, such as some parasitoids, the foundress can facultatively determine its off-spring's sex, using cues (host size, superparasitism). In many circum-stances, however, sex is determined later by some environmental influence. Environmental sex determination (ESD) is well known to occur in reptiles, where temperature plays an important role in determining the sex of the developing egg (for a recent review, see Shine, 1999). Analogous to the condition-dependent sex allocation proposed by Charnov (1979) and Bull (1981), ESD is favoured when an individual's fitness is strongly influenced by environmental conditions, where an offspring will enter an environment away from the parent and where the individual has little control over the environment it will experience (Charnov and Bull, 1977). ESD, as either the major or the secondary determining influence enabling facultative control of sex ratio, is probably widespread in both gono-choristic and hermaphroditic parasites (Charnov and Bull, 1977). ESD is known to occur in parasitic copepods and mermithid nematodes, where adults are free-living and the larval forms grow and attain sexual maturity within their host. In these cases, sex is determined according to the number of coinfecting larvae (Blackmore and Charnov, 1989; Charnov, 1993). Similarly, both trematode and cestode platyhelminths adjust their male : female investment according to the number of coinfecting individuals (Didyk and Burt, 1998; Trouvé *et al.*, 1999). In both cases, LMC and probably host quality play a part. In addition, ESD is important when reductions in host quality affect reproductive success *per se* – that is, when mating assurance is jeopardized. In the examples given above, this is probably not the case. However, examples from parasitic protozoa, discussed later, suggest that mating assurance can play a major role and that immune responses targeting sexual reproduction provide a selective force for the evolution of ESD.

Sex Determination, Phenotype Plasticity and Host Heterogeneity

Variability either in host quality or in the intensity of coinfection is expected to select for great plasticity in sex allocation. Although the theories have been largely based on parasitoid systems, it is clear that sex determination in other parasite systems follows the adaptive predictions from resource-allocation theory. The extension of this framework to parasite–vertebrate host systems offers great potential, because it intro-duces further elements, most notably those based on the relative longevity of the host–parasite interactions and the increased importance of host immunity and parasite epidemiology. This is of particular consequence for vector-borne protozoan parasites, where the transmission season (and hence sexual reproduction) can vary considerably for a single host–parasite species interaction. The complex nature of such host–parasite systems, where the host is of highly variable quality (e.g. degree of immu-nity) and the epidemiology often unstable, provides the parasite with a challenging diversity of circumstance, where no single sex-allocation strategy may be appropriate. To what extent sex determination can be adaptive when faced with such uncertainty will be addressed with reference to one such complex but well-studied system – malaria and related parasites in their vertebrate hosts.

Malaria and related parasites – a worked example

Malaria parasites belong to the haemosporidian apicomplexan protozoans, which share a similar life cycle (Box 10.1), proliferating asexually within their vertebrate host and transmitting via an insect vector. At some point during infection, a proportion of the parasites switches from asexual- to sexual-stage production. In addition to commitment to sexual differentiation (gametocytogenesis), malaria parasites determine the sex of their gametocytes (male or female). During its entire cycle in the vertebrate host, *Plasmodium* is haploid and there are no sex chromosomes. Each clone is capable of self-fertilization, producing both male and female gametocytes, i.e. it is a simultaneous hermaphrodite. *In vitro* culture of *Plasmodium falciparum* has shown that the proportion of male and female gametocytes is fixed, clone-specific and therefore genetically determined (Ranford-Cartwright *et al.*, 1993). How sex is determined is unknown, but it appears, at least for *P. falciparum*, to occur at the moment

---

Box 10.1. Malaria-parasite life cycle (Carter and Graves, 1988; Sinden *et al.*, 1996)

At blood-feeding, an infected mosquito injects parasite sporozoite stages, which invade vertebrate host cells. These sporozoites undergo asexual proliferation in their host cells, producing many hundreds of thousands of merozoite-stage parasites. These merozoites invade erythrocytes, where they grow, divide asexually to produce further merozoites and burst out of the cell to invade further erythrocytes, a cycle occurring every 24, 36 or 48 h, according to the *Plasmodium* species. At some point during the course of the infection, most notably when such asexual proliferation is slowed, the merozoite stages grow but do not divide, and thus produce the sexual stages, the gametocytes, which are gamete precursors. Transmission from the vertebrate host to the mosquito vector is mediated solely by these sexual stages of the parasite, which are distinguishable as males and females. When taken up in the blood meal by another female mosquito, these gametocytes transform into gametes: each male gametocyte undergoes exflagellation, by which process up to eight male gametes are produced; each female gametocyte produces only one female gamete. Such gametogenesis occurs within 10–15 min following uptake in the blood meal, and within 30 min the male must actively swim to find and fertilize the female gamete. The subsequent zygote transforms into a mobile ookinete, which penetrates the mosquito midgut, where it encysts. Eight to 15 days later (depending on the *Plasmodium* species), this mature oocyst bursts, releasing several thousand sporozoites, which invade the salivary glands of the mosquito and are injected into the vertebrate host during its next blood meal.

of or shortly after commitment to becoming a sexual form (Bruce *et al.*, 1990; Smith *et al.*, 2000).

LMC and Sex Determination

Several authors have suggested that the life cycle of malaria parasites lends itself to LMC (Ghiselin, 1974; Read *et al.*, 1992). *Plasmodium* spp. undergo obligate sexual reproduction within their mosquito vectors during transmission. Fertilization therefore occurs between the gametes produced from those gametocytes taken up in the mosquito blood meal (~3 µl). Although cross-fertilization between different clones occurs in the laboratory (Walliker *et al.*, 1987), as well as in nature (Babiker *et al.*, 1994; Paul *et al.*, 1995), *Plasmodium* can also effectively have a clonal mode of reproduction. The propensity to self-fertilize will be, to a large extent, determined by the number of overlapping infections in the human host. The mean number of different parasite clones per infected person varies considerably, according to the transmission intensity of the region. In regions of intermediate transmission intensity, such as Papua New Guinea (PNG) (Paul *et al.*, 1995), the number of clones per infected person is lower (range 1–3, mean 1.8) than in regions of high transmission intensity (range 1–6, mean 3.2), such as Tanzania (Babiker *et al.*, 1994). Therefore the number of infected people harbouring only a single parasite clone, and hence the likelihood of self-fertilization, will be higher in regions of low or intermediate transmission than in regions of high transmission. When applying LMC to such blood parasites, Hamilton's basic equation can be rearranged and formulated in terms of the selfing rate, *s* (Read *et al.*, 1992), which has been shown to be the equivalent of Wright's inbreeding coefficient *F* (Dye and Godfray, 1993). Thus the optimum gametocyte sex ratio, *r*, relates to the selfing rate, *s*, by $r = (1 - s)/2$, and where *s* is related to the number of clones per host, *n*, by $s = 1/n$.

Examination of gametocyte sex ratios in PNG showed a female bias and predicted an inbreeding coefficient, Wright's *F*, of from 0.64 to 1 (Read *et al.*, 1992). Genetic analyses of oocysts (zygotes), which contain the haploid products of meiosis, from regions of high (Tanzania) and intermediate (PNG) transmission intensity found that there was a significant reduction in degree of heterozygosity from that expected under random mating and that it was more extreme in the region of lower (PNG, *F* = 0.9: Paul *et al.*, 1995) than in that of higher (Tanzania, *F* = 0.3: Hill *et al.*, 1995) transmission intensity. The high incidence of self-fertilization in PNG therefore confirmed the hypothesis, based on gametocyte sex ratios, of high inbreeding in PNG malaria-parasite populations (Read *et al.*, 1992). The positive relationship between sex ratio and transmission intensity (Read *et al.*, 1992; Robert *et al.*, 1996) suggests that the sex ratio is adaptive and responds to the population genetic structure of the parasite and hence inbreeding rate (LMC). Evidence supporting LMC has been found in the sex ratios of some related haemosporidians (e.g. *Leucocytozoon* parasites of birds), but not in others (i.e. *Haemoproteus* parasites of birds). *Leucocytozoon* spp. were found to produce sex ratios optimal for the local inbreeding probabilities based on the prevalence of infection, which reflects the transmission intensity and hence the distribution of coinfections (Read *et al.*, 1995). *Haemoproteus* spp., however, which are vectored by *Culicoides* spp. (Diptera: Heleidae), were found to produce fewer female-biased sex ratios than predicted (Shutler *et al.*, 1995), which was subsequently suggested to be the result of the need to produce insurance males, because of the small size of the vector-species blood meal compared with that taken by mosquitoes (Shutler and Read, 1998), i.e. the insurance males compensate for the lower probability of gametocytes being present in the diminutive blood meal. In situations where there may be low numbers of gametocytes in the blood meal, LMC predictions for the optimal sex ratio can be dramatically altered, reflecting the influence of mating assurance (West *et al.*, 2002; Fig. 10.2).

Further evidence that LMC may operate on the sex determination of *Plasmodium* and related parasites has recently been found in a comparative analysis of apicomplexan parasites that have syzygy. Syzygy is the process by which a single male gametocyte pairs together with a single female gametocyte, either in host cells or in the lumen of host organs, just prior to gametogenesis (Barta, 1999) and therefore gametes from a single

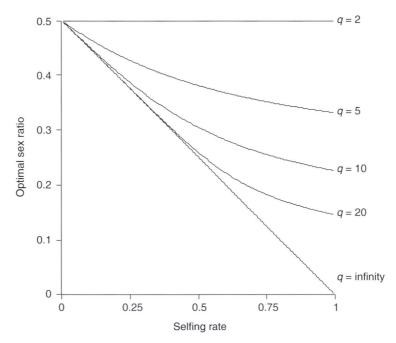

Fig. 10.2. The optimal sex ratio in relation to selfing rate for various gametocyte numbers in the mating pool, where *q* is the number of gametes able to interact. Derived from West *et al.* (2002) with permission from the *Journal of Parasitology*.

male are only able to fertilize the gamete from their paired female. Syzygy therefore removes the factors favouring female-biased sex ratios under LMC: increasing female number will not reduce competition among male gametes from the same gametocyte and will not allow an increase in fertilization success by a single male gametocyte. As predicted, species with syzygy (adelorin species, including the hepatozoans) tended to have an even sex ratio (West *et al.*, 2000).

There is therefore some field evidence that LMC shapes malaria-parasite sex determination and that optimal sex ratios will vary according to the parasite population structure and hence the transmission intensity of the parasite. However, a cross-sectional study found that, although the modal sex ratio was female-biased, transmission success increased with sex ratio, i.e. the sex ratio that maximized mosquito infection rates was neither the most common nor that expected under LMC (Robert *et al.*, 1996). This apparent paradox may simply reflect more complex sex-allocation strategies: in contrast to *in vitro* cultures, the *Plasmodium* sex ratio *in vivo* is not fixed but can vary during the course of infection. It has been known since the mid-1900s (James, 1931; Shute and Maryon, 1951) that the gametocyte sex ratio could vary quite dramatically during the course of a single infection, although with little clear significance for transmission success. Of what adaptive value, if any, therefore is such a fluctuating sex ratio? There is considerable, albeit anecdotal, evidence that males are a limiting factor. Boyd (1949) first noticed the importance of the number of male gametocytes and suggested that the number of males rather than females determine infection success. In addition, a comparison of clones from a single isolate adapted to *in vitro* culture showed that the less female-biased sex-ratio clones were more infectious to mosquitoes (Burkot *et al.*, 1984). Furthermore, the importance of insurance males was suggested as an explanation for the more even sex ratios in *Haemoproteus*, where small blood-meal size would decrease the probability of taking up a male. In malaria parasites, although LMC may operate, the need for insurance males may alter the optimal sex ratio, but does so only under specific conditions – i.e. when the habitat becomes deleterious for reproduction and insurance males are beneficial. Such flexibility in sex allocation would require the ability to respond to the environment and facultatively alter sex ratio accordingly. Two possible, but not exclusive, explanations are therefore considered: (i) *Plasmodium* can alter its sex ratio facultatively according to the number of coinfecting clones (facultative LMC); and (ii) variable host quality jeopardizes reproduction and selects for insurance males.

Facultative LMC

One interpretation of these fluctuating sex ratios in *Plasmodium* has been proposed from longitudinal gametocyte sex-ratio data in lizard populations (Pickering *et al.*, 2000). Here, the sex ratio was found to be positively correlated with the gametocyte density, which the authors interpreted in terms of facultative LMC and which relies on one major assumption: that gametocyte densities increase with the number of coinfecting clones. When there are multiple clones present, the parasite clones are expected to adjust their sex ratio in response to local conditions: single clone infections produce low gametocytaemia and LMC predicts the maintenance of a female bias; multiple clone infections result in a higher gametocytaemia and there is a facultative switch to a more equal sex ratio. This requires that coinfecting clones are able to estimate the number of coinfecting clones, which they could do using any number of cues, as long as the cues vary in intensity according to the number of clones present. In mouse malaria, coinfection resulted in significantly increased anaemia independent of parasitaemia in mixed rather than single infections, with an accompanying increase in gametocytaemia and infectiousness to mosquitoes (Taylor *et al.*, 1997, 1998). Here clones may be responding to haematological cues, which vary according to coinfecting clone number. In contrast, mixed-strain infections of *Plasmodium gallinaceum* in their chicken hosts did not result in an increase in anaemia, gametocytaemia or sex ratio. Rather, the gametocytaemia reflected the asexual reproduction rate, which differed between the individual strains, and the sex ratio, as discussed below, altered with the host's haematological response to the infection. Despite these apparently contradictory results, which probably reflect differences in the host–parasite systems and, most notably, the differing immune systems of the vertebrate hosts, all suggest that changes in the blood environment play a significant role in parasite sexuality (production of gametocytes or sex ratio) and hence transmission. Both asexual growth rates and anaemia are considerably influenced by the host immune response to infection. Although the former reflects the parasite-virulence/host-resistance relationship, the role of immunity in malaria-induced anaemia is poorly understood. Anaemia is not only the result of red blood-cell haemolysis by parasites, but is also a mechanism (perhaps autoimmune), at least in some host–parasite interactions, by which the host controls the infection: many malaria parasite species are only able to infect mature red blood cells and not young ones (reticulocytes). The intricate relationship between haematology and immunology will determine how host quality varies. How multiple clones affect each other via their individual effects on host quality is implicit to malaria research and yet remains poorly understood. That parasites do use haematological cues to increase reproductive success does, however, seem to be the case, as discussed next.

Host quality and mating assurance

The second selection-based explanation for fluctuating sex ratios concerns the effect of a varying habitat (blood) on reproductive success –

i.e. mating assurance. During the course of a malaria infection in the vertebrate host, the blood environment becomes increasingly deleterious for the parasite: accompanying the increasing anaemia (caused in part by parasite-induced haemolysis), there is a developing immune response against the parasite. Until recently, the effect of the host's immune responses on parasite sexual development was thought to be restricted to stimulation of gametocytogenesis (Carter and Graves, 1988). However, it has been recently demonstrated that such changes in the blood environment may also alter parasite sex determination. In animal models, the proportion of gametocytes that were male was significantly elevated when the vertebrate host was in a state of increased erythropoiesis (red blood-cell production) and that the hormone responsible for triggering erythropoiesis, erythropoietin, was implicated in this increased allocation of male gametocytes (Paul *et al.*, 2000). Erythropoiesis is induced in response to increased anaemia, such as that resulting from *Plasmodium* asexual-stage proliferation. Sexual stages are produced from these asexual stages and therefore the immune and haematological environment within which the sexual stages occur will worsen over time, largely due to their asexual progenitors. The immune system also responds to the sexual stages themselves, and this response is actually effective against the gametes once they are formed within the mosquito blood meal (Carter *et al.*, 1979). Essentially, an antibody response agglutinates the gametes and slows down the ability of the male gametes (equivalent to actively searching sperm) to find a female gamete, which they must do within 30 min to achieve successful fertilization and hence infection of the mosquito. Therefore both erythropoiesis (reflecting the extent of parasite-induced anaemia) and the immune response to gametocytes (produced from these asexual stages) increase simultaneously (Fig. 10.3a). Intuitively, if males become individually less efficient, producing more males would compensate for this. In the highly charged blood environment during a malaria infection, fertilization is actually very inefficient: very few gametocytes become zygotes. Can subtle changes in sex ratio make a difference? Using a simple heuristic model of fertilization (where two random clouds of males and females unite), producing compensatory males can, in theory, have a dramatic effect on zygote production. Figure 10.3b shows the mean oocyst density found in mosquitoes over the course of an actual vertebrate infection where the sex ratio becomes increasingly male. Using the fertilization model, a dramatic decrease in zygotes is predicted if the sex ratio had remained female-biased, purely as a result of the physics of fertilization (Paul *et al.*, 1999a). Direct evidence that sex allocation is important and probably adaptive for mating assurance was provided by manipulating the sex ratio experimentally. The sex ratio was increased precociously by artificially elevating the erythropoietin levels during the host infection at a time when asexual parasite damage and the host's immune response were both minimal, and this increase in sex ratio was accompanied by a drop in mosquito infectivity rate (Paul *et al.*, 2000). Why? Simply because the

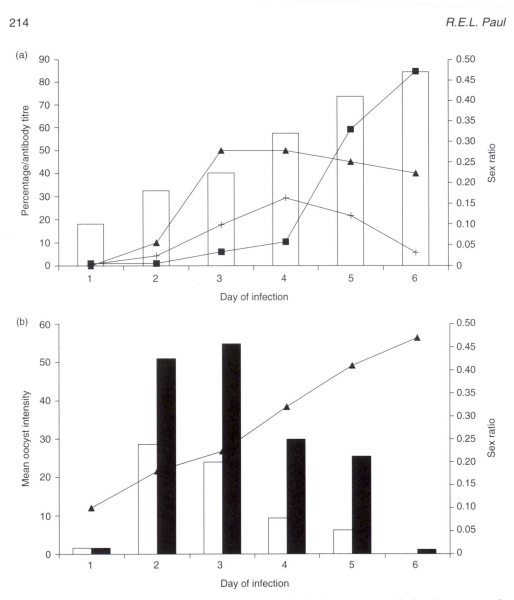

Fig. 10.3. (a) Parasite, haematological and immunological parameters during the course of a single infection of *Plasmodium gallinaceum* in its chicken host. Gametocyte sex ratio (open bars), % of red blood cells which are reticulocytes, indicating erythropoiesis (■), agglutinating antibody titre (▲), parasitaemia (percentage of red blood cells infected) (+). (Data adapted from Carter *et al.*, 1979, and Paul *et al.*, 2000). (b). Comparison of observed (solid bars) and expected (open bars) parasite oocyst loads in mosquitoes under observed (line) and fixed female-biased (4 : 1) sex ratios during the course of a *Plasmodium gallinaceum* infection in its chicken host. Observed infection loads are based on the oocyst density from 30 mosquitoes gorged on the infected chicken each day. Expected values are estimated as follows. By inputting the observed mean oocyst loads into the fertilization model (details in Paul *et al.*, 1999a), the number of effective male gametes produced per male gametocyte (i.e. resulting from the negative effect of the agglutinating antibody on male efficacy) can be calculated for the observed gametocytaemias and sex ratio. Then, changing only the gametocyte sex ratio to a constant 4 : 1, the model will produce the new expected mean oocyst loads.

parasite had been tricked into allocating more of its gametocytes into males at a time when the male gametes were able to search for females unimpeded and that therefore there was a shortage of females. Thus sex allocation appears to be adaptive and is able to compensate for the decreasing efficacy of the male gametes (due to host immunity) by producing more male 'insurance' gametocytes in response to erythropoietin, induced by the anaemia provoked by the asexual parasite proliferation. However, in human and lizard malaria parasites, there seems to be no clear correlation between sex ratio and transmission success (James, 1931; Schall, 2000). Although there may be considerable differences between malaria species, most notably the host immunological response to infection, further consideration of parasite life histories, with special reference to the human malaria *P. falciparum*, may elucidate this apparent paradox.

In regions of endemic malaria, where most of the population has some immunity, the majority of sporozoite inoculations result in a low asymptomatic parasitaemia, which may last for several months (Greenwood, 1987). Such chronic infections are significant for parasite persistence and the infectious reservoir (Jeffery and Eyles, 1955), especially in regions with a very short transmission season, where parasites must survive for up to 10 months without transmission (Arnot, 1998). It is assumed that parasites restart transmission upon the arrival of the mosquitoes but must achieve fertilization with a very low gametocyte density, which demands an extremely efficient fertilization process, opposite to that seen during the acute phase of infection. As can be deduced from a simple mechanistic model of fertilization, one way to ensure fertilization success when increasing gametocyte density is not an option would be to increase the proportion of males (Paul *et al.*, 1999a). Hence a clone can find itself in hosts of widely ranging quality (immune response) or in a condition (high, acute vs. low, chronic density) determining its reproductive capacity. In such 'low-quality' conditions, mating may no longer be assured. Malaria-parasite clones will often find themselves alone in a host and, because sexual reproduction is a prerequisite of transmission, clones must therefore self-fertilize to ensure transmission success. This selects not only for the maintenance of a hermaphroditic state, but also for an ability to adjust its male : female function investment to ensure fertilization as host quality varies. Thus, although parasite population structure (number of parasite clones per person) may promote sex allocation under LMC, variable host quality, which influences asexual parasite density (i.e. foundress condition) and hence gametocyte numbers, can also significantly affect mating assurance and so influence the optimal sex-allocation strategy.

Concluding Remarks

Parasites must exist in discrete habitats of highly variable quality, between which they must transmit. In addition to developing strategies to exploit their hosts and evade host defence systems, sexually reproducing parasites must further divert resources to sexual reproduction, thus introducing a further subdivision of investment into males and females. Whether sex-allocation strategies can evolve depends on how sex is determined. In parasitoids, the foundress determines the sex of its offspring. For hermaphrodites, the adult can invest differentially in male/female function in response to environmental stimuli. Although chromosomal sex determination reduces genetic variance in sex ratio and will tend to restrain the evolution of an allocation strategy, environmental factors can strongly influence the sex ratio, whether primary or secondary. In short, parasites exhibit a tremendous range of sex-determination mechanisms that enable the evolution of strategies. Such strategies tend to reflect three basic principles: mating assurance in low-density conditions; resource optimization reflecting coinfection probabilities; and condition-dependent fitness differences in the sexes. Such principles do appear to explain the wide variety of sex ratios observed in parasitic systems.

Our comprehension of how malaria parasites modulate their resource allocation to maximize reproduction remains limited. For malaria parasites in particular, where transmission and sexual reproduction are simultaneous, strategies that ensure fertilization are expected to be under very strong selection. The tremendous variability in host quality is expected to select for a very plastic resource-allocation strategy, and further cues of habitat quality in both malaria and other parasites are anticipated to exist. How long-term evolutionary selection, such as that observed under LMC, can further influence parasite sex-determining behaviour and resource allocation in the complex environment of the vertebrate host remains a relatively unexplored realm with exciting possibilities. As well as evaluating, for example, classical sex-allocation theory, one interesting novelty considers the value of such plastic ESD not only in ensuring fertilization but also in promoting cross-fertilization. If clones vary in their response to haematological cues, for example, coinfecting clones can have different sex ratios in the same host environment. Sexual-stage proteins are notably conserved and hence host immunological responses will be shared across clones. If males are indeed the limiting sex, then a clone poorly responding to cues will transmit poorly and be outcompeted by a good responder. However, when the two clones (poor and good responder) are coinfecting, their differing sex ratios will lead to a relative increase in the rate of cross- vs. self-fertilization. Cross-fertilization not only leads to recombination, but also guarantees the presence of both clones in the same mosquito – that is, co-transmission. The majority of infected mosquitoes in the field contain one or very few oocysts, each the product of a single fertilization event, and therefore cotransmission is strongly determined by cross-fertilization frequencies. Although recombinant parasites may have an advantage over parental clones if immunity is clone-specific, cotransmission may be advantageous in itself. There is some evidence that coinoculated clones have a greater success rate in initiating an infection in humans (by antagonizing the host immune response (Gilbert *et al.*, 1998)) and that there is greater than expected cotransmission in low transmission-intensity human malaria foci (Paul *et al.*, 1999b). Plasticity in sex determination may therefore be additonally maintained by the weak advantages of cross-fertilization. The uncertain relevance of such 'down-stream' effects on sex allocation and the widely variable nature of the host–parasite interaction present novel challenges for sex-allocation theory, with potentially important significance for the fight against the malaria parasite.

Until recently, research concerning parasites of medical and veterinary importance has been largely the domain of specialists in the field and there has been an absence of an evolutionary approach to essentially practical issues. The dawn of Darwinian medicine (Williams and Nesse, 1991), whose premise is the application of adaptationist argumentation to infectious diseases, has encouraged evolutionary biologists to take an active role in infectious-disease research. The apparently successful application of sex-allocation theory to malaria and related parasites (Read *et al.*, 1992, 1995; West *et al.*, 2000) provides optimistic grounds for expanding adaptationist reasoning to more complex and perhaps more pertinent medical phenotypes, such as virulence (Pickering *et al.*, 2000). As well as evolutionary biology making a significant contribution to infectious-disease research, model systems such as malaria provide an increasing opportunity for developing our understanding of behavioural ecology. The medical importance of malaria has resulted in one of the most well-developed and best-documented systems, which is now amenable to laboratory manipulation, as well as field experimentation. The variability in the host environment, whether as a result of host immunology or of parasite epidemiology, presents a challenging range of circumstances for parasite evolution. The rapid expansion in our understanding of the mechanisms involved in parasite–host interactions, whether host immunological responses to the parasite or the molecular subtleties of the parasite life cycle, will enable a more detailed investigation of the adaptive nature of observed parasite behaviours. Application of such adaptationist approaches to life-history traits must, however, proceed with caution. At present, such methodology assumes equilibrium states that are not at all certain in host–parasite systems, which can display quite variable and even chaotic dynamics. In such systems, there is the danger of stretching adaptive theories too far and generating 'just-so' stories. On the other hand, application of adaptationist ideas to more complex systems does offer the possibility of elaborating evolutionary theories, as well as making a significant contribution to serious public-health issues.

References

Arnot, D. (1998) Clone multiplicity of *Plasmodium falciparum* infections in individuals exposed to variable levels of disease transmission. *Transactions of the Royal Society of Tropical Medicine and Hygiene* 92, 580–585.

Babiker, H.A., Ranford-Cartwright, L.C., Currie, D., Charlwood, J.D., Billingsley, P., Teuscher, T. and Walliker, D. (1994) Random mating in a natural population of the malaria parasite *Plasmodium falciparum*. *Parasitology* 109, 413–421.

Barta, J.R. (1999) Suborder Adeleorina Leger, 1911. In: Lee, J.J., Leedale, G.F., Patterson, D.J. and Bradbury, P.C. (eds) *Illustrated Guide to the Protozoa*. Society of Protozoologists, Lawrence, Kansas, pp. 70–107.

Bell, G. (1982) *The Masterpiece of Nature: the Evolution and Genetics of Sexuality*. University of California Press, Berkeley, 635 pp.

Blackmore, M. and Charnov, E.L. (1989) Adaptive variation in environmental sex determination in a nematode. *American Naturalist* 134, 817–823.

Boyd, M.F. (1949) Epidemiology: factors related to the definitive host. In: Boyd, M.F. (ed.) *Malariology*. W.B. Saunders, London, pp. 608–697.

Bruce, M.C., Alano, P., Duthie, S. and Carter, R. (1990) Commitment of the malaria parasite *Plasmodium falciparum* to sexual and asexual development. *Parasitology* 100, 191–200.

Brusca, R.C. and Brusca, G.J. (1990) *Invertebrates*. Sinauer, Sunderland, Massachusetts, 922 pp.

Bull, J.J. (1981) Sex ratio evolution when fitness varies. *Heredity* 46, 9–26.

Burkot, T., Williams, J.L. and Schneider, I. (1984) Infectivity to mosquitoes of *Plasmodium falciparum* clones grown *in vitro* from the same isolate. *Transactions of the Royal Society of Tropical Medicine and Hygiene* 78, 339–341.

Carter, R. and Graves, P.M. (1988) Gametocytes. In: Wernsdorfer, W. and McGregor, I. (eds) *Malaria: Principles and Practice of Malariology*, Vol. 1. Churchill Livingstone, London, pp. 253–306.

Carter, R., Gwadz, R.W. and Green, I. (1979) *Plasmodium gallinaceum*: transmission-blocking immunity in chickens. II. The effect of antigamete antibodies *in vitro* and *in vivo* and their elaboration during infection. *Experimental Parasitology* 47, 194–208.

Charlesworth, D. and Jarne, P. (1993) The evolution of the selfing rate in functionally hermaphrodite plants and animals. *Annual Review of Ecology and Systematics* 24, 441–466.

Charnov, E.L. (1979) The genetical evolution of patterns of sexuality: Darwinian fitness. *American Naturalist* 113, 465–480.

Charnov, E.L. (1982) *The Theory of Sex Allocation*. Princeton University Press, Princeton, New Jersey, 355 pp.

Charnov, E.L. (1993) *Life History Invariants*. Oxford University Press, Oxford, 167 pp.

Charnov, E.L. and Bull, J. (1977) When is sex environmentally determined? *Nature* 266, 828–830.

Didyk, A.S. and Burt, M.D. (1998) Geographical, seasonal, and sex dynamics of *Shipleya inermis* (Cestoidea: Dioecocestidae) in *Limnodromus griseus* Gmelin (Aves: Charadriiformes). *Journal of Parasitology* 84, 931–934.

Dye, C. and Godfray, H.C.G. (1993) On sex ratio and inbreeding in malaria parasite populations. *Journal of Theoretical Biology* 161, 131–134.

Eshel, I. and Feldman, M.W. (1982) On evolutionary genetic stability of the sex ratio. *Theoretical Population Biology* 21, 430–439.

Falconer, D.S. (1981) *Introduction to Quantitative Genetics*, 2nd edn. Longman, London, 340 pp.

Fisher, R.A. (1930) *The Genetical Theory of Natural Selection*. Dover, New York, 291 pp.

Fried, B. (1986) Chemical communication in hermaphroditic digenetic trematodes. *Journal of Chemical Ecology* 12, 1659–1677.

Gemmill, A., Viney, M.E. and Read, A.F. (1997) Host immune status determines sexuality in parasitic nematodes. *Evolution* 51, 393–401.

Ghiselin, M.T. (1969) The evolution of hermaphroditism among animals. *Quarterly Review of Biology* 44, 189–208.

Ghiselin, M.T. (1974) *The Economy of Nature and the Evolution of Sex*. University of California Press, Berkeley, 346 pp.

Gibson, W. and Garside, L. (1991) Genetic exchange in *Trypanosoma brucei brucei*: variable chromosomal location of housekeeping genes in different trypanosome stocks. *Molecular and Biochemical Parasitology* 45, 77–89.

Gilbert, S.C., Plebanski, M., Gupta, S., Morris, J., Cox, M., Aidoo, M., Kwiatkowski, D., Greenwood, B.M., Whittle, H.C. and Hill, A.V. (1998) Association of malaria parasite population structure, HLA, and immunological antagonism. *Science* 279, 1173–1177.

Godfray, H.C.J. (1994) *Parasitoids: Behavioral and Evolutionary Ecology*. Princeton University Press, Princeton, New Jersey, 473 pp.

Greenwood, B.M. (1987) Asymptomatic malaria infections – do they matter? *Parasitology Today* 3, 206–214.

Griffiths, N. and Godfray, H.C.J. (1988) Local mate competition, sex ratio and clutch size in bethylid wasps. *Behavioural Ecology and Sociobiology* 22, 211–217.

Hamilton, W. (1967) Extraordinary sex ratios. *Science* 156, 477–488.

Hamilton, W.D. (1979) Wingless and fighting males in fig wasps and other insects. In: Blum, M.S. and Blum, N.A. (eds) *Sexual Selection and Reproductive Competition in Insects*. Academic Press, London, pp. 167–220.

Hamilton, W.D. (1980) Sex vs. non-sex vs. parasite. *Oikos* 35, 282–290.

Hamilton, W.D. (1993) Haploid dynamic polymorphism in a host with matching parasites: effects of mutation/subdivision, linkage, and patterns of selection. *Journal of Heredity* 84, 328–338.

Hamilton, W.D., Axelrod, R. and Tanese, R. (1990) Sexual reproduction as an adaptation to resist parasites (a review). *Proceedings of the National Academy of Sciences, USA* 87, 3566–3573.

Hartl, D.L. (1971) Some aspects of natural selection in arrhenotokous populations. *American Zoologist* 11, 309–325.

Hill, W.G., Babiker, H.A., Ranford-Cartwright, L.C. and Walliker, D. (1995) Estimation of inbreeding coefficients from genotypic data on multiple alleles, and application to estimation of clonality in malaria parasites. *Genetical Research* 65, 53–61.

Jaenike, J. (1978) An hypothesis to account for the maintenace of sex within populations. *Evolutionary Theory* 3, 191–194.

James, S.P. (1931) Some general results of a study of induced malaria in England. *Transactions of the Royal Society of Tropical Medicine and Hygiene* 24, 477–538.

Jeffery, G.M. and Eyles, D.E. (1955) Infectivity to mosquitoes of *Plasmodium falciparum* as related to gametocyte density and duration of infection. *American Journal of Tropical Medicine and Hygiene* 4, 781–789.

King, B.H. (1987) Offspring sex ratios in parasitoid wasps. *Quarterly Review of Biology* 62, 367–396.

King, B.H. (1989) Host-size dependent sex ratios among parasitoid wasps: does host growth matter? *Oecologia* 78, 420–426.

Ladle, R.J. (1992) Parasites and sex: catching the red queen. *Trends in Ecology and Evolution* 7, 405–408.

Lively, C. M. (1987) Evidence from a New Zealand snail for the maintenance of sex by parasitism. *Nature* 328, 519–521.

Marin, I. and Baker, B.S. (1998) The evolutionary dynamics of sex determination. *Science* 281, 1990–1994.

Maynard Smith, J. (1978) *The Evolution of Sex*. Cambridge University Press, Cambridge, 222 pp.

Maynard-Smith, J. and Price, G.R. (1973) The logic of animal conflict. *Nature* 246, 15–18.

Michod, R.E. and Levin, B.R. (1988) *The Evolution of Sex*. Sinauer, Sunderland, Massachusetts, 342 pp.

Paul, R.E.L., Packer, M.J., Walmsley, M., Lagog, M., Ranford-Cartwright, L.C., Paru, R. and Day, K.P. (1995) Mating patterns in malaria parasite populations of Papua New Guinea. *Science* 269, 1709–1711.

Paul, R.E.L., Raibaud, A. and Brey, P.T. (1999a) Sex ratio adjustment in *Plasmodium gallinaceum*. *Parassitologia* 41, 153–158.

Paul, R.E., Brockman, A., Price, R.N., Luxemburger, C., White, N.J., Looareesuwan, S., Nosten, F. and Day, K.P. (1999b) Genetic analysis of *Plasmodium falciparum* infections on the north-western border of Thailand. *Transactions of the Royal Society of Tropical Medicine and Hygiene* 93, 587–593.

Paul, R.E.L., Coulson, T.N., Raibaud, A. and Brey, P.T. (2000) Sex determination in malaria parasites. *Science* 287, 128–131.

Pickering, J., Read, A.F., Guerrero, S. and West, S.A. (2000) Sex ratio and virulence in two species of lizard malaria parasites. *Evolutionary Ecology Research* 2, 171–184.

Price, P.W. (1980) *Evolutionary Biology of Parasites*. Princeton University Press, Princeton, 237 pp.

Ranford-Cartwright, L.C., Balfe, P., Carter, R. and Walliker, D. (1993) Frequency of cross-fertilization in the human malaria parasite *Plasmodium falciparum*. *Parasitology* 107, 11–18.

Read, A.F., Narara, A., Nee, S., Keymer, A.E. and Day, K.P. (1992) Gametocyte sex ratios as indirect measures of outcrossing rates in malaria. *Parasitology* 104, 387–395.

Read, A.F., Anwar, M., Shutler, D. and Nees, S. (1995) Sex allocation and population structure in malaria and related parasitic protozoa. *Proceedings of the Royal Society London Series B* 260, 359–363.

Robert, V., Read, A.F., Essong, J., Tchuinkam, T., Mulder, B., Verhave, J.P. and Carnevale, P. (1996) Effect of gametocyte sex ratio on infectivity of *Plasmodium falciparum* to *Anopheles gambiae*. *Transactions of the Royal Society of Tropical Medicine and Hygiene* 90, 621–624.

Schall, J.J. (2000) Transmission success of the malaria parasite *Plasmodium mexicanum* into its vector: role of gametocyte density and sex ratio. *Parasitology* 121, 575–580.

Seger, J. and Hamilton, W.D. (1988) Parasites and sex. In: Michod, R.E. and Levins, B.R. (eds) *The Evolution of Sex.* Sinauer, Sunderland, Massachusetts, pp. 176–193.

Shine, R. (1999) Why is sex determined by nest temperature in many reptiles? *Trends in Ecology and Evolution* 14, 186–189.

Shute, P.G. and Maryon, M. (1951) A study of gametocytes in a West African strain of *Plasmodium falciparum. Transactions of the Royal Society of Tropical Medicine and Hygiene* 44, 421–438.

Shutler, D. and Read, A.F. (1998) Local mate competition, and extraordinary and ordinary blood parasite sex ratios. *Oikos* 82, 417–426.

Shutler, D., Bennett, G.F. and Mullie, A. (1995) Sex proportions of *Haemoproteus* blood parasites and local mate competition. *Proceedings of the National Academy of Sciences, USA* 92, 6748–6752.

Sinden, R.E., Butcher, G.A., Billker, O. and Fleck, S.L. (1996) Regulation of infectivity of *Plasmodium* to the mosquito vector. *Advances in Parasitology* 38, 53–117.

Smith, T.G., Lourenco, P., Carter, R., Walliker, D. and Ranford-Cartwright, L.C. (2000) Commitment to sexual differentiation in the human malaria parasite, *Plasmodium falciparum. Parasitology* 121, 127–133.

Stearns, S.C. (1987) *The Evolution of Sex and its Consequences.* Birkhauser, Basle, 403 pp.

Stouthamer, R., Luck, R.F. and Hamilton, W.D. (1990) Antibiotics cause partheno-genetic *Trichogramma* (Hymenoptera: Trichogrammatidae) to revert to sex. *Proceedings of the Natural Academy of Sciences, USA* 87, 2424–2427.

Taylor, L.H., Walliker, D. and Read, A.F. (1997) Mixed-genotype infections of the rodent malaria *Plasmodium chabaudi* are more infectious to mosquitoes than single-genotype infections. *Parasitology* 115, 121–132.

Taylor, L.H., Mackinnon, M.J. and Read, A.F. (1998) Virulence of mixed clone and single clone infections of the rodent malaria *Plasmodium chabaudi. Evolution* 52, 489–497.

Tibayrenc, M., Kjellberg, F. and Ayala, F.J. (1990) A clonal theory of parasitic protozoa: the population structures of *Entamoeba, Giardia, Leishmania, Naegleria, Plasmodium, Trichomonas and Trypanosoma* and their medical and taxonomical consequences. *Proceedings of the National Academy of Sciences, USA* 87, 2414–2418.

Trivers, R.L. and Willard, D.E. (1973) Natural selection of parental ability to vary the sex ratio of offspring. *Science* 179, 90–92.

Trouvé, S. and Coustau, C. (1999) Chemical communication and mate attraction in echinostomes. *International Journal of Parasitology* 29, 1425–1432.

Trouvé, S., Jourdane, J., Renaud, F., Durand, P. and Morand, S. (1999) Adaptive sex allocation in a simultaneous hermaphrodite. *Evolution* 53, 1599–1604.

Viney, M.E. (1999) Exploiting the life cycle of *Strongyloides ratti. Trends in Parasitology* 15, 231–235.

Waage, J.K. (1982) Sib-mating and sex ratio strategies in scelionid wasps. *Ecological Entomology* 7, 103–112.

Walliker, D., Quakyi, I.A., Wellems, T.E., McCutchan, T.F., Szarfman, A., London, W.T., Corcoran, L.M., Burkot, T.R. and Carter, R. (1987) Genetic analysis of the human malaria parasite *Plasmodium falciparum. Science* 236, 1661–1666.

Werren, J.H. (1983) Sex ratio evolution under local mate competition in a parasitic wasp. *Evolution* 37, 116–124.

West, S.A., Lively, C.M. and Read, A.F. (1999) A pluralist approach to sex and recombination. *Journal of Evolutionary Biology* 12, 1003–1012.

West, S.A., Smith, T.G. and Read, A.F. (2000) Sex allocation and population structure in apicomplexan (protozoa) parasites. *Proceedings of the Royal Society London Series B* 267, 257–263.

West, S.A., Smith, T.G., Nee, S. and Read, A.F. (2002) Fertility insurance and the sex ratios of malaria and related haemosporor in blood parasites. *Journal of Parasitology* (in press).

Williams, G.C. (1975) *Sex and Evolution.* Monographs in Population Biology, Princeton University Press, Princeton, New Jersey, 200 pp.

Williams, G.C. and Nesse, R.M. (1991) The dawn of Darwinian medicine. *Quarterly Review of Biology* 66, 1–22.

Interactions between 11
Intestinal Nematodes and
Vertebrate Hosts

M.V.K. Sukhdeo,[1] S.C. Sukhdeo[2] and A.D. Bansemir[3]

*[1]Department of Ecology, Evolution and Natural Resources, Rutgers
University, 14 College Farm Road, New Brunswick, NJ 08901, USA;
[2]Department of Ecology, Evolution and Natural Resources, Rutgers
University, 84 Lipman Drive, New Brunswick, NJ 08901, USA;
[3]Division of Life Sciences, Rutgers University, 604 Allison Road,
Piscataway, NJ 08904, USA*

Introduction

The phylum Nematoda (roundworms) is one of the largest groups in the animal kingdom, and some estimates suggest that four out of every five metazoans are nematodes (Bongers and Ferris, 1999). Gastrointestinal nematodes represent only a tiny subset of this vast assemblage, but, because of the severe medical and economic consequences of human infection (> 3 billion people infected (Montresor *et al.*, 1998)), we know a lot about these worms. Members of this group are obligate parasites that spend all, or part, of their lives in the guts and associated organs of their hosts. Entry into the vertebrate host is usually via oral ingestion or skin (percutaneous) penetration, and the worms often travel long distances in their hosts to reach their adult sites. Adult worms typically end up in narrow and precise microhabitats within the gut, even though most of the intestinal habitat tends to be vacant of competing parasite species (Rohde and Hobbs, 1986). Interspecific competition and/or resource-dependent mechanisms have been proposed to explain this phenomenon (Holmes, 1973; Rohde, 1979, 1992), but there have been few rigorous tests of these ideas. In addition, parasite decisions on the best locations within the gut are not independent of higher-level effects of life history and the constraints of phylogeny. This chapter will focus on the basic strategies used by gastrointestinal nematodes to get to their habitats in the gut of vertebrate hosts. Studies on the proximate mechanisms of habitat selection behaviour and the role of resources (food, attachment and mates) in parasite strategic decisions will be used to re-examine some traditional hypotheses on the behaviour of these worms.

Why do Nematodes Select Narrow Microhabitats in the Host?

Nematode site (microhabitat) selection is so predictable and so precise that the location of a species within the host gut is a robust taxonomic character. These worms are found in all regions of the gut, from nose to anus, but their specific distributions are typically very narrow, a phenomenon generally referred to as niche restriction (Crompton, 1973; Holmes, 1973; Rohde, 1979). Even within discrete organs in the gut, the worms are very selective (Wertheim, 1970). For example, pinworm species were long thought to share the same habitat within the rectum, but, in a study where eight cohabiting species of pinworms were partitioned both linearly and radially in the rectum of their turtle host, there was little overlap among the species (Schad, 1963).

One of the earliest explanations for the precision of habitat selection was proffered by Holmes (1961, 1973). He felt that niche restriction was a legacy of past interspecific competition. In this scenario, two parasite species initially competing for the same habitats would, over time, segregate into distinct niches (microhabitats) through ecological and physiological specialization (Hair and Holmes, 1975). Eventually, as competing species became extinct for one reason or other, the surviving worms would remain in their narrow niches. There were several criticisms of this model. Apart from the difficulties of dealing empirically with the 'ghost of competition past', it was not clear why narrow habitats did not expand as the competitors became extinct. Studies of habitat selection in both guts and gills suggested that many potential parasite habitats were unoccupied (Rohde, 1979; Rohde and Hobbs, 1986).

Rohde argued that competition for food and space did not adequately explain habitat specificity, because of the many available habitats (Rohde, 1981; Rohde and Hobbs, 1986), and suggested that the reason for the restriction to narrow niches was to increase intraspecific contact and facilitate mating (Rohde, 1977, 1994). Clearly, at the low densities typical of most parasite infections, this behaviour would be adaptive, because it would ensure that the sexes got close enough for sexual attraction to occur. However, there are also some problems with this hypothesis. For example, it predicts that dioecious worms will be more restricted in their habitats than hermaphroditic worms, and this is not necessarily so (Adamson and Caira, 1994). In addition, some parasitic nematodes mate before reaching their final habitats. For example, female *Dracunculus medinensis* (human guinea-worm) mate with their males before they migrate to their specific habitats in the subcutaneous tissue of the leg (Roberts and Janovy, 2000). In this parasite, the final site is determined by the need to shed infective propagules through a cutaneous ulcer in the human foot, and not by the need to encounter mates.

There have been additional suggestions that parasite habitat selection might be driven by many other factors, including the evolution of parasite community dynamics (Holmes and Price, 1986; Price, 1986) and phylo-genetic constraints (Brooks and McLennan, 1993; Adamson and Caira, 1994; Sukhdeo *et al.*, 1997). Space prevents detailed treatment of the many ideas. Nevertheless, although the situation remains unresolved, it is likely that all or several of these mechanisms shape the decision-making strategies of parasites as they progress through their life cycles.

Laboratory Models

Nematodes have similar developmental patterns; after hatching from the egg, usually as first-stage larvae (L1), the worms undergo four moults to adulthood. In almost every parasitic nematode, whether the L1 is contained within an egg or a sheath, the infective stage is usually the L3 larva. Eggs offer good protection, but they must be passively dispersed to their hosts. Ensheathment of the infective stage combines mobility with protection against desiccation and other adverse environmental extremes. There is usually a large active component in the host-finding strategies of ensheathed species. In these worms, the unshed cuticle of the previous moult covers the infective L3 like a protective sheath, but the worms cannot feed and must use their stored energy for all activities during finding and infecting the host (Medica and Sukhdeo, 1997).

Gastrointestinal nematodes have been recovered from every vertebrate host examined (Anderson, 2000), but for most of these species we know little beyond their descriptive taxonomy and some anecdotal observations. Most of our detailed knowledge comes from a handful of species that are amenable to experimentation because of their ease of maintenance and infection in the laboratory. Much of the work discussed in this chapter will focus on three of the most popular laboratory models. *Trichinella spiralis* is the pig parasite that is responsible for the human disease trichinellosis and is easily maintained in mice and rats. Infection occurs through the ingestion of raw infected meat, and the worms establish in the anterior small intestine, mature and produce larvae, which migrate to skeletal muscles and encyst. *Heligmosomoides polygyrus* is a luminal parasite of murine hosts and is used as a model for nematode diseases of domestic animals. Infection is through the ingestion of infective L3 larvae, and the adult worms establish in the lumen of the small intestine. Eggs are passed in the faeces and develop into infective L3 larvae. *Nippostrongylus brasiliensis* is a rodent parasite that is used as a model for hookworm disease. Infection occurs when the larvae penetrate the skin of the host and migrate through the body to the gut. These worms are also lumen dwellers and, although generally restricted to the anterior small intestine, their site fidelity is somewhat lower (Croll and Smith, 1977). Eggs are passed in the faeces and develop into mobile L3 larvae.

Living in the Gut

For gastrointestinal nematodes, the most popular route of host entry is through ingestion with food or water. After infection, these nematodes tend to travel passively along with the flow of food in the gut to get to their sites. The second route most typically begins with penetration of the skin by infective larvae (as in hookworms) and requires an obligate migration through the host to arrive at the gut. The infecting larvae migrate to the lungs, where they are coughed up and swallowed, and thus they also enter the gut orally and travel passively to their sites. There is an odd variation (e.g. *Strongylus* species in horses) where the worms infect orally, migrate around the body and then return to the gut where they began. *Ascaris lumbricoides*, the giant roundworm that infects almost a billion humans, provides a good example of this strategy. Orally infecting eggs hatch in the small intestine and the larvae penetrate the gut wall and migrate to the lungs, where they break through the alveoli and are coughed up and swallowed. This 'superfluous' migration through the body can be very costly, since most of the migrating worms are killed by the host's immune responses (Duncan, 1972; McCraw and Slocombe, 1976).

After arrival in the gut, there are two major obstacles to the successful establishment and survival of intestinal nematodes: intestinal peristalsis and the host immune response. Often the worms' solutions to these two problems are not mutually exclusive. For example, adult *T. spiralis* have an intracellular habitat within the epithelial cells of the small intestine (Despommier *et al.*, 1978) and thus the worms are spared the expulsive effects of intestinal peristalsis. However, this tissue habitat is readily accessible to the host's immune responses, and the worms are typically expelled within 7–9 days after infection (Larsh and Race, 1954; Sukhdeo and Meerovitch, 1977). The ovoviviparous females are able to produce all of their progeny in this short time, and their tissue habitat makes it easy to disperse the larvae into blood-vessels, from where they are passively transported to the musculature (Sukhdeo and Meerovitch, 1980).

In contrast, *H. polygyrus*, which lives in the lumen of the small intestine, may survive for 10–12 months in the gut. The host still mounts an immune response to these worms, but it is much less severe and not sterilizing (Urban *et al.*, 2000). In vertebrate hosts, more than half of the humoral immune responses are concentrated in the gut to control the intestine's bacterial flora (Roitt, 1977), so few worms can fully escape them, regardless of the habitat they choose. A highly detailed under-standing of the complex nature of immune responses to these worms is available (see Callard *et al.*, 1996; Lenschow *et al.*, 1996; Ekkens *et al.*, 2000; Urban *et al.*, 2000), but this topic is beyond the scope of this chapter. However, it appears that successful parasites are frequently invisible to their host's immune responses.

Lumen-dwelling nematodes may escape the brunt of the immune response, but they must deal with the inexorable expulsive effects of peristalsis. Several holdfast mechanisms have evolved to deal with gut flow. Some nematodes, such as the human hookworms *Necator americanus* and *Ancylostoma duodenale*, have armed buccal cavities, which allow them to simultaneously attach and extract blood meals. Several of the nematodes of cattle and sheep, such as *Ostertagia ostertagi* and *Haemonchus contortus*, similarly attach by embedding their anterior ends in the tissue or in pits and crypts of the gut. Large ascarid worms actively swim against the flow, as do some of the small nematodes, such as *N. brasiliensis*. Adult *H. polygyrus* attach by wrapping their body coils around villi in the duodenum.

Infecting L3 larvae tend to have the most difficulty with holdfasts, because of their small size. A typical strategy during infection is for the infecting larvae to penetrate into the tissue of the gut, where they are safe from gut flow, develop to adulthood within the tissue and then emerge as adults into the lumen. In *H. polygyrus* infections, the larvae penetrate into the small intestine within minutes of infection, develop to adults by 7–10 days and re-enter the lumen just before the specific immune response gets going (Sukhdeo *et al.*, 1984). A similar strategy is seen in several of the nematodes infecting cattle and sheep. In *O. ostertagi* and *H. contortus*, the infecting worms penetrate the gut wall and often arrest their development within the tissue until an appropriate time. These arrested larvae can survive for long periods, because they are metabolically inert and therefore invisible to the host's immune response. Hormonal responses during lactation in the host triggers a 'spring rise', whereupon the worms develop into adults, enter the lumen and produce their eggs just in time to be available to the weanlings in the spring (Roberts and Janovy, 2000).

Proximate Behaviours in Habitat Selection

Free-living nematodes and the free-living stages of parasitic nematodes have a large repertoire (> 20 classes) of distinct behaviour patterns (e.g. locomotion, swimming, nictation, jumping, oviposition) and complex orientation responses to temperature, and a wide variety of simple and complex molecules have been identified (Croll and Sukhdeo, 1981). In contrast, while adult parasitic nematodes from freshly dead intestines are always extremely active, with a lot of twisting and wriggling, it is not clear how these worms behave in the living hosts. Intestinal nematodes have extensive arrays of sensory organs and possess intricate nervous circuitries, which are as complex as those of their free-living relatives (McLaren, 1976; Sukhdeo and Sukhdeo, 1994; Halton *et al.*, 1998; Reuter *et al.*, 1998). They may also possess an equally impressive suite of behaviours, but we have been able to identify only two types of behaviours that operate within the intestine: orientation responses (attraction/repulsion) and releaser responses (fixed action patterns).

Intestinal nematodes orientate towards (are attracted to) members of the opposite sex. This was long suspected. In hookworm infections, when the sexes are unbalanced, with more males than females, there is an increase in the laceration of the mucosa and subsequent blood loss as the males migrate more actively to find the females (Beaver et al., 1964). Males and females of several species will find each other, even if inoculated at separate locations in the gut (Beaver, 1955; Roche, 1966; Sukhdeo and Meerovitch, 1977). Laboratory studies using specially designed orientation chambers have conclusively demonstrated that the sexes are attracted to each other (Bonner and Etgers, 1967; Anya, 1976; Bone and Shorey, 1977; Belosevic and Dick, 1980).

Clearly, these worms possess the mechanisms for orientation, and these orientation responses are undoubtedly adaptive in mate finding over short distances in the gut. (Turbulence from peristalsis would make gradients unlikely over large distances.) Surprisingly, however, orientation responses have not been demonstrated in the habitat-selection strategy of parasites: that is, parasitic worms do not orientate to any signal within their hosts except those coming from potential mates (Sukhdeo and Sukhdeo, 1994; Sukhdeo, 1997). This is odd because the classic mechanistic explanation for why parasite habitat selection is so precise is that the worms are attracted to specific signals emanating from their habitats (Bone, 1981). Yet, despite intense efforts to identify these attractive signals during habitat selection, none have been found in parasites (Sukhdeo, 1990; Sukhdeo and Sukhdeo, 2002).

Releaser Responses

On entering the intestines of their hosts, nematode infective stages are usually carried passively to their sites of penetration. The encysted stages of T. spiralis are digested in the stomach and, stimulated by pepsin, the worms whip their tails around to help break out of the cysts (Fig. 11.1a). These tail-whipping worms are then carried passively with the stomach contents into the small intestine, where bile is secreted into the very first section (duodenum). Bile triggers an instantaneous change in the worms' behaviour and they enter into a frenetic sinusoidal behaviour pattern (Fig. 11.1b). This behaviour is characteristic of genetically fixed behaviour patterns called releaser responses (Lorenz and Tinbergen, 1957), and it makes the worms automatically migrate out of the gut contents and penetrate into the gut wall (Sukhdeo, 1990). The worms will even penetrate abnormal habitats in the large intestine when triggered by bile. Once triggered, this response will proceed unabated for hours, until the worms run out of energy and die (Sukhdeo and Croll, 1981b). At this stage, there is no need for the worms to conserve energy because, if they do not successfully penetrate, they will be swept out with the faeces.

Releaser responses are fixed stereotyped activities (occur in the same way every time) that are triggered by sign stimuli, in this case, bile. These responses generally evolve in response to environmental or behavioural conditions that are predictable or constant, and they are triggered by sign stimuli that are unambiguous (Lorenz and Tinbergen, 1957). For example,

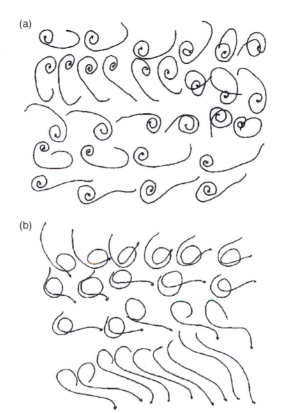

Fig. 11.1. Frame-by-frame analysis of the movement patterns of infective larvae of *Trichinella spiralis*. The sequence moves from top left to bottom right, the anterior end of the worm is depicted by an arrowhead and the elapsed time between each frame is 0.017 s. (a) Movement pattern of a larva in the stomach showing coiling and uncoiling of the tail. (b) Movement pattern of a larva that has just been triggered by bile, as it changes from the coiling pattern into a rapid sinusoidal pattern.

the dot on a parent bird's beak releases a fixed begging behaviour from the hatchling, or the red colour of a male adversary in some fish elicit a fixed fighting response (Alcock, 1998). Host intestines are extremely predictable environments to an invading parasite because, in every individual of the same host species, intestines are functionally and physically identical. Hosts have predictable diets, moulded by natural selection, and the design of the gut is intimately connected to the predictable process of digesting these foodstuffs. Under these predictable conditions, releaser responses may provide the optimal solutions.

The infecting worms travel passively in the gut and thus the locations where the worms penetrate are a function of how fast the gut is flowing when they are triggered. If intestinal transit is artificially increased, the worms will establish significantly more posteriorly than normal controls and, when gut peristalsis is slowed down, the worms establish more anteriorly (Sukhdeo and Croll, 1981b). Habitat selection by the infective larvae of *H. polygyrus* is similar to that in *T. spiralis*, and the worms also penetrate the gut when triggered by bile. If the entry of bile into the intestine is surgically relocated to more posterior regions of the intestine, the worms also move posteriorly to establish at the location of the 'new' bile-duct (Sukhdeo and Croll, 1981a). In normal hosts, bile is a reliable signal that the worms have arrived in the small intestine.

Bile is not the only signal used by parasites in the gut, but, while several components of the gut, including pepsin, trypsin and pH, have also been shown to trigger these responses, bile appears to be the most widely used (Lackie, 1975). Activation responses to bile have been seen in hundreds of diverse parasites of the small intestine (Lackie, 1975), and it has been experimentally demonstrated that bile is a critical requirement for successful infection in numerous species (Sukhdeo and Mettrick, 1984, 1986). Nevertheless, these parasites do not orientate to bile, but simply use it as a sign stimulus.

Fixed behaviours are also important in the habitat selection of several cestodes and trematodes (Sukhdeo, 1990; Sukhdeo and Sukhdeo, 2002). Even in apparently complex migrations, parasitic worms do not appear to utilize orientation mechanisms, but, rather, take advantage of predictable body architecture (Sukhdeo and Mettrick, 1986) or the unidirectional flows of blood in the host. Skin-penetrating nematodes appear to have the most complex migrations through the host body to get to the gut, but in reality they need only migrate to the lungs, where they are coughed up and swallowed. It appears that, for most of their migration through the body, they are just carried passively through the pipes and tubes of the circulatory system to the lungs. The migration of live skin-penetrating larvae of *N. brasiliensis* to the lungs was no different from the migration of killed worms injected into small veins (Croll, 1972a,b), suggesting that penetrating worms might require only a single behaviour that allows them to burrow through tissue until they hit a vein, and from there they are passively carried to the lungs, where they get trapped in the pulmonary capillary beds (Bone, 1981).

Theoretical Models of Habitat Selection

The most widely used theoretical model of habitat selection in animals is the ideal free distribution (IFD) model, which was developed to explain bird behaviour (Fretwell and Lucas, 1970). Subsequently, there have been numerous tweakings and modifications of this basic model (Fretwell, 1972; Gillis and Kramer, 1987; Weber, 1998), but the underlying assumptions of optimality have not been altered (van der Steen, 1998; Weber, 1998). Modern behaviourists often claim that the IFD was designed to be a null model on to which more realistic factors could be added, and they tacitly and explicitly argue that the assumptions of the model are too simplistic for all but the most basic or artificial sets of conditions.

Nevertheless, it is still important for its heuristic value in defining the way in which questions on habitat selection should be approached. However, it is doubtful that flawed models can teach us how to think about nature, and clinging to accepted wisdom is often the most important obstacle to the development of new ideas and models.

IFD assumes that, in any environment, potential habitats vary in resource quality, and it predicts how animals in the population (competitors) will choose among the habitats. In this model, competitors are 'ideal' in that they have perfect knowledge of their environment and know which habitats have the highest gains, and they are 'free' to enter into any habitat because the resources are not defended in any way. The model predicts that competitors will distribute themselves in proportion to habitat profitabilities such that their distribution in the environment will be aggregated (more competitors in some habitats than others) but the fitness of each competitor will be similar (Fretwell and Lucas, 1970). In a very fundamental way, this theoretical model applies directly to parasite habitat selection. Parasites have an ideal knowledge of their living environments, which developed over long evolutionary periods of intimate host/parasite contact. In parasites, habitat-selection decisions are so fine-tuned that the responses have become genetically fixed. Clearly, then, these fixed behaviour patterns should represent optimal solutions to the problem of habitat selection. However, parasites in the small intestine do not distribute themselves in accordance with the IFD model (Sukhdeo, 1991, 2000; Sukhdeo and Bansemir, 1996).

In *T. spiralis* infections, the adults remain at the site of larval establishment in the anterior small intestine (Sukhdeo and Meerovitch, 1980). The distribution of these worms in the gut is aggregated (Tyzzer and Honeij, 1916; Roth, 1938; Gursch, 1949), but it is not an IFD, specifically because individual reproductive fitness (fecundity) is not similar at all locations (Sukhdeo, 1991). Worms inhabiting the poorer habitats should move to the better habitats until density-dependent effects reduce the fitness of these good sites to that in the poorer sites. The non-IFD distributions in *T. spiralis* may be due to constraints on the parasite's ability to move freely, because it is an intracellular parasite. However, similar non-IFD distributions are seen in the habitat selection behaviour of *H. polygyrus* (Bansemir and Sukhdeo, 1996). These worms are mobile and easily migrate around in the gut; worms transplanted into the most terminal ileal regions can return upstream to the duodenum within 24 h, a distance of more than 30 cm (Sukhdeo and Bansemir, 1996). Yet, in low-level infections where there are few competitors, worms will often establish in suboptimal locations, even when the best habitats (those associated with highest fecundity) are completely vacant (Sukhdeo, 2000). Such behaviours are not easy to explain, because there are obvious penalties on reproductive success for choosing poor locations.

IFD and other models based on optimal foraging (MacArthur and Pianka, 1966) have not fared well in studies in the free-living world either (Seger and Stubblefield, 1996; van der Steen, 1998). The problem may lie in the notion of optimality. The idea of optimal behaviour was based on a 1960s concept of the gene as an orderly library of adaptively constructed genetic information, under which optimal responses could evolve. However, the modern genome, with its junk, non-coding, repetitive and selfish DNA, is nothing like the old perception. It is difficult to imagine that these imperfect genes can lead to optimal behaviour, and yet the idea of optimal behaviour still permeates our thinking about and our modelling of animal responses. The assumptions of optimality are rarely questioned. Failure of the model (or its many modifications) in the real world is explained by the presumption that animals in the experiments were not perfectly adapted to their environments or that there was a lack of complete information about the environment and the additional costs and benefits that compete with the ones being investigated (see Krebs and Davies, 1978; Begon *et al.*, 1996). Optimality models are mere caricatures of nature and should not be promoted for their value as yardsticks against which to compare nature. These models are simple and easy to program into computers, are relevant to many classes of ecological problems and are more easily testable than any other ecological model (Weber, 1998); nevertheless, mere convenience should not be a justification for sticking with a flawed paradigm, especially if it precludes fresh new ideas.

Resources

In all classical hypotheses on parasite habitat selection (Holmes, 1973; Hair and Holmes, 1975; Rohde, 1981; Price, 1986; Rohde and Hobbs, 1986; Brooks and McLennan, 1993), the role of resources is considered to be the most critical parameter. However, there have been few efforts to identify these putative resources in parasite systems. Resources are defined through their use by individuals and their effects on individual fitness (Wiens, 1984). For gastrointestinal nematodes, we generally consider three categories of resources: mates, attachment and food. Thus, we would need to know that changes in mates, attachment or food will influence their reproductive fitness.

Food resources are probably the most important determinant of habitat suitability (Partridge, 1978; Dill, 1987). The food resource of intestinal nematodes may come from one of three compartments: host ingesta, host tissue or host blood. However, for most parasites, we know little of what they eat and much of what we think is based on circumstantial evidence (Bansemir and Sukhdeo, 1996). For instance, examples of nematodes that feed on host ingesta have always included *H. polygyrus* (Bawden, 1969) and *N. brasiliensis* (Croll, 1976). In fact, utilizing techniques that differentially label the food compartments, it has been shown that both of these parasites feed exclusively on host epithelial tissue (Bansemir and Sukhdeo, 1994, 2002). Similarly, the spearlike stylet in the mouth of the human caecal whipworm *Trichuris trichiura* suggested that these worms were blood feeders (Li, 1933; Chitwood and

Chitwood, 1937), but, in fact, their mouth-parts are never near any blood source (Lee and Wright, 1978). Clearly, we need to be more rigorous in our assumptions about how parasites feed, especially since this behaviour can have significant impacts on their habitat selection.

Altering food availability in these parasite can be difficult. When the host is fasted, there is a significant reduction in the individual fecundities of the female worms of *H. polygyrus* (Sukhdeo and Bansemir, 1996) and there is also a significant change in their intestinal distribution patterns (Bansemir and Sukhdeo, 1996). These worms use intestinal villi both for attachment (wrapping around) and as food, and the length of the villi is the most important determinant of their habitat selection (Bansemir and Sukhdeo, 1996). Thus, it is difficult to separate the effects of food reduction and attachment effort in these worms, because starving the host causes shortening of the villi and the worms respond by migrating to the longest villi with the better holdfasts. Nevertheless, attachment resources appear to be more important than mates in this parasite because, when males and females are surgically transplanted to the posterior small intestine (short villi), the smaller males leave the larger females behind as they scramble towards the long villi in the duodenum (Sukhdeo and Bansemir, 1996). Eventually the females do catch up, but this response suggests that, in these long-lived worms, secure holdfasts take precedence over mates. In short-lived worms, these decisions may be different. For example, in both *T. spiralis* and *N. brasiliensis*, mates are very important, and the male worms will actively search for their females, regardless of other conditions related to attachment or food (Sukhdeo and Croll, 1981a; Bansemir and Sukhdeo, 2002).

Clearly, future models of habitat selection in intestinal nematodes will have to recognize that each species has distinct priorities and that their behavioural strategies must be considered in light of the constraints imposed by their life histories.

Life-history Constraints

It is not always intuitive how life history may shape habitat selection behaviour in intestinal nematodes. At a gross level, it seems clear that the manner in which the parasites transmit themselves to the next host can have a significant effect on habitat selection. For example, *H. polygyrus*'s transmission strategy involves shedding infective stages into the faecal stream; thus a luminal habitat may be more appropriate than a tissue habitat, such as that of *T. spiralis*, which is transmitted by carnivory. In fact, life-history constraints may exert effects at even finer scales. In these nematodes, habitats are very narrow, and significant differences in female fitness can occur over very short distances within the gut (Sukhdeo, 1991; Bansemir and Sukhdeo, 1996). Several theoretical models suggest that a trade-off might exist between a parasite's virulence (a direct effect of using host resources for parasite reproduction) and its transmission success

(Anderson and May, 1991). Small changes in microhabitat choice might mediate this process, but this is unlikely. There are easier ways to achieve this, e.g. altering the reproductive apparatus. Among closely related species of *Trichinella* that differ in their virulence, there is a significant positive correlation between uterus size and the number of infective stages produced (Sukhdeo and Meerovitch, 1980).

Much of the way we think about host/parasite interactions comes from the gene-for-gene (GFG) hypothesis, which originated in studies of plant pathogens (Flor, 1956). There are actually now two hypotheses (variants) of GFG interactions, and geneticists are debating over them because there is not enough evidence to unequivocally support either (Newton and Andrivon, 1995). The original interpretation is that, for each gene that conditions resistance in the host, there is a corresponding gene that conditions virulence (pathogenicity) in the parasite (Vanderplank, 1991). The second and more popular interpretation is that the parasite gene conditions avirulence (Kerr, 1987). The first implies a constant escalation between the partners, while the second considers the fact that it might not necessarily be advantageous for a pathogen to be virulent at the expense of its host.

GFG is a simplistic explanation for how hosts and parasites interact with each other, and it is now acknowledged that GFG interactions, if they exist, are often complex polymorphisms, with large numbers of resistance and virulence genes involved (Leonard, 1993). In animal-parasite studies, there does not seem to be such an emphasis on GFG, but concepts such as the 'red queen' and 'arms race' are modifications of GFG (Lively, 1996). Parasitologists appear to be much more comfortable discussing an arms race because it does not imply that there are specific rigid gene products that control resistance and virulence.

In GFG scenarios, the immune response is thought to be the major player. There are several examples of long-lived nematode infections that are not actively controlled by host resistance and are slow to elicit acquired responses, despite repeated exposure. For example, the human hookworm *Necator americanus* can live for 17 years in the host (Behnke, 1987) and the filarial nematode *Onchocerca volvulus*, which causes tropical river blindness in humans, can live for 18 years (Plaiser *et al.*, 1991). It is not clear why there is no escalation by the host to kill the parasite in these situations. It may be that host immune responses have evolved to restrict infections within narrow tolerable bounds (narrow niches), rather than totally eliminate infections. This may explain the scarcity of sterilizing immunity among these worms (Behnke *et al.*, 1992).

Several theoretical and empirical studies now suggest that parasite transmission and its need to produce infective stages at the cost of host resources are directly coupled to the evolution of virulence in parasites (Anderson and May, 1991; Ewald, 1994; Lenski and May, 1994). This mechanism might clearly operate in microparasites that multiply within their hosts, e.g. bacteria, viruses and protozoa (Bull, 1994; Read, 1994). However, intestinal nematodes (and other macroparasites) differ fundamentally from microparasites. Microparasites destroy tissue during replication, and thus their virulence is directly related to replication. In macroparasites, virulence is directly related to the density of worms in the gut and, in high-density infections with high virulence, density-dependent effects reduce egg production per worm (Fleming, 1988), so there is no intrinsic benefit to increased virulence (Anderson and May, 1991). Indeed, virulence may be decoupled from transmission under certain conditions. For example, in *Strongylus* species, nematodes that infect orally before migrating through the body, virulence is not caused by the adults but by the migrating larvae damaging tissue before they return to the gut to reproduce. In these nematodes, there is no clear relation between transmission and virulence (Medica and Sukhdeo, 2001). However, in trichostrongyles, such as *H. contortus*, where adult worms cause damage as they feed on tissue and reproduce, there is a positive relationship between virulence and transmission rate (Medica and Sukhdeo, 2001).

Phylogenetic Constraints

Phylogeny undoubtedly constrains several aspects of nematode biology, including habitat selection. However, while behavioural and life-history characters are typically mapped on to existing phylogenies, they are rarely included as character states in phylogenetic inferences. One of the most perplexing aspects of nematode habitat selection is the superfluous tissue migration before re-entering the gut. In *Strongylus vulgaris*, the virulent caecal worm of horses, infection is oral, but the worms penetrate the gut, migrate through the circulatory system and then return to the intestines as adult worms (McCraw and Slocombe, 1976). At the heart of this odd migration is the question of how parasitic nematodes evolved.

Nematode parasitism probably did not evolve until animals invaded land, about 430 million years ago (Chabaud, 1954, 1955; Anderson, 1984). Thus, there are almost no nematode parasites of marine groups, such as molluscs, polychaetes and crustaceans, but a rich nematode fauna in terrestrial groups, such as earthworms, insects and terrestrial molluscs (Anderson, 2000). The entire group of monoxenic (one host) intestinal nematodes are thought to have evolved from free-living, bacteria-feeding soil nematodes, and the first mode of transmission was not by oral ingestion but through skin penetration (Fülleborn, 1929; Adamson, 1986, 1989). In this scenario, nematode parasitism began with the accidental penetration of the moist skin of an early amphibian host (Fig. 11.2). The free-living nematodes probably already used the skin of an animal as a mechanism to avoid dry periods in the environment and as transportation between suitable habitats (Dougherty, 1951; Chabaud, 1982). After penetrating the skin, the worms had to make the obligatory migration to the gut because they were bacteria feeders, and the bacteria in the host are found in the gut. Subsequently, large herbivores evolved and they

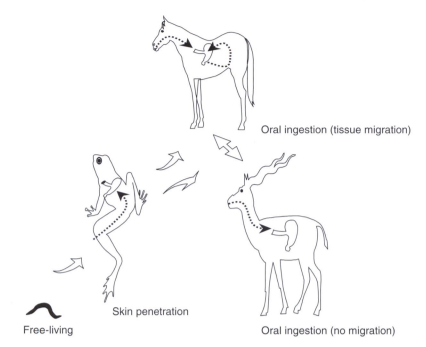

Oral ingestion (tissue migration)

Skin penetration

Free-living   Oral ingestion (no migration)

Fig. 11.2. The evolution of transmission strategies in intestinal nematodes (Strongylida). Free-living soil nematodes first invaded amphibian hosts via skin penetration. Subsequently, with the evolution of large herbivores, these skin penetrators were ingested and the oral routes evolved. (After Sukhdeo *et al.*, 1997.)

accidentally ingested the skin penetrators, who became oral infectors but retained their migration through the host, and thus the odd migration was explained (Durette-Desset, 1985; Adamson, 1989). This hypothesis was tested in related species representing oral infectors, skin penetrators and oral infectors that migrate in the host, using sequences of part of the mitochondrial DNA (mtDNA) gene encoding cytochrome c oxidase subunit I. These phylogenetic analyses supported the hypothesis that the most primitive nematode parasites were skin penetrators (Sukhdeo *et al.*, 1997). The data suggest that the odd migration of *S. vulgaris* is a legacy of a past transmission strategy, and reminds us that evolution does not always produce the optimal solutions.

These examples represent the effects of phylogeny at a gross level in the habitat-selection behaviour of nematodes, but there are many presumed effects at finer scales that have not been tested. For example, a common assumption is that the restriction of the skin penetrators to the gut may have been the result of their bacteria-feeding preference (Durette-Desset, 1985; Adamson, 1989). However, there is now some evidence suggesting that the earliest skin penetrators lived as adults in the lungs and not the intestines (S.C. Sukhdeo, unpublished). Clearly, much more work is needed, and the growing accessibility of molecular and other analytical tools portends an exciting future.

References

Adamson, M.L. (1986) Modes of transmission and evolution of life histories in zooparasitic nematodes. *Canadian Journal of Zoology* 64, 1375–1384.

Adamson, M.L. (1989) Constraints in the evolution of life histories in zooparasitic Nematoda. In: Ko, R.C. (ed.) *Current Concepts in Parasitology*. University Press, Hong Kong, pp. 221–253.

Adamson, M.L. and Caira, J.N. (1994) Evolutionary factors influencing the nature of parasite specificity. *Parasitology* 109 (suppl.), S85–S96.

Alcock, J. (1998) *Animal Behavior: an Evolutionary Approach*, 6th edn. Sinauer Associates, Sunderland, Massachusetts, 640 pp.

Anderson, R.C. (1984) The origins of zooparasitic nematodes. *Canadian Journal of Zoology* 62, 317–328.

Anderson, R.C. (2000) *Nematode Parasites of Vertebrates*. CAB International, Wallingford, 650 pp.

Anderson, R.M. and May, R.M. (1991) *Infectious Diseases of Humans: Dynamics and Control*. Oxford University Press, New York, 375 pp.

Anya, A.O. (1976) Studies on the reproductive physiology of nematodes: the phenomenon of sexual attraction and the origin of the attractants in *Aspiculuris tetraptera*. *International Journal for Parasitology* 6, 173–177.

Bansemir, A.D. and Sukhdeo, M.V.K. (1994) The food resource of adult *Heligmosomoides polygyrus* in the small intestine. *Journal of Parasitology* 80, 24–28.

Bansemir, A.D. and Sukhdeo, M.V.K. (1996) Habitat selection by a gastrointestinal parasite: proximal cues involved in decision making. *Parasitology* 113, 311–316.

Bansemir, A.D. and Sukhdeo, M.V.K. (2002) The food resource of *Nippostrongylus brasiliensis*. *Journal of Parasitology* 87, 1470–1472.

Bawden, R.J. (1969) Some effects of the diet of mice on *Nematospiroides dubius* (Nematoda). *Parasitology* 59, 203–213.

Beaver, P.C. (1955) Observations on *Necator* infections resulting from exposure to three larvae. *Revista Ibero Parasitologica Tomo Extranjera* 713, 721.

Beaver, P.C., Yoshida, Y. and Ash, L.R. (1964) Mating of *Ancylostoma caninum* in relation to blood loss in the host. *Journal of Parasitology* 50, 286–293.

Begon, M., Harper, J.L. and Townsend, C.R. (1996) *Ecology: Individuals, Populations, and Communities*, 3rd edn. Blackwell Science, Oxford, 1068 pp.

Behnke, J.M. (1987) Evasion of immunity by nematode parasites causing chronic infections. *Advances in Parasitology* 26, 1–71.

Behnke, J.M., Barnard, C.J. and Wakelin, D. (1992) Understanding chronic nematode infections: evolutionary considerations, current hypotheses and the way forward. *International Journal for Parasitology* 2, 861–907.

Belosevic, M. and Dick, T.A. (1980) Chemical attraction in the genus *Trichinella*. *Journal of Parasitology* 66, 88–93.

Bone, L.W (1981) Chemotaxis of parasitic nematodes. In: Bailey, W.S. (ed.) *Cues that Influence the Behavior of Internal Parasites*. US Department of Agriculture, New Orleans, pp. 52–66.

Bone, L.W. and Shorey, H.H. (1977) Interactive influences of male and female produced pheromones on male attraction to female *Nippostrongylus brasiliensis*.
Journal of Parasitology 63, 843–848.

Bongers, T. and Ferris, H. (1999) Nematode community structure as a bioindicator
in environmental monitoring. *Trends in Ecology and Evolution* 14, 224–228.

Bonner, T.P. and Etges, F.J. (1967) Chemically mediated sexual attraction in
Trichinella spiralis. *Experimental Parasitology* 21, 53–60.

Brooks, D.R. and McLennan, D.A. (1993) *Parascript: Parasites and the Language of
Evolution*. Smithsonian Institution Press, Washington, DC, 429 pp.

Bull, J.J. (1994) Virulence. *Evolution* 48, 1423–1437.

Callard, R.E., Matthews, D.J. and Hibbert, L. (1996) IL-4 and IL-13 receptors: are
they one and the same? *Immunology Today* 17, 108–110.

Chabaud, A.G. (1954) Sur le cycle évolutif des spirurides et des nématodes ayant
une biologie comparable: valeur systématique des caractères biologiques.
Annales de Parasitologie Humaine et Comparée 29, 42–88.

Chabaud, A.G. (1955) Essai d'interprétation phylétique des cycles évolutifs chez
les Nématodes, parasites de Vertébrés: conclusions taxonomiques. *Annales de
Parasitologie Humaine et Comparée* 30, 83–126.

Chabaud, A.G. (1982) Evolution et taxonomie des Nématodes – revue. In: Mettrick,
D.F. and Desser, S.S. (eds) *Parasites: Their World and Ours. Fifth International Congress of Parasitologists*. Elsevier Biomedical, New York,
pp. 216–221.

Chitwood, B.G. and Chitwood, M.B. (1937) The histological anatomy of
Cephalobellus papiliger Cobb, 1920. VIII. The esophagi of representatives
of the Enoplida. *Zeitschrift Zellforschung* 27, 517–531.

Croll, N.A. and Smith, J.M. (1977) The location of parasites within their hosts:
the behaviour of *Nippostrongylus brasiliensis* in the anaesthetized rat.
International Journal for Parasitology 7, 195–200.

Croll, N.A. and Sukhdeo, M.V.K. (1981) Hierarchies of nematode behaviour. In:
Zuckerman, B.M. (ed.) *Plant Parasitic Nematodes*, Vol. 3. Academic Press,
New York, pp. 227–251.

Croll, N.C. (1972a) Behaviour of larval nematodes. In: Canning, E.U. and Wright,
C.A. (eds) *Behavioural Aspects of Parasite Transmission*. Academic Press,
London, pp. 31–52.

Croll, N.C. (1972b) Behavioural activities of nematodes. *Helminthological
Abstracts* 41A, 359–377.

Croll, N.C. (1976) The location of parasites within their hosts: the influence of host
feeding and diet on the dispersion of adults of *Nippostrongylus brasiliensis* in
the intestine of the rat. *International Journal for Parasitology* 6, 441–448.

Crompton, D.W.T. (1973) The sites occupied by some parasite helminths in the
alimentary tract of vertebrates. *Biological Reviews* 47, 27–83.

Despommier, D.D., Sukhdeo, M.V.K. and Meerovitch, E. (1978) *Trichinella
spiralis*: site selection of the larva during its enteral phase of infection in mice.
Experimental Parasitology 44, 209–215.

Dill, L.M. (1987) Animal decision making and ecological consequences: the future
of aquatic ecology and behaviour. *Canadian Journal of Zoology* 65, 803–811.

Dougherty, E.C. (1951) Evolution of zooparasitic groups in the phylum Nematoda,
with special reference to host-distribution. *Journal of Parasitology* 37,
353–378.

Duncan, J.L. (1972) The life cycle, pathogenesis and epidemiology of *Strongylus
vulgaris* in the horse. *Equine Veterinary Journal* 5, 20–25.

Durette-Desset, M.-C. (1985) Trichostrongyloid nematodes and their vertebrate hosts: reconstruction of the phylogeny of a parasitic group. *Advances in Parasitology* 24, 239–306.

Ekkens, M., Liu, Q., Fang, H., Byrd, C., Sharpe, A.H., Urban, J. and Gausel, W.C. (2000) OX40L is not required for the development of a B7-dependent type 2 immune response to the nematode parasite, *Heligmosomoides polygyrus*. *FASEB Journal* 14, A954.

Ewald, P.W. (1994) *Evolution of Infectious Disease*. Oxford University Press, Oxford, 298 pp.

Fleming, M.W. (1988) Size of inoculum dose regulates in part worm burdens, fecundity, and lengths in ovine *Haemonchus contortus* infections. *Journal of Parasitology* 74, 975–978.

Flor, H.H. (1956) The complementary genic systems in flax and flax rust. *Advances in Genetics* 8, 29–54.

Fretwell, S.D. (1972) *Populations in a Seasonal Environment*. Princeton University Press, Princeton, New Jersey, 217 pp.

Fretwell, S.D. and Lucas, H.L. (1970) On territorial behavior and other factors influencing habitat distribution in birds. *Acta Biotheoretica* 19, 1–36.

Fülleborn, F. (1929) On the larval migration of some parasitic nematodes in the body of the host and its biological significance. *Journal of Helminthology* 7, 15–26.

Gillis, D.M. and Kramer, D.L. (1987) Ideal interference distributions: population density and patch use by zebrafish. *Animal Behaviour* 35, 1875–1882.

Gursch, O.F. (1949) Effect of digestion and refrigeration on the ability of *Trichinella spiralis* to infect rats. *Journal of Parasitology* 34, 394–395.

Hair, J.D. and Holmes, J.C. (1975) The usefulness of measures of diversity, niche width and niche overlap in the analysis of helminth communities in waterfowl. *Acta Parasitologica Polonica* 23, 253–269.

Halton, D.W., Maule, A.G., Mair, G.R. and Shaw, C. (1998) Monogenean neuromusculature: some structural and functional correlates. *International Journal for Parasitology* 28, 1609–1623.

Holmes, J.C. (1961) Effects of concurrent infections on *Hymenolepis diminuta* (Cestoda) and *Moniliformis dubius* (Acanthocephala). I. General effects and comparison with crowding. *Journal of Parasitology* 47, 209–216.

Holmes, J.C. (1973) Site segregation by parasitic helminths: interspecific interactions, site segregation, and their importance to the development of helminth communities. *Canadian Journal of Zoology* 51, 333–347.

Holmes, J.C. and Price, P.W. (1986) Communities of parasites. In: Anderson, D.J. and Kikkawa, A.J. (eds) *Community Ecology: Pattern and Process*. Blackwell Scientific Publications, Oxford, pp. 187–213.

Kerr, A. (1987) The impact of molecular genetics on plant pathology. *Annual Review of Phytopathology* 25, 87–110.

Krebs, J.R. and Davies, N.B. (1978) *Behavioral Ecology, an Evolutionary Approach*. Sinauer Associates, Sunderland.

Lackie, A.M. (1975) The activation of infective stages of endoparasites of vertebrates. *Biological Reviews* 50, 285–323.

Larsh, J.E., Jr and Race, G.J. (1954) A histopathologic study of the anterior small intestine of immunized and nonimmunized mice infected with *Trichinella spiralis*. *Journal of Infectious Diseases* 94, 262–272.

Lee, T.D.G. and Wright, K.A. (1978) The morphology of the attachment and probable feeding site of the nematode *Trichuris muris* (Schrank, 1788) Hall, 1916. *Canadian Journal of Zoology* 56, 1889–1905.

Lenschow, D.J., Walunas, T.L. and Bluestone, J.A. (1996) CD28/B7 system of T cell costimulation. *Annual Review of Immunology* 14, 233–258.

Lenski, R.E. and May, R.M. (1994) The evolution of virulence in parasites and pathogens: reconciliation between two competing hypotheses. *Journal of Theoretical Biology* 169, 253–265.

Leonard, K.J. (1993) Stability of equilibria in a gene-for-gene coevolution model of host–parasite interactions. *Phytopathology* 84, 70–77.

Li, H.C. (1933) On the mouth-spear of *Trichocephalus trichuria* and of a *Trichocephalus* sp. from monkey, *Macacus rhesus. Chinese Medical Journal* 47, 1343–1346.

Lively, C.M. (1996) Host–parasite coevolution and sex. *BioScience* 46, 107–114.

Lorenz, K.Z. and Tinbergen, N. (1957) Taxis and instinct. In: Schiller, C.H. (ed.) *Instinctive Behavior: the Development of a Modern Concept.* International Universities Press, New York, pp. 176–208.

MacArthur, R.H. and Pianka, E.R. (1966) On optimal use of a patchy environment. *American Naturalist* 100, 603–609.

McCraw, B.M. and Slocombe, J.O.D. (1976) *Strongylus vulgaris* in the horse: a review. *Canadian Veterinary Journal* 17, 150–157.

McLaren, D.J. (1976) Sense organs and their secretions. In: Croll, N.A. (ed.) *The Organization of Nematodes.* Academic Press, London, pp. 139–162.

Medica, D.L. and Sukhdeo, M.V.K. (1997) Role of lipids in the transmission of the infective stage (L3) of *Strongylus vulgaris* (Nematoda: Strongylida). *Journal of Parasitology* 83, 775–779.

Medica, D.L. and Sukhdeo, M.V.K. (2001) Estimating transmission potential in gastrointestinal nematodes (Order Strongylida). *Journal of Parasitology* 87, 442–445.

Montresor, A., Crompton, D.W.T., Hall, A., Bundy, D.A.P. and Savioli, L. (1998) Guidelines for the Evaluation of Soil-transmitted Helminthiasis and Schisto-somiasis at Community Level. WHO/CTD/SIP 98.1, WHO, Geneva.

Newton, A.C. and Andrivon, D. (1995) Assumptions and implications of current gene-for-gene hypotheses. *Plant Pathology* 44, 607–618.

Partridge, L. (1978) Habitat selection. In: Krebs, J.R. and Davies, N.B. (eds) *Behavioural Ecology: an Evolutionary Approach.* Sinauer Associates, Sunderland, Massachusetts, pp. 351–376.

Plaiser, A.P., van Oortmarssen, G.J., Remme, J. and Habbema, J.D.F. (1991) The reproductive lifespan of *Onchocerca volvulus* in West African savanna. *Acta Tropica* 48, 271–284.

Price, P.W. (1986) Evolution in parasite communities. In: Howell, M.J. (ed.) *Parasitology – Quo Vadit? Proceedings of the Sixth International Congress of Parasitology.* Australian Academy of Science, Canberra, pp. 209–214.

Read, A.F. (1994) The evolution of virulence. *Trends in Microbiology* 2, 73–76.

Reuter, M., Mantyla, K. and Gustafsson, M.K.S. (1998) Organization of the orthogon-main and minor nerve cords. *Hydrobiologia* 383, 175–182.

Roberts, L.S. and Janovy, J., Jr (2000) *Foundations of Parasitology,* 6th edn. McGraw-Hill, Boston, 658 pp.

Roche, M. (1966) Influence of male and female *Ancylostoma caninum* on each other's distribution in the intestine of the dog. *Experimental Parasitology* 19, 327–331.

Rohde, K. (1977) A non-competitive mechanism responsible for restricting niches. *Zoologischer Anzeiger* 199, 164–172.

Rohde, K. (1979) A critical evaluation of intrinsic and extrinsic factors responsible for niche restriction in parasites. *American Naturalist* 114, 648–671.

Rohde, K. (1981) Niche width of parasites in species-rich and species-poor communities. *Experientia* 37, 359–361.

Rohde, K. (1992) Latitudinal gradients in species diversity: the search for the primary cause. *Oikos* 65, 514–527.

Rohde, K. (1994) Niche restriction in parasites: proximate and ultimate causes. *Parasitology* 109, S69-S84.

Rohde, K. and Hobbs, R.P. (1986) Species segregation: competition or reinforcement of reproductive barriers? In: Cremin, C., Dobson, C. and Moorhouse, D.E. (eds) *Parasite Lives. Papers on Parasites, their Hosts and their Association to Honor J.F.A. Sprent*. University of Queensland Press, Brisbane, Queensland, pp. 189–199.

Roitt, I. (1977) *Essential Immunology*, 3rd edn. Blackwell Scientific Publications, Oxford, 324 pp.

Roth, H. (1938) On the localization of adult trichinae in the intestine. *Journal of Parasitology* 24, 225–231.

Schad, G.A. (1963) The ecology of co-occurring congeneric pinworms in the tortoise *Testudo graeca*. In: *Proceedings XVI International Congress of Zoology*, Vol. 1, pp. 223–224.

Seger, J. and Stubblefield, J.W. (1996) Optimization and adaptation. In: Rose, M.R. and Lauder, G.V. (eds) *Adaptation*. Academic Press, New York, pp. 93–123.

Sukhdeo, M.V.K. (1990) Habitat selection by helminths: a hypothesis. *Parasitology Today* 6, 234–237.

Sukhdeo, M.V.K. (1991) The relationship between intestinal location and fecundity in *Trichinella spiralis*. *International Journal for Parasitology* 21, 855–858.

Sukhdeo, M.V.K. (1997) Earth's third environment: a worm's eye view! *Bioscience* 47, 141–149.

Sukhdeo, M.V.K. (2000) Inside the vertebrate host: ecological strategies by parasites living in the third environment. In: Poulin, R., Morand, S. and Skorping, A. (eds) *Evolutionary Biology of Host–Parasite Relationships*. Development in Animal and Veterinary Sciences 32, Elsevier, New York, pp. 43–62.

Sukhdeo, M.V.K. and Bansemir, A.D. (1996) Critical resources that influence habitat selection decisions by gastrointestinal helminth parasites. *International Journal for Parasitology* 26, 483–498.

Sukhdeo, M.V.K. and Croll, N.A. (1981a) The location of parasites in their hosts: bile and site selection behaviour in *Nematospiroides dubius*. *International Journal for Parasitology* 11, 157–162.

Sukhdeo, M.V.K. and Croll, N.A. (1981b) The location of parasites in their hosts: the effects of physiological factors on site selection by *Trichinella spiralis*. *International Journal for Parasitology* 11, 163–168.

Sukhdeo, M.V.K. and Meerovitch, E. (1977) Comparison of three geographical isolates of *Trichinella*. *Canadian Journal of Zoology* 55, 2060–2064.

Sukhdeo, M.V.K. and Meerovitch, E. (1980) A biological explanation for lower infectivities in some *Trichinella* isolates. *Canadian Journal of Zoology* 58, 1227–1231.

Sukhdeo, M.V.K. and Mettrick, D.F. (1984) *Heligmosomoides polygyrus* (Nematoda): distribution and net fluxes of glucose, H_2O, Na^+, K^+, and Cl^- in the mouse small intestine. *Canadian Journal of Zoology* 62, 37–40.

Sukhdeo, M.V.K. and Mettrick, D.F. (1986) The behavior of juvenile *Fasciola hepatica*. *Journal of Parasitology* 72, 492–497.

Sukhdeo, M.V.K. and Sukhdeo, S.C. (2002) Fixed behaviours and migration in parasitic flatworms. *International Journal for Parasitology* 32, 329–342.

Sukhdeo, M.V.K., O'Grady, R. and Hsu, S.C. (1984) The site selected by the larvae of *Heligmosomoides polygyrus*. *Journal of Helminthology* 58, 19–23.

Sukhdeo, S.C. and Sukhdeo, M.V.K. (1994) Mesenchyme cells in *Fasciola hepatica* (Platyhelminthes): primitive glia? *Tissue and Cell* 26, 123–131.

Sukhdeo, S.C., Sukhdeo, M.V.K., Black, M. and Vrijenhoek, R. (1997) The evolution of tissue-migration in parasitic nematodes (Nematoda: Strongylida) inferred from a protein-coding mitochondrial gene. *Biological Journal of the Linnean Society* 61, 281–298.

Tyzzer, E.E. and Honeij, J.A. (1916) The effects of radiation on the development of *Trichinella spiralis*. *Journal of Parasitology* 3, 43–56.

Urban, J., Fang, H., Liu, Q., Ekkens, M., Chen, S.J., Nguyen, D., Mitro, V., Donaldson, D.D., Byrd, C., Peach, R., Morris, S.C., Finkelman, F.D., Schopf, L. and Gausel, W.C. (2000) IL-13-mediated worm expulsion is B7 independent and IFN-gamma sensitive. *Journal of Immunology* 164, 4250–4256.

Vanderplank, J.E. (1991) The two gene-for-gene hypotheses and a test to distinguish them. *Plant Pathology* 40, 1–3.

van der Steen, W.J. (1998) Methodological problems in evolutionary biology. XI. Optimal foraging theory revisited. *Acta Biotheoretica* 46, 321–336.

Weber, T.P. (1998) News from the realm of the ideal free distribution. *Trends in Ecology and Evolution* 13, 89–90.

Wertheim, G. (1970) Experimental concurrent infections with *Strongyloides ratti* and *S. venezuelensis* in laboratory rats. *Parasitology* 61, 389–395.

Wiens, J.A. (1984) Resource systems, populations, and communities. In: Price, P.W., Slobodchikoff, C.N. and Gaud, W.S. (eds) *A New Ecology: Novel Approaches to Interactive Systems*. John Wiley & Sons, New York, pp. 397–436.

Parasite Manipulation of Host 12 Behaviour

Robert Poulin

*Department of Zoology, University of Otago, PO Box 56, Dunedin,
New Zealand*

Introduction

Parasites of all kinds are known to modify the behaviour of their hosts in
ways that appear to enhance the parasite's chances of completing its life
cycle (Holmes and Bethel, 1972; Moore and Gotelli, 1990; Poulin, 1995,
1998a). This phenomenon can be seen as the extended phenotype (*sensu*
Dawkins, 1982) of the parasite, i.e. as parasite genes being expressed in
host phenotypes. Because it is the product of natural selection and not
mere pathology, parasite manipulation of host behaviour should be a
good area of focus for behavioural ecologists. A simple definition of
behavioural ecology is that it is the study of the evolution of behaviour in
relation to ecology. Textbooks of behavioural ecology (e.g. Krebs and
Davies, 1997) do not specifically address host manipulation by parasites,
but they present the theoretical framework necessary to approach this
phenomenon with more rigour than it has received in the past.

Consider the following examples. An orb-weaving spider suddenly
begins to build a strange web hours before the parasitic wasp larva it has
harboured for several days emerges from the spider; the strange web
will serve to protect the wasp larva from being swept away by heavy rain
after the spider's death (Eberhard, 2000). Male mayflies infected by a
mermithid nematode develop female morphological traits and adopt a
female behaviour: they return to water to oviposit, but release only the
eggs of the parasite they carry (Vance, 1996). Rats infected by the protozoan parasite *Toxoplasma gondii* show an imprudent attraction to the
odours of cats, the parasite's definitive host, and not the innate aversion to
such odours typical of uninfected rats (Berdoy *et al.*, 2000). In each case it
has been shown that the parasite causes the behavioural change. We can
then ask the question: why does the parasite alter the behaviour of its
host?

One of the founding fathers of behavioural ecology, Niko Tinbergen (1963), proposed that there are four ways of answering the question 'why?' in such cases. The first way is in terms of function: the parasite alters host behaviour because it increases the parasite's chances of completing its life cycle. This answer emphasizes the adaptive nature of the phenomenon. The second way is in terms of causation: the parasite alters host behaviour because it interferes with the biochemistry or physiology of its host in ways that have repercussions on host behaviour. This answer shifts the focus toward the mechanisms of host manipulation. The third way is in terms of evolutionary history: the parasite alters host behaviour because it inherited this ability from its ancestors and now shares it with related parasite species. The fourth way to answer the 'why' question would be in terms of development and would involve the learning processes necessary for a parasite to achieve host manipulation. Since parasites have no opportunity to learn about manipulation (they do it as part of their genetic programming or they die), this fourth type of answer is not truly applicable to the host-manipulation phenomenon.

After an early enthusiasm for the functional aspects of behaviour, the recent trends in behavioural ecology suggest a return to an integrated approach linking function with mechanisms and evolution (Krebs and Davies, 1997). When studying the behavioural ecology of parasites, it is only fitting to adopt the perspective of a behavioural ecologist. Thus, here I shall summarize recent advances in the study of host manipulation by parasites at the functional, evolutionary and mechanistic levels, and try to link these levels. I shall not present a comprehensive review of all examples documented in the literature; these are available elsewhere (e.g. Moore and Gotelli, 1990; Moore, 1993; Poulin, 1995, 1998a). Instead, I shall try to provide the conceptual and experimental framework necessary for the rigorous study of parasite manipulation of host behaviour.

Manipulation as a Functional Adaptation

Changes in host behaviour following parasitic infection are not necessarily beneficial for the parasite. They can be responses of the host aimed at eliminating the parasite or compensating for its effects. For example, behavioural fever allows parasitized insects to raise their body temperature and kill their parasites (Boorstein and Ewald, 1987). Alternatively, they can be pathological side-effects with few or no consequences for parasite fitness. Here I shall focus on behavioural changes in infected hosts thought to be cases of adaptive manipulation by the parasite. In these situations, alterations in host behaviour apparently lead to an increase in the probability of successful parasite transmission and completion of the life cycle. Such adaptive manipulation can take many forms. For instance, parasitoid wasps can change the behaviour of their insect host to ensure their own survival following their emergence from the host (Brodeur and Vet, 1994; Eberhard, 2000). Parasitic fungi can make their insect hosts go to places where the release and dispersal of fungal spores will be facilitated (Evans, 1988; Maitland, 1994). Mermithid nematodes can force their terrestrial insect hosts to enter water, where the parasite and/or its eggs must be released (Maeyama *et al.*, 1994; Vance, 1996). Malaria and other vector-borne protozoans can modify the flight and probing behaviour of mosquitoes to enhance their transmission to new hosts (Rossignol *et al.*, 1984; Rowland and Boersma, 1988). The majority of examples, however, involve helminth parasites with complex life cycles, in which transmission from an intermediate host to the next host occurs via predation. Many parasites with such life cycles are known to alter the behaviour of their hosts in ways that increase their susceptibility to predation by the next host in the cycle. The following discussion bears mainly on these parasites.

Three points need to be made about the available empirical support for the functional role of host manipulation by parasites. First, the vast majority of studies made specific predictions about the type of behavioural changes they expected parasites to induce in their hosts, but not about their magnitude. Without quantitative predictions, it is difficult to assess how much these changes in host behaviour are the product of cost-effective natural selection. Secondly, the magnitude of published estimates of changes in host behaviour induced by parasites has been steadily decreasing over the years (Poulin, 2000). The existence of the phenomenon of host manipulation by parasites is not in doubt; however, changes in host behaviour resulting from infection appear much more subtle on average than the first, and still widely cited, examples to be published. Thirdly, small changes in host behaviour can result in substantial increases in parasite transmission. For instance, Lafferty and Morris (1996) showed that larval trematodes that induced a fourfold change in fish behaviour benefited from a 30-fold increase in their transmission rate (by predation) to bird definitive hosts. In general, published estimates of parasite-mediated behavioural changes tend to be rather small, whereas estimates of how these changes translate into enhanced transmission success provide more striking results (Fig. 12.1). Whereas most studies focus on proximate changes in host behaviour induced by parasites, it is their ultimate effect on parasite transmission that is under selection.

Behavioural ecology provides some powerful conceptual tools, allowing rigorous, quantitative predictions to be made and tested in respect of when a parasite should manipulate its host and how strongly. Here, I shall discuss how the three theoretical frameworks forming the cornerstones of behavioural ecology (optimality theory, game theory and evolutionarily stable strategies, and kin selection) can be used to study host manipulation by parasites.

Optimality models are based on the premise that natural selection is an optimizing agent, favouring behavioural strategies that best promote an individual's reproductive success. Taking an approach borrowed from economics, optimality models weight the various strategies available to an organism in terms of their costs and benefits, and make predictions

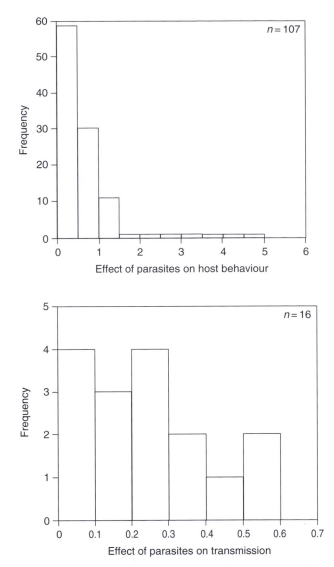

Fig. 12.1.   Frequency distributions of effect sizes of helminth parasites on the behaviour of their intermediate host (top) and of effects of helminths on predation of the intermediate host by the definitive host, or transmission success (bottom). Data are derived from published values. Effect sizes of helminths on host behaviour are the absolute value of the difference in mean behaviour between infected and uninfected hosts, corrected for sample size and variability; values smaller than 0.5 are considered relatively small effects (Cohen, 1988). The effects of helminths on predation rate on the intermediate host by the definitive host are computed as the proportion of infected intermediate hosts eaten minus the proportion of uninfected hosts eaten; with a maximum value of 1, any value greater than 0.2 indicates a substantial benefit for the parasite. (Data are from Poulin, 2000.)

about the combinations likely to maximize the net reproductive output. This approach has been particularly fruitful in the area of foraging ecology (Stephens and Krebs, 1986). In the context of host manipulation by parasites, and assuming that the currency to be maximized is the transmission success of the parasite to its definitive host, optimality theory allows us to predict the optimal investment into manipulation of the intermediate host that a parasite should make (Poulin, 1994; Brown, 1999). From both the benefits and the costs of manipulation, the optimal level of manipulation can be derived as the one that achieves the greatest net benefits (benefits minus costs).

For instance, let us assume that the energy invested into manipulation by a parasite is variable within a population and under genetic control. This is most probably true of most cases. We might expect that the more a parasite invests into manipulation of its intermediate host, the more it increases its probability of transmission to the definitive host beyond what it was initially. Without manipulation, the parasite might still be transmitted, but with a lower probability, P, corresponding to passive transmission. Manipulation enhances this probability, but with diminishing returns, i.e. small investments in manipulation yield greater returns per unit investment than larger investments (Fig. 12.2). Beyond a certain level of investment into manipulation, the probability of transmission approaches 1. At the same time, the costs of manipulation are also likely to increase in proportion to the investment in manipulation. Little is known about how costly it is for parasites to control the behaviour of their hosts. Two examples, however, suggest that the cost may sometimes be of an all-or-nothing nature. Metacercariae of the trematode *Dicrocoelium dendriticum* alter the behaviour of their ant intermediate host to increase their transmission to their sheep definitive host. Typically, of the many metacercariae inside one ant, only one migrates to the ant's brain to induce the behavioural change: this metacercaria normally does not infect the mammalian host and dies (Wickler, 1976). Similarly, metacercariae of *Microphallus papillorobustus* alter the behaviour of their amphipod host to facilitate their transmission to bird hosts. Those that induce the manipulation are the ones that encyst in the brain of the amphipod; many others encyst in the abdomen and cause no altered behaviour. Frequently metacercariae in the brain are encapsulated and melanized by the immune system of the amphipod host, whereas abdominal parasites are very rarely attacked (Thomas *et al.*, 2000). Both these examples suggest that the probability of dying as a result of attempting to manipulate the intermediate host can increase drastically. For this reason, the shape of the cost function may be close to a sigmoidal curve, with costs rising sharply with only modest investments in manipulation (Fig. 12.2). Other functions and curve shapes are also possible, of course.

The optimal investment in manipulation, then, is the investment level that achieves the greatest positive difference between the benefits and the costs in terms of transmission success (Fig. 12.2). A more rigorous

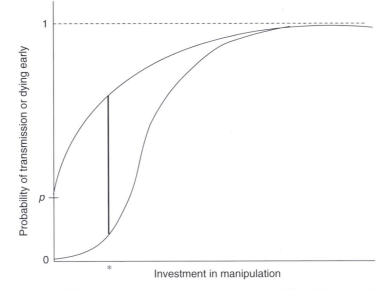

Fig. 12.2. Probability of parasite transmission and probability of the parasite dying early as a function of investment in the manipulation of the intermediate host. Without any investment in manipulation, the parasite has a passive transmission rate (p) that is greater than zero; increasing investment in manipulation yields higher transmission probabilities but with diminishing returns (top curve). At the same time, the cost of manipulation, or the probability of dying early, increases with the level of investment, following a sigmoidal function in this hypothetical example (bottom curve). The optimal investment in manipulation (*) is the level at which the net gain (benefits minus costs) in transmission probability is maximized.

mathematical treatment is given in Brown (1999) and more detailed disussion is presented elsewhere (Poulin, 1994, 1998b), but the superficial and qualitative treatment given above is sufficient to illustrate the optimality approach. Many other ecological parameters are likely to influence the optimal level of investment in manipulation, such as the likelihood that a parasite will share an intermediate host with conspecifics and the longevity of both the parasite and the intermediate host following infection (Poulin, 1994). These can also be incorporated in a quantitative mathematical model to obtain more refined predictions. There are then two ways of testing these predictions. First, as suggested by Poulin (1994) and Brown (1999), comparisons can be made across species that differ in respect of passive transmission rate, mean number of parasites per intermediate host or longevity of the intermediate host. Secondly, using model systems where some of these variables can be manipulated, certain predictions can also be tested experimentally. The advantage of the optimality approach is that it generates specific quantitative predictions. Given that the bulk of the published empirical work on host manipulation by parasites consists of investigations into whether a particular parasite shows signs of being a manipulator or not, optimality theory could at least serve to focus efforts on quantitative hypothesis testing, rather than simply adding to existing lists of manipulating species.

The second cornerstone of modern behavioural ecology is the concept of evolutionarily stable strategies (ESS), derived from game theory (Maynard Smith, 1982). It states that the optimal strategy for an individual in any given situation depends on the strategies adopted by other individuals in the population. The ESS are the mixture of strategies in the population where the net reproductive pay-off to all strategies is the same. The ESS can be a stable ratio of different genotypes, or they can consist of each individual adopting the different strategies for stable proportions of their time. The outcome is that no other feasible alternative strategy can invade an evolutionarily stable population.

This concept has applications in the area of host manipulation. In certain host-parasite systems, there is usually only one parasite per host, and that parasite must manipulate the behaviour of its host on its own. When more than one individual parasite ends up in the same intermediate host, they may all incur additional costs because of limited space or other resources in the host (e.g. Dezfuli *et al.*, 2001). In other systems, however, several conspecific parasites normally share the same intermediate host. In these cases, the optimal group investment in manipulation can be modelled using an optimality approach (Brown, 1999). We might expect the cost of manipulation to be shared by all co-occurring parasites, such that the investment of any individual is inversely proportional to the size of the group it belongs to. Large numbers of parasites in the same intermediate host also introduce the possibility of cheating. In a large group, a parasite that invests less in manipulation than its expected share can reap all the benefits of manipulation without paying the full costs (Poulin, 1994). As seen above for the trematodes *D. dendriticum* and *M. papillorobustus*, one parasite can pay the full costs and receive no benefits, while the others get all the benefits at no cost. Not all parasites in a population adopt the same strategy. An ESS approach is ideal for studying the ratio of the two strategies, honest (manipulative) and cheating (non-manipulative), in the parasite population. When cheaters are more common than their ESS proportion, their relative fitness would decrease because there are not enough manipulators present to guarantee an increase in the transmission rate, and their frequency will return to the ESS. If the proportion of manipulators increases above the ESS, the few cheaters left achieve greater fitness by paying even lower costs and their success leads to their proportion returning to the ESS. Thus we may expect a stable equilibrium ratio of honest and cheating parasites. Accurate quantitative predictions are difficult to make but should be possible when knowledge of the costs in particular systems is available.

Interestingly, a form of cheating has been documented at the inter-specific level: it appears that some non-manipulative parasite species try to 'hitch' a ride with a manipulative parasite species with which they share intermediate and definitive hosts (Thomas *et al.*, 1997, 1998; Lafferty *et al.*, 2000). These hitchhikers obtain the benefits of manipulation, i.e. increased transmission rate to the definitive host, but let the other species assume the full costs. Using simple models, it is possible to predict which conditions should favour the evolution of hitchhiking (Thomas *et al.*, 1998) and, even though there are two parasite species involved, it should be possible to derive the ESS or the optimal ratio of manipulators and non-manipulators.

Finally, the concept of kin selection and its associated notion of inclusive fitness have shed a new light on cooperation and conflict among individuals (Hamilton, 1964). Any organism shares alleles with its relatives, and its inclusive fitness depends on its own success at reproduction (direct fitness) and on how much it raises the reproductive success of its relatives by helping them in any way (indirect fitness). In the examples discussed above where several conspecific parasites occur in the same intermediate host, what happens if the parasite group consists mostly or entirely of close kin? This is the case in the trematode *D. dendriticum*: because of the way in which they are acquired by their ant intermediate host, all metacercariae inside one ant are likely to be derived from a single egg by asexual reproduction in their first host, a snail (Wickler, 1976). The one metacercaria that induces the manipulation of ant behaviour and then dies thus benefits in terms of inclusive fitness: some of its genes are transmitted via its kin. Presumably, having the full costs incurred by a single individual rather than shared among relatives yields greater overall benefits. Because *D. dendriticum* metacercariae are clones, it may be that it is more advantageous for one to die than for all to suffer from a slightly lower fitness. Systems similar to this one may be relatively common. Because of patchiness in the distribution of eggs from different individuals and because of the amplification of this patchiness by asexual multiplication of larval stages in trematodes, it may be that manipulative parasites sharing an intermediate host with conspecifics may in fact be surrounded by relatives. Kin-selection theory would allow various predictions to be made about the optimal partitioning of investments in manipulation among co-occurring parasites once their genetic relatedness is determined.

My aim in this section has been to illustrate how some of the powerful conceptual tools developed by behavioural ecologists can be applied to the study of host manipulation by parasites. Other theoretical frameworks have been used to generate predictions regarding whether or not parasites should manipulate host behaviour and how strong the manipulation should be. These include population-dynamics models (Dobson, 1988) and verbal models based on life history and ecological classification of parasites (Kuris, 1997). The many predictions derived from these theoretical explorations remain mostly untested to this date. At present, our knowledge of the phenomenon of host manipulation by parasites remains superficial, consisting mainly of a check-list of known systems in which manipulation has been documented. We now have the tools to generate testable, quantitative predictions that would shed light on how the phenomenon has evolved; the time has come to use them.

Phylogenetic Evolution of Manipulation

The most informative way to examine the evolutionary history of host manipulation by parasites, like any other trait, is to view it in a phylogenetic context (Moore and Gotelli, 1990). This is the only way that allows one to determine how often and in what circumstances the trait has evolved. In certain taxa, the ability to manipulate hosts may be an ancestral trait inherited by all living species and would thus have evolved only once. This appears to be the case for acanthocephalans (Moore, 1984). In other cases, it may have evolved repeatedly in different lineages exposed to similar conditions. Only by mapping the ability to manipulate on to a parasite phylogeny can we begin to address these issues.

There has been only one rigorous attempt to achieve this. Moore and Gotelli (1996) studied the effects of the acanthocephalan *Moniliformis moniliformis* on the behaviour of seven species of cockroaches, all suitable intermediate hosts for this parasite. In laboratory experiments, the behaviour of infected cockroaches of all seven species was compared with that of uninfected conspecifics. The behavioural measures scored were all likely to influence a cockroach's susceptibility to predation by rats, the acanthocephalan's definitive host. There was no strong concordance between the distribution of observed alterations in particular behaviours among the seven host species and a cockroach phylogeny based on morphological features (Moore and Gotelli, 1996). This finding suggests that the effect of the parasite is independent of phylogenetic constraints and has evolved separately in each host–parasite combination. Unfortunately, the weakness of this study is that certain of these seven host–parasite associations do not occur in nature, even if all cockroach species proved to be suitable as hosts in the laboratory. Nevertheless, this study shows the power of phylogenetic approaches to evaluate whether particular feats of manipulation are unique or inherited.

Another approach is to examine how phylogenetically unrelated parasite taxa have solved similar or even identical transmission problems (Poulin, 1995). Convergent evolution is likely to have favoured similar types of manipulation in different taxa that have intermediate and definitive hosts involved in similar predator–prey relationships. For instance, two completely unrelated parasite species, the acanthocephalan *Polymorphus paradoxus* (Holmes and Bethel, 1972) and the trematode *M. papillorobustus* (Helluy, 1984), both use littoral amphipods (*Gammarus* spp.) as intermediate hosts and the birds that feed on them as definitive hosts. Both parasites induce a very similar photophilic behaviour in their amphipod hosts, an analogous trait that arose twice independently by convergent evolution. Another example comes from a comparison of nematomorphs (hair-worms) and mermithid nematodes, two unrelated taxa that both evolved similar life cycles. Juveniles of both groups develop inside terrestrial insects until they become adults, when they must finish their life in fresh water. Although predation by a definitive host is not involved, the parasites still need to manipulate the behaviour of their hosts to induce them to throw themselves in water. Indeed, both nematomorphs and mermithid nematodes are known to trigger a marked hydrophilia in their insect hosts, which results in the parasites emerging into a suitable aquatic habitat (Poinar, 1991; Maeyama et al., 1994). Again, this suggests convergent evolution by different parasites facing similar difficulties.

In other cases, related parasites in similar situations appear to use different approaches to solve the same problem. For instance, two gymnophallid trematodes that must be transmitted by predation from bivalves to shore birds (oyster-catchers, gulls) have chosen different manipulation of the bivalve host to achieve a greater transmission success. Meiogymnophallus fossarum encysts in the mantle of its bivalve host and causes it to reverse its position in the mud, so that the shell opening faces upward, and to remain closer to the surface than uninfected bivalves (Bartoli, 1974). In contrast, Parvatrema affinis encysts anywhere in its bivalve host, but somehow causes it to leave noticeable tracks on the surface of tidal flats that indicate where it has burrowed (Swennen, 1969; but see Mouritsen, 1997). Clearly, the problem of getting birds to feed on bivalves has many solutions, and the two gymnophallid trematodes above have independently solved that problem. There are other solutions as well that have been adopted independently by echinostomatid trematodes with identical life cycles: some of these species encyst in the foot of their bivalve host, greatly impairing its function and leaving heavily infected bivalves incapable of burrowing into the sediments (Lauckner, 1984; Thomas and Poulin, 1998).

The impression one gets from the above examples is that host manipulation by parasites has evolved independently on numerous occasions. It is too early to estimate just how often this trait evolved; however, based on the range of parasite taxa in which transmission via predation occurs (Kuris, 1997) and on the wide number of taxa in which manipulation has been documented (Moore and Gotelli, 1990), the actual number must be much greater than 20 times. Because of convergent evolution, manipulations of host behaviour in different systems often appear superficially similar, as though there were only a few solutions to a particular transmission problem. However, no doubt the true diversity of approaches used by manipulating parasites will become apparent once we uncover the mechanisms that different parasites may use to induce seemingly identical alterations in their hosts' behaviour.

Mechanisms of Manipulation

Surely one of the most convincing pieces of evidence that a parasite is truly manipulating the behaviour of its host, and not merely causing a non-adaptive pathological behaviour, would be finding a parasite organ whose sole purpose is the induction of the behavioural change in the host (Poulin, 1995). The existence of such an organ could only be accounted for by natural selection favouring individuals possessing the organ over ones that do not. There is a growing body of information on the nature of the mechanisms used by parasites to alter the behaviour of their hosts (reviewed by Holmes and Zohar, 1990; Hurd, 1990; Beckage, 1991; Thompson and Kavaliers, 1994; Kavaliers *et al.*, 2000). Knowledge of the mechanisms involved, however, is still limited to a few host–parasite systems, and it lags behind knowledge of the actual effects on host behaviour.

In some systems, chemical modulation of host behaviour is the method employed by the parasite (Hurd, 1990; Beckage, 1991). This is particularly well documented in interactions between parasitoids and their insect hosts (Beckage, 1985, 1993; Eberhard, 2000), but is also known from systems involving helminths. For instance, in the case of the previously mentioned acanthocephalan *P. paradoxus* and the photophilic behaviour it induces in its amphipod host, Helluy and Holmes (1990) have shown that an identical response can be elicited by chemicals without the need for infection. Thus the way in which the parasite achieves manipulation is by altering host chemistry through its own secretions. The trematode *M. papillorobustus*, which causes a similar photophilia in its amphipod host, apparently also achieves this via chemicals, though the evidence is more circumstantial (Thomas *et al.*, 2000).

The production of modulating chemicals or any other active alteration of host physiology is not always necessary for a parasite to modify host behaviour. Certain parasites have adopted a potentially cheaper method, consisting simply of encysting in a particular tissue of the intermediate host. For instance, the presence of larval parasites in either the nervous system (Crowden and Broom, 1980; Lafferty and Morris, 1996; Barber and Crompton, 1997), lungs (Rau and Caron, 1979) or muscles (Rau and Putter, 1984; Thomas and Poulin, 1998) of certain intermediate hosts is sufficient to change their behaviour in ways that render them more susceptible to capture by the definitive host. In these systems, natural selection has increased the site selectivity of the parasite to ensure it benefits from higher transmission success at virtually no cost beyond those associated with reaching the infection site.

A complex phenomenon such as host manipulation by parasites must be investigated at both the proximate and ultimate levels. In other words, the phenomenon can only be truly understood if we know the mechanisms that lead to a change in host behaviour as well as the fitness consequences for the parasite. For some reason, we know which host behaviours are changed in a large number of host–parasite associations, but we only know how these changes are mediated in a few systems. Recent trends in behavioural ecology suggest that functional and mechanistic approaches to the study of behaviour are increasingly being merged into a more holistic approach (Krebs and Davies, 1997). Hopefully this new way of thinking will soon reach the area of parasite manipulation of host behaviour.

Conclusions

The study of host manipulation by parasites presents a special complication for behavioural ecologists: the trait of interest is the product of the interaction between the genotypes of two different organisms. The behaviour of a parasitized host is the simultaneous expression of host and parasite genes, and investigating the function, evolution and proximate basis of a change in host behaviour can be very challenging. In this chapter, I have considered only cases in which the modified host behaviour appeared to be mainly the product of parasite genes and therefore of benefit to the parasite. However, without proper study of both the immediate causes of the change in host behaviour and of its net fitness consequences for the parasite, it is premature to consider it as a parasite adaptation. Behavioural ecologists have been accused in the past of adopting a Panglossian view of nature, in which every feature of an organism is always beneficial (Gould and Lewontin, 1979). Although changes in host behaviour are often detrimental to the host, they are not necessarily always advantageous for the parasite. And, even if they are advantageous for the parasite, they may not be true adaptations. Given the tools now available, the next step in this field of research should be to integrate investigations into the functional, causal and historical aspects of host manipulation. First, clear, quantitative predictions derived from optimality theory or game theory need to be tested in experimental systems or in comparative analyses to provide a deeper understanding of when the ability to manipulate hosts evolves and how strongly it manifests itself. Secondly, the proximal cause of behavioural changes needs to be elucidated in more than a handful of systems. Thirdly, information on both the magnitude and type of manipulation induced by parasites and on the physiological mechanisms used to induce them needs to be mapped on parasite phylogenies to provide insight into their evolutionary history. There have been recent developments in the broader implications of host manipulation by parasites, such as its influence on the development of parasite communities, on the evolution of other sympatric parasites and on the evolution of hosts themselves (see Poulin and Thomas, 1999; Lafferty *et al.*, 2000). To better understand the phenomenon itself, however, the three steps outlined above will be necessary if we are to go beyond the mere natural-history perspective.

References

Barber, I. and Crompton, D.W.T. (1997) The distribution of the metacercariae of *Diplostomum phoxini* in the brain of minnows, *Phoxinus phoxinus. Folia Parasitologica* 44, 19–25.

Bartoli, P. (1974) Recherches sur les Gymnophallidae F.N. Morozov, 1955 (Digenea), parasites d'oiseaux des côtes de Camargue: systématique, biologie et écologie. PhD thesis, Université d'Aix-Marseille, France.

Beckage, N.E. (1985) Endocrine interactions between endoparasitic insects and their hosts. *Annual Review of Entomology* 30, 371–413.

Beckage, N.E. (1991) Host–parasite hormonal relationships: a common theme? *Experimental Parasitology* 72, 332–338.

Beckage, N.E. (1993) Games parasites play: the dynamic roles of proteins and peptides in the relationship between parasite and host. In: Beckage, N.E., Thompson, S.N. and Federici, B.A. (eds) *Parasites and Pathogens of Insects*, Vol. 1. Academic Press, New York, pp. 25–57.

Berdoy, M., Webster, J.P. and Macdonald, D.W. (2000) Fatal attraction in rats infected with *Toxoplasma gondii. Proceedings of the Royal Society of London B* 267, 1591–1594.

Boorstein, S.M. and Ewald, P.W. (1987) Costs and benefits of behavioral fever in *Melanoplus sanguinipes* infected by *Nosema acridophagus. Physiological Zoology* 60, 586–595.

Brodeur, J. and Vet, L.E.M. (1994) Usurpation of host behaviour by a parasitic wasp. *Animal Behaviour* 48, 187–192.

Brown, S.P. (1999) Cooperation and conflict in host-manipulating parasites. *Proceedings of the Royal Society of London B* 266, 1899–1904.

Cohen, J. (1988) *Statistical Power Analysis for the Behavioral Sciences*, 2nd edn. L. Erlbaum Associates, Hillsdale, New Jersey, 567 pp.

Crowden, A.E. and Broom, D.M. (1980) Effects of the eyefluke, *Diplostomum spathaceum*, on the behaviour of dace (*Leuciscus leuciscus*). *Animal Behaviour* 28, 287–294.

Dawkins, R. (1982) *The Extended Phenotype*. Oxford University Press, Oxford, 307 pp.

Dezfuli, B.S., Giari, L. and Poulin, R. (2001) Costs of intraspecific and interspecific host sharing in acanthocephalan cystacanths. *Parasitology* 122, 483–489.

Dobson, A.P. (1988) The population biology of parasite-induced changes in host behavior. *Quarterly Review of Biology* 63, 139–165.

Eberhard, W.G. (2000) Spider manipulation by a wasp larva. *Nature* 406, 255–256.

Evans, H.C. (1988) Coevolution of entomogenous fungi and their insect hosts. In: Pirozynski, K.A. and Hawksworth, D.L. (eds) *Coevolution of Fungi with Plants and Animals*. Academic Press, London, pp. 149–171.

Gould, S.J. and Lewontin, R.C. (1979) The spandrels of San Marco and the Panglossian paradigm: a critique of the adaptationist programme. *Proceedings of the Royal Society of London B* 205, 581–598.

Hamilton, W.D. (1964) The genetical evolution of social behaviour, parts I and II. *Journal of Theoretical Biology* 7, 1–52.

Helluy, S. (1984) Relations hôtes–parasites du trématode *Microphallus papillorobustus* (Rankin, 1940). III. Facteurs impliqués dans les modifications du comportement des *Gammarus* hôtes intermédiaires et tests de prédation. *Annales de Parasitologie Humaine et Comparée* 59, 41–56.

Helluy, S. and Holmes, J.C. (1990) Serotonin, octopamine, and the clinging behavior induced by the parasite *Polymorphus paradoxus* (Acanthocephala) in *Gammarus lacustris* (Crustacea). *Canadian Journal of Zoology* 68, 1214–1220.

Holmes, J.C. and Bethel, W.M. (1972) Modification of intermediate host behaviour by parasites. In: Canning, E.U. and Wright, C.A. (eds) *Behavioural Aspects of Parasite Transmission*. Academic Press, London, pp. 123–149.

Holmes, J.C. and Zohar, S. (1990) Pathology and host behaviour. In: Barnard, C.J. and Behnke, J.M. (eds) *Parasitism and Host Behaviour*. Taylor and Francis, London, pp. 34–64.

Hurd, H. (1990) Physiological and behavioural interactions between parasites and invertebrate hosts. *Advances in Parasitology* 29, 271–318.

Kavaliers, M., Colwell, D.D. and Choleris, E. (2000) Parasites and behaviour: an ethopharmacological perspective. *Parasitology Today* 16, 464–468.

Krebs, J.R. and Davies, N.B. (1997) *Behavioural Ecology: an Evolutionary Approach*, 4th edn. Blackwell Science, Oxford, 456 pp.

Kuris, A.M. (1997) Host behavior modification: an evolutionary perspective. In: Beckage, N.E. (ed.) *Parasites and Pathogens: Effects on Host Hormones and Behavior*. Chapman & Hall, New York, pp. 293–315.

Lafferty, K.D. and Morris, A.K. (1996) Altered behavior of parasitized killifish increases susceptibility to predation by bird final hosts. *Ecology* 77, 1390–1397.

Lafferty, K.D., Thomas, F. and Poulin, R. (2000) Evolution of host phenotype manipulation by parasites and its consequences. In: Poulin, R., Morand, S. and Skorping, A. (eds) *Evolutionary Biology of Host–Parasite Relationships: Theory Meets Reality*. Elsevier Science, Amsterdam, pp. 117–127.

Lauckner, G. (1984) Impact of trematode parasitism on the fauna of a North Sea tidal flat. *Helgoländer Meeresuntersuchungen* 37, 185–199.

Maeyama, T., Terayama, M. and Matsumoto, T. (1994) The abnormal behavior of *Colobopsis* sp. (Hymenoptera: Formicidae) parasitized by *Mermis* (Nematoda) in Papua New Guinea. *Sociobiology* 24, 115–119.

Maitland, D.P. (1994) A parasitic fungus infecting yellow dungflies manipulates host perching behaviour. *Proceedings of the Royal Society of London B* 258, 187–193.

Maynard Smith, J. (1982) *Evolution and the Theory of Games*. Cambridge University Press, Cambridge, 224 pp.

Moore, J. (1984) Altered behavioral responses in intermediate hosts: an acanthocephalan parasite strategy. *American Naturalist* 123, 572–577.

Moore, J. (1993) Parasites and the behavior of biting flies. *Journal of Parasitology* 79, 1–16.

Moore, J. and Gotelli, N.J. (1990) A phylogenetic perspective on the evolution of altered host behaviours: a critical look at the manipulation hypothesis. In: Barnard, C.J. and Behnke, J.M. (eds) *Parasitism and Host Behaviour*. Taylor and Francis, London, pp. 193–233.

Moore, J. and Gotelli, N.J. (1996) Evolutionary patterns of altered behavior and susceptibility in parasitized hosts. *Evolution* 50, 807–819.

Mouritsen, K.N. (1997) Crawling behaviour in the bivalve *Macoma balthica*: the parasite manipulation hypothesis revisited. *Oikos* 79, 513–520.

Poinar, G.O. (1991) Nematoda and Nematomorpha. In: Thorp, J.H. and Covich, A.P. (eds) *Ecology and Classification of North American Freshwater Invertebrates*. Academic Press, New York, pp. 249–283.

Poulin, R. (1994) The evolution of parasite manipulation of host behaviour: a theoretical analysis. *Parasitology* 109, S109–S118.

Poulin, R. (1995) 'Adaptive' changes in the behaviour of parasitized animals: a critical review. *International Journal for Parasitology* 25, 1371–1383.

Poulin, R. (1998a) Evolution and phylogeny of behavioural manipulation of insect hosts by parasites. *Parasitology* 116, S3–S11.

Poulin, R. (1998b) *Evolutionary Ecology of Parasites: From Individuals to Communities*. Chapman & Hall, London, 212 pp.

Poulin, R. (2000) Manipulation of host behaviour by parasites: a weakening paradigm? *Proceedings of the Royal Society of London B* 267, 787–792.

Poulin, R. and Thomas, F. (1999) Phenotypic variability induced by parasites: extent and evolutionary implications. *Parasitology Today* 15, 28–32.

Rau, M.E. and Caron, F.R. (1979) Parasite-induced susceptibility of moose to hunting. *Canadian Journal of Zoology* 57, 2466–2478.

Rau, M.E. and Putter, L. (1984) Running responses of *Trichinella spiralis*-infected CD-1 mice. *Parasitology* 89, 579–583.

Rossignol, P.A., Ribeiro, J.M.C. and Spielman, A. (1984) Increased intradermal probing time in sporozoite-infected mosquitoes. *American Journal of Tropical Medicine and Hygiene* 33, 17–20.

Rowland, M. and Boersma, E. (1988) Changes in the spontaneous flight activity of the mosquito *Anopheles stephensi* by parasitization with the rodent malaria *Plasmodium yoelli*. *Parasitology* 97, 221–227.

Stephens, D.W. and Krebs, J.R. (1986) *Foraging Theory*. Princeton University Press, Princeton, New Jersey, 247 pp.

Swennen, C. (1969) Crawling-tracks of trematode infected *Macoma balthica*. *Netherlands Journal of Sea Research* 4, 376–379.

Thomas, F. and Poulin, R. (1998) Manipulation of a mollusc by a trophically transmitted parasite: convergent evolution or phylogenetic inheritance? *Parasitology* 116, 431–436.

Thomas, F., Mete, K., Helluy, S., Santalla, F., Verneau, O., De Meeüs, T., Cézilly, F. and Renaud, F. (1997) Hitch-hiker parasites or how to benefit from the strategy of another parasite. *Evolution* 51, 1316–1318.

Thomas, F., Renaud, F. and Poulin, R. (1998) Exploitation of manipulators: 'hitch-hiking' as a parasite transmission strategy. *Animal Behaviour* 56, 199–206.

Thomas, F., Guldner, E. and Renaud, F. (2000) Differential parasite (Trematoda) encapsulation in *Gammarus aequicauda* (Amphipoda). *Journal of Parasitology* 86, 650–654.

Thompson, S.N. and Kavaliers, M. (1994) Physiological bases for parasite-induced alterations of host behaviour. *Parasitology* 109, S119–S138.

Tinbergen, N. (1963) On aims and methods of ethology. *Zeitschrift für Tierpsychologie* 20, 410–433.

Vance, S.A. (1996) Morphological and behavioural sex reversal in mermithid-infected mayflies. *Proceedings of the Royal Society of London B* 263, 907–912.

Wickler, W. (1976) Evolution-oriented ethology, kin selection, and altruistic parasites. *Zeitschrift für Tierpsychologie* 42, 206–214.

Parasite Manipulation of **13** Vector Behaviour

J.G.C. Hamilton and H. Hurd

Centre for Applied Entomology and Parasitology, School of Life Sciences, Keele University, Keele, Staffordshire ST5 5BG, UK

Introduction

One of the major differences between a parasitic and a free-living lifestyle is the requirement that a parasite must periodically pass from host to host. This transmission phase of its life cycle may involve active host-seeking stages or passive transfer via ingestion of the current host by the next one. However, many parasites dwelling in the circulatory system or superficially, in the skin, utilize a third mode of transfer, that of vector transmission. In this case a haematophagous, or blood-sucking, arthropod transports the parasite to another host. Many micro- and macroparasites of medical or veterinary importance, such as arboviruses, bacteria, rickettsia, protozoa and nematodes, are carried between hosts in this way (Lehane, 1991).

The parasite may have very transient contact with its transport vehicle, by passing from one host to the next on contaminated vector mouth-parts or by being ingested but passing through the gut unchanged and exiting in contaminated faeces. This type of transmission is known as 'mechanical'. Alternatively, 'biological' transmission occurs when the parasite undergoes a distinct life-cycle phase within the vector and engages in a parasitic association with it. This can just involve a phase of parasite growth and differentiation, as seen in filarial worms, such as *Wuchereria bancrofti*, where exsheathed microfilariae penetrate the midgut and undergo two moults within the thoracic flight muscles (Kettle, 1990). Alternatively, pathogens such as the plague bacillus, *Yersinia pestis*, and many of the arboviruses multiply in the vector but do not change in form. Finally, the malaria parasites and many of the trypanosomes and leishmanias both multiply and change to new forms within the vector.

Mechanical transmission is, in effect, passive and no opportunity is created for the parasite or pathogen to interact with the vector. For mechanical passage via contaminated mouth-parts to be successful, rapid biting of successive new hosts must occur, because vector-transmitted parasites are short-lived in the environment. This mode of transmission is probably quite rare. Examples include parasites transmitted by fleas, such as the myxoma virus, which, in Britain, is carried from rabbit to rabbit by the flea *Spilopsyllus cuniculi*. In addition, parasite density in the host's blood meal must be great for the tiny volumes of blood that remain on most mouth-parts to contain sufficient parasites to infect a new host. In this respect, tabanid flies, with their spongy mouth-parts, which soak up blood, could be expected to be capable of mechanical transmission. *Tabanus fuscicostatus* transmits *Trypanosoma evansi*, the causative agent of surra in horses and camels, in this manner.

In contrast, if the vector also acts as a 'host' to parasite life stages, complex interactions between arthropod and parasite are likely to evolve. Evolutionary pressure will drive parasites to enhance their transmission success. In the case of vector transmission, it is reasonable to assume that the basic case reproduction number, R_0 (the number of new infections that arise from a single current infection), in a defined population of susceptibles (Macdonald, 1957) will be increased if more vector–host contact occurs and if transfer between vector and host becomes more efficient. This is particularly so as transmission dynamics of vector-borne diseases are very different from those of directly transmitted diseases, because R_0 can be an order of magnitude greater. The basic equations used to model R_0 for vector-transmitted diseases incorporate the vectorial capacity (C) of the bloodsucking insect population. Vectorial capacity is the daily rate at which future inoculations arise from a current infective case (Garrett-Jones, 1964). Thus, in these cases, $R_0 = C/r$ (where r is the rate of recovery from infectiousness) (Garrett-Jones and Shidrawi, 1969). An estimate of C is derived from eight components (Dye, 1990); however, for each of these components assumptions are made that do not always hold (Dye, 1990; Rogers and Packer, 1993). For example, host selection by vectors is not always random and may be influenced by the disease status of the host (Day and Edman, 1983) and mosquito biting behaviour changes when infected with malaria (Anderson *et al.*, 1999). It is almost inevitable that adaptive changes that enhance parasite transmission will cause changes to vector life-history traits, because the interests of the parasite are intricately linked with that of the vector via blood feeding (Hurd *et al.*, 1995; Klowden, 1995); thus C will be underestimated.

Blood feeding is essential to the fitness of the vector, as it provides the female with a proteinaceous meal for egg production, and it is essential to the parasite, as host contact provides an opportunity for transmission. But blood feeding can be risky, due to host defensive behaviour (Day and Edman, 1984a; Randolph *et al.*, 1992). Increased host contact increases the chances of vector mortality. Thus many vectors face trade-offs in terms of feeding strategies between carbohydrate (nectar) meals, which allow maintenance metabolism, and blood meals, which will increase reproductive output but may shorten lifespan (Koella, 1999). This is exemplified by mosquitoes, where feeding drive will be matched to the minimum meal size required to mature a batch of eggs (Clements, 1992) and trade-off decisions will be made with regard to egg-batch size/mortality risk (Anderson and Roitberg, 1999). For the parasite too, a conflict exists between maximizing host contact by increased vector biting and minimizing vector mortality by decreased biting. The parasite's success is thus constrained by a conflict between increasing transmission by increasing biting and decreasing mortality by decreasing biting. However, the optimum balance between host contact (biting) and mortality risk may not be the same for vector and parasite. This concept has been explored in a model of the transmission success of malaria sporozoites as a function of mosquito biting rate (Koella, 1999; Schwartz and Koella, 2001). Koella's model predicts that selection pressure will increase parasite transmission by changing the compromise between vector biting rate and vector reproductive output such that parasite success is maximized. If the trade-off positions postulated in this model also pertain to other haematophagous insects, we should expect many vector-transmitted parasites to evolve to a situation where they alter some aspect of biting behaviour. Evidence will be presented below to demonstrate that there are indeed many examples of parasites that do change both the host-seeking and the biting behaviour aspects of the blood-feeding behaviour of their vectors in ways that intuitively suggest that transmission to the vertebrate host will be enhanced.

Despite the obvious importance of vector blood-feeding behaviour to parasite-transmission dynamics, there is a major lack of quantitative data on the consequence of parasitic-induced alteration of this life-history trait. In particular, few studies have demonstrated that changes in vector feeding actually do increase parasite transmission. Yet verification of increased parasite success is fundamental to our understanding of the evolutionary significance of parasite-induced alteration of vector blood feeding.

If and when changes in behaviour are observed, are these non-adaptive pathological consequences of infection or has the parasite directly manipulated the host? Many authors have argued the importance of distinguishing between these options if we are to understand the evolutionary significance of host behavioural changes (Minchella, 1985; Moore and Gotelli, 1990; Horton and Moore, 1993; Hurd, 1998). Poulin (1995) identified key indicators of adaptive manipulation. Paramount among these was the need to demonstrate fitness benefits for the parasite. Natural selection will clearly favour a parasite that is able to alter the behaviour of its vector such that its transmission success is enhanced compared with that of its conspecifics. In addition, Poulin (1995) suggested that parasite-induced changes in host behaviour that are adaptive are likely to be complex, to function precisely to enhance transmission, to have arisen by chance and to have evolved several times in different taxa (see also Moore and Gotelli, 1990; Poulin, 1998).

True vector manipulation will occur via an extension of the pheno-type of the parasite (Dawkins, 1982, 1990) such that manipulator molecules are produced that directly affect vector behaviour patterns (Hurd, 1998). Manipulative effort is costly for the parasite; thus the degree of manipulation is likely to be constrained and manipulation will be optimized, not maximized (Poulin, 1994). A survey of the current literature on vector–parasite interactions suggests that we appear to be a very long way from identifying manipulator molecules in any vector-transmitted parasite. Thus, at the present time, the issue of pathology versus adaptive manipulation is difficult to assess and will remain so until substantive studies of several associations provide evidence with which to weigh the alternative hypotheses.

Blood feeding provides the point of contact between the vector and the next host, but this is not the only factor that is a candidate for adaptive manipulation. The activity and longevity of the vector will also affect chances of parasite success. Vectors with reduced lifespan may have fewer encounters with their hosts and thus fewer chances for trans-mission. This is particularly so if the parasite has a long developmental or extrinsic latent period in the host prior to becoming a patent infection, as do malaria parasites and filarial worms. If vector lifespan is compromised by infection during this developmental phase, there will be no trans-mission. It is difficult to generalize concerning the effect of parasites on vector lifespan, because conclusions resulting from survivorship studies are contradictory, even when the same parasite–vector association is being assessed (Lines *et al.*, 1991; Lyimo and Koella, 1992). In addition, studies need to be conducted in the field, where stressors in addition to infection operate and parasites are associated with their natural vectors. One of the major problems that we face here is the difficulty in deter-mining the age of insect vectors. Methods for assessing age or, more importantly, parous status (number of blood meals taken) that are simpler, quicker and more reliable than the present age-grading techniques (Detinova, 1962; Sokolova, 1994) need to be devised.

Parasites that develop and reproduce within the vector inevitably utilize host resources. If the vector does not replenish these resources more often than an uninfected organism, activity and/or longevity are likely to be compromised. In addition, many parasites evoke a defence response in their vector that, even if it does not eliminate the parasite, will be costly to the vector in metabolic terms (Ferdig *et al.*, 1993; Richman *et al.*, 1997; Luckhart *et al.*, 1998). There is growing evidence to support the view that vector resource allocation may be altered by infection such that the balance between reproduction and growth and maintenance is changed, due to a curtailment of reproductive effort (Hurd, 2001). Evolutionary theory suggests that delayed or depressed reproduction will result in increased lifespan (Price, 1980); thus, if fecundity is reduced, activity levels and longevity may remain unchanged by the demands imposed by the infection, and fecundity reduction may be a strategy that minimizes the effects of infection (Hurd, 2001). Although parasites from many taxa are known to adversely affect the egg production of their vectors (Hurd, 1990), studies that provide data with which to assess the evolutionary history of this manipulation are rare.

Changes in vector behaviour must have a physiological, biochemical and molecular basis; thus it is as important to consider these aspects of manipulation as to try to determine whether behavioural changes actually enhance parasite success. A few studies are beginning to unravel the mechanisms underlying the curtailment of vector reproductive success. However, our understanding of the biochemical and molecular bases of the various aspects of haematophagous behaviour is negligible (apart from the endocrine control of the responsiveness of *Aedes aegypti* lactic acid receptors (Davis, 1984; Klowden *et al.*, 1987)).

Blood-feeding Behaviour – Haematophagy

The acquisition of a blood meal by temporary ectoparasites depends upon the realization of a series of behavioural and non-behavioural steps leading from a state of hunger to imbibing blood. This process is not strictly sequential, thus allowing the insect a more flexible approach to the circumstances in which it finds itself (Lehane, 1991). Sutcliffe (1987) described the initial stages of host location; however, the description stops when the insect reaches the host's general vicinity. Takken (1991) described the host-finding process as ending when the insect alights on the host. However, there is clearly more to the acquisition of a blood meal than host location, and the insect must not only alight on the host but also choose a feeding site and then commence probing and imbibing blood.

A general and more comprehensive model of blood-feeding behaviour has four steps: (i) the appetitive search; (ii) activation and orientation; (iii) attraction; and (iv) landing and probing. These steps include host location, host acceptance and the initiation of blood feeding. The initiation of any or all of these stages is dependent on endogenous factors, such as the age, host preference, nutritional and reproductive state of the insect, whether or not the insect is crepuscular, nocturnal or diurnal, and their interaction. Exogenous factors, such as temperature, humidity, wind speed, light intensity and the availability of hosts, are also likely to affect the outcome of the blood-meal-seeking process (Fig. 13.1). All the endogenous factors and the insect's response to the exogenous factors may change because of infection by parasites. In addition, vector behaviour may be altered by proxy, i.e. the behaviour of the vector is not altered by direct action on the vector but by action instead on the host. For example, a parasite might modify its host's odour, thus making it more attractive to the vector; this may then have a beneficial outcome for the parasite in terms of an improved chance of transmission. We shall examine the potential role of the parasite in manipulating the process to its advantage and examine each of the steps outlined above for any evidence of parasite manipulation.

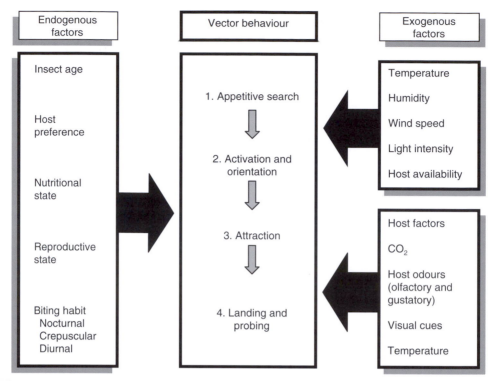

Fig. 13.1. Factors affecting the blood-meal-seeking process.

The appetitive search

After emergence, or after taking a blood meal, an adult female mosquito goes through a latent period when it will not blood-feed. In *Anopheles gambiae*, this period of inhibition of response to human odour lasts for at least 40 h after a blood meal (Takken *et al.*, 2001). Similar inhibition of activity is seen in other mosquito species (see references in Davis *et al.*, 1987), and in *A. aegypti* inhibition lasts until after oviposition (Klowden, 1987).

However, as the female matures or the blood meal is digested and eggs mature, the imperative for the female to acquire a blood meal becomes increasingly urgent. Thus the female undertakes periodic, non-orientated flight activity to increase her chances of encountering host-derived cues and obtaining a blood meal (Lehane, 1991). This behavioural pattern has been recognized in other Diptera apart from mosquitoes, e.g. ranging behaviour in tsetse-flies (Vale, 1980). The questing behaviour exhibited by ixodid ticks is believed to perform a similar function (Waladde and Rice, 1982). Undoubtedly this behaviour occurs in many groups of blood-feeding arthropods.

No work has been undertaken that we are aware of to specifically address the effect of parasitic infection on appetitive search behaviour.

However, the effect may be inferred from experiments designed to determine the effect of infection on activity levels, although one significant problem with this approach is that human odour from the experimenter was unlikely to have been excluded from these experiments and therefore its potentially confounding influence cannot be ignored. In laboratory experiments, Berry *et al.* (1988) showed that *Aedes trivittatus* infected with *Dirofilaria immitis* had greatly increased flight activity. *D. immitis* is the common heart-worm of dogs, cats, foxes and wolves in the USA, where it occurs most frequently in the southern states. *Aedes*, *Culex* and *Anopheles* species of mosquitoes are all competent vectors of this parasite and *A. trivittatus* has been found to be naturally infected with *D. immitis*. Within 24–36 h after taking an infected blood meal, larvae are found in the Malpighian tubules of the mosquito vector. They reside there and develop into the infective L3 stage, which then migrate to the haemocoel. Eleven to 12 days postinfection, they are found in and among the fat bodies in the lower half of the thorax, from where they finally migrate to the mouth-parts. Berry *et al.* (1988) fed *A. trivittatus* with either uninfected blood or with blood infected with *D. immitis* 3 days prior to the start of the experiment and then monitored activity patterns for 15 days in acoustic chambers. During this time, mosquitoes would have been likely to be conducting appetitive search behaviour. Flight activity increased within and outside normal flight-activity periods as the parasitic infection developed and from 8 days postinfection was significantly greater than in the uninfected controls. A greater number of flights of shorter duration were made, a behavioural pattern that was particularly marked in older mosquitoes with a heavier parasite burden. Flight activity was significantly suppressed at specific times when parasites were believed to be moulting or migrating to the mouth-parts. It is interesting to note that the vector's behaviour was not modified for 8 days after infection and was dependent on the parasitic burden. Increased flight activity as the parasite approached maturity could increase the vector's chance of encountering a host and thus improve the chances of the parasite being successfully transmitted. Although there may be several explanations to account for these observations, the increased activity is in line with predictions made by Koella *et al.* (1998a).

Although general activity levels cannot be specifically related to ranging behaviour, it is interesting to note that Ryan (1984) observed that *Glossina longipalpis* in the field and naturally infected with trypanosomes were probably more active than uninfected flies.

In contrast, when *D. immitis* was present in a non-natural vector, *A. aegypti* flight activity was reduced (Berry *et al.*, 1987). Similar loss of flight activity was seen when *A. aegypti* (a non-natural vector) was infected with *Brugia pahangi* (Berry *et al.*, 1986; Rowland and Lindsay, 1986) (see also review by Moore, 1993). Loss of flight activity was also seen when *Anopheles stephensi* was infected with the rodent malaria *Plasmodium yoelii* (Rowland and Boersma, 1988). As *A. stephensi* is not a natural vector of *P. yoelii*, perhaps this illustrates that there is a conflict between studies conducted on natural and non-natural combinations of vectors and parasites. For the purposes of this review, we would consider vectors infected by a parasite with which they may become naturally infected in the wild as being a natural combination and vectors infected with parasites that they are unlikely to become infected with in the wild as being a non-natural combination. Of course, in the natural situation, there are likely to be degrees of association between vector and parasite. In those with a long evolutionary association, the effect of the parasite may be significantly different from that in a recently developed association. Further problems may occur with laboratory infections, which may be excessively heavy or light and thus not reflect the true situation encountered by either parasite or vector. Thus the loss of flight activity in *A. aegypti* infected with *B. pahangi* occurs when the filarial worm invades flight muscle, causing a loss of flight activity associated with a specific tissue pathology (Townson, 1970; Paige and Craig, 1975). Although cells of the Malpighian tubules of *A. trivittatus* infected with *D. immitis* are ruptured as L3 larvae migrate to the mouth-parts, there have been no other pathological effects reported to account for the increased flight activity.

Activation and orientation

Activation occurs when the insect comes into contact with a suitable signal from a potential host and switches from appetitive non-directional flight to directed host-location behaviour. Orientation is the tendency for the insect to fly upwind after activation. Activation itself may not result in a behavioural change but, rather, may prime the insect for acceptance of a subsequent orientational cue. These cues are generally olfactory, but visual cues are also important, especially in tsetse- and blackflies. The role of CO_2 in activating and inducing upwind flight is well known in mosquitoes (Laarman, 1958; Clements, 1963), and Takken (1991) considered that it might possibly trigger the flight response in hungry mosquitoes. In tsetse-flies, activation and orientation are also caused by CO_2.

No experiments have been specifically designed or conducted to investigate the effect of parasitic infection on vector behaviour in this aspect of host blood-feeding behaviour.

Attraction

Attraction is the process by which the vector is drawn from a distance to the host. Host odours (other than CO_2), visual cues, heat and water vapour all play a role in attraction to the host. Odours may play a more prominent role as long-range attractants, with visual, heat and water-vapour cues becoming more important factors at shorter range. There may also be significant additive and synergistic interactions between these cues. These interactions have been largely unexplored in blood-feeding arthropods.

Attraction of a vector to a host depends on two separate factors: one is the inherent attractiveness of the host and the other is the extent to which the vector responds. Therefore a parasite may potentially alter either aspect of the attraction equation, depending on its stage of development and location (in host or vector). For example, a parasite infecting a vector might be able to alter the sensitivity of the vector's host-odour receptors, making the vector more or less responsive to host factors. Again, this could be dependent on the stage of development of the parasite within the vector. In the alternative scenario, a parasite infecting the host could make the host more attractive by producing a greater quantity of attractive chemicals, inducing the production of new attractive chemicals or possibly reducing the repellence of an individual, thus indirectly modifying the vector's behaviour and making the host more likely to be fed upon by the vector. In any scenario, the parasite would presumably improve its chances of transmission. There is very little direct evidence for any of these proposed manipulations and yet their significance in the epidemiology of disease transmission must be underestimated.

Parasitized-host attractiveness

Infection of a host may modify its odour profile, a fact that has been recognized by physicians for hundreds of years (Penn and Potts, 1998). These changes have been associated with exhaled breath (Davies *et al.*, 1997; Smith *et al.*, 1998) or may be associated with changes in epidermal microbial flora (Braks *et al.*, 1999). Little evidence exists for the role of parasites in altering host odour. There is further confusion as to whether any alteration is due to direct physiological change or an alternative scenario where the odour is altered as a secondary consequence of infection, e.g. the sick animal may groom less, which leads to increased bacterial activity, or alternatively pyrexia increases the volatilization of host odour compounds. However, if the parasite were altering the attractiveness of the host in a manner that would increase parasite transmission, we might expect to see increased attraction at periods associated with the maturation of the infective stages of the parasite.

Laboratory-based experiments describe the preferential selection of rodents infected with malaria above non-infected rodents by vectors seeking a blood meal (Day and Edman, 1983; Day *et al.*, 1983). In these experiments, rats were infected with either *Plasmodium berghei*, *Plasmodium chabaudi* or *P. yoelii* and experiments were conducted at periods of peak gametocytaemia (gametocytes are the malarial stage that can infect the vector). However, studies by Burkot *et al.* (1989) showed that at a field site in Madang Province, Papua New Guinea, *A. punctulatus* mosquitoes did not preferentially select malaria- or filaria-infected humans. This field site is hyperendemic for malaria, with *Plasmodium falciparum*, *Plasmodium malariae*, *Plasmodium vivax* and *Plasmodium ovale* all present. In none of the households examined were infected individuals preferentially selected by *Anopheles punctulatus*. In this natural situation, the peak gametocytaemia rarely exceeded 0.4% for

P. falciparum and, for over 80% of the infected population, gameto-cytaemia was under 0.0016% (Burkot *et al.*, 1989). Burkot *et al.* (1989) suggested that the increased attraction in laboratory experiments (Day and Edman, 1983; Day *et al.*, 1983) was due to the excessively high levels of parasitaemia in these animals.

Freier and Friedman (1976) demonstrated that *A. aegypti* probed less often and attempted to feed less often on chickens infected with *Plasmodium gallinaceum* than on uninfected chickens. Their experiments showed that this differential response was not due to temperature variation between infected and uninfected chickens. Mahon and Gibbs (1982) showed that arbovirus-infected hens attracted more mosquitoes than uninfected hens.

Only a small amount of work has been done to explore these possible parasite influences in other vector–host associations. Coleman *et al.* (1988) reported that *Leishmania*-infected dogs were more attractive to sandflies than uninfected dogs. Baylis and Mbwabi (1995) obtained field evidence from Kenya that oxen infected with *Trypanosoma congolense* were more attractive to *Glossina pallidipes* than uninfected oxen or oxen infected with *Trypanosoma vivax*. Although the results approached significance, it is worth remembering that oxen are not the natural host of either *T. congolense* or *T. vivax* and that *G. pallidipes* may prefer to feed on other bovids, such as bushbuck (Leak, 1999), and thus any observed effect may or may not mimic the natural situation.

None of these experiments were specifically designed to measure the effect of infection on the attractiveness of the host; specifically, they did not measure attraction over distance nor did their experiments take into account differences in the innate attractiveness of individual hosts prior to infection or the stage of parasite development.

Three recent sets of experiments have attempted to address these issues using a dual-port wind-tunnel olfactometer to demonstrate attraction over 1 m. In the first series of experiments the odour of Syrian hamsters infected with *Leishmania infantum* (*chagasi*) was shown to be more attractive than non-infected hamster odour to female *Lutzomyia longipalpis* (the natural vector of *L. infantum* in South America) (Rebollar-Téllez, 1999). Female *L. longipalpis* were given a choice between infected hamster odour and uninfected hamster odour; a significantly greater number of female *L. longipalpis* were attracted to the odour of *L. infantum*-infected hamsters. Subsequently, O'Shea *et al.* (2001) repeated and extended these observations and confirmed the results obtained by Rebollar-Téllez (1999). O'Shea *et al.* (2001) also collected the odours from infected and non-infected hamsters by air entrainment; when these odours were tested in the olfactometer the same results were obtained. Taylor (2001) showed that, given a choice of infected mouse odour and uninfected mouse odour, a significantly greater number of females were attracted to the odour of mice infected with *P. yoelii nigeriensis*. These preliminary experiments also showed that attractiveness varied throughout the cycle of malaria infection. Mice that were recently infected were no more attractive than uninfected mice; however, after 4 days, infected mice were significantly more attractive. This increase in attraction corresponded with increasing parasitaemia (Taylor, 2001), as suggested by Kingsolver (1987), but as yet cannot be related directly to gametocyte levels.

These experiments exclude the effect of host activity, temperature and haemostasis mechanisms in the blood-feeding process and suggest that parasitic infection may enhance odour-mediated host attraction of at least some parasite–host–vector complexes at a period of infection likely to enhance parasite transmission. Increasing attractiveness of the host at a time when infective parasitic stages are present would be adaptively favourable to the parasite by improving transmission prospects. It is possible – although this hypothesis has not been tested – that host attractiveness may even decrease during periods when non-infective stages of the parasite are present. Clearly, this is an area that warrants further investigation.

It has been suggested that the fever induced by parasitic infection may make infected individuals more attractive than uninfected individuals (Gillett and Connor, 1976; Turell *et al.*, 1984). This would be an important factor in the epidemiology of disease transmission; however, the role of heat as an attractant remains to be established. The raised temperature of malaria-infected mice has no differential attractive effect on mosquitoes (Day and Edman, 1984b) and the experiments described above would suggest that heat is not important over even relatively short distances for sandflies and mosquitoes.

Parasitized vector attraction to host odour

No work has been undertaken in this area as far as we are aware, although a significant amount of work has been done to examine the changes in landing and probing behaviour of parasitized vectors.

Landing and probing

Landing is believed to be controlled by stimuli acting over a short range, including many of the same stimuli that are responsible for other stages of the attraction sequence (e.g. odour, visual, heat and moisture cues). It is at this stage that the decision to alight, or not, on the potential host is made. This stage is highly regulated by the host and its defensive mechanisms and by the persistence and density of the vector (Edman, 1989). Probing, the active movement of the mouth-parts into the surface of the host, the uptake of blood and cessation of feeding are dependent on the interaction between the host and the vector. The parasite, whether infecting the host or the vector and depending on its developmental stage, may manipulate the process to its own advantage. Schwartz and Koella (2001) postulated that the malaria parasite should try to manipulate the mosquito's feeding behaviour and life history to increase its transmission and that these changes would be dependent on the developmental stage of the parasite.

Landing and probing on a parasitized host

An important factor influencing host selection is host defensive behaviour, which occurs during the landing and probing stage (Burkot, 1988). Day and Edman (1983) and Day *et al.* (1983) showed that animals experimentally infected with rodent malaria, *P. berghei*, *P chabaudi* and *P. yoelii*, exhibited reduced defensive behaviour and were consequently preferentially fed upon by mosquitoes. This reduced host activity coincided with peaks of gametocytaemia and would result in increased transmission.

Parasite infection of the host may also lower the quality of the blood meal by lowering the haematocrit. However, the consequence is more freely flowing blood, so the mosquito spends less time on the host probing for a blood meal (Daniel and Kingsolver, 1983; Rossignol *et al.*, 1985; Shieh and Rossignol, 1992). Recent work by Taylor and Hurd (2001) showed that, as a result of the slight anaemia early in the infection of mice with *P. yoelii nigeriensis*, *A. stephensi* took a larger blood meal than from uninfected mice and hence a greater number of erythrocytes infected with gametocytes. A further consideration was that a blood meal taken at this early anaemic/gametocyte stage of infection would provide sufficient protein to mature a batch of eggs; therefore the mosquito would not require a further blood meal in the current gonotrophic cycle. This in turn would have a beneficial outcome in terms of mosquito fitness by reducing risky host contact and hence improving the chances of the parasite surviving to produce sporozoites. It is likely, however, that different transmission profiles will occur in other parasite–host–vector complexes. A field study by Baylis and Mbwabi (1995) reported that the success of the tsetse-fly *G. pallidipes* in obtaining a blood meal from oxen infected with *T. congolense* was 60% greater than when feeding on *T. vivax*-infected oxen. They suggested that vasodilatation, caused by *T. congolense* attaching with their flagella to the walls of the host microvasculature, might account for this difference, rather than levels of parasitaemia, anaemia or skin temperature. Similar results were obtained in the laboratory by Moloo *et al.* (2000), working with combinations of *T. congolense*, *T. vivax*, *G. pallidipes* and *Glossina morsitans centralis*. The results were more convincing for *T. congolense* than for *T. vivax* infections. These results taken together would suggest that parasites do indeed modify their host to improve the feeding success of their vector and thus improve their chances of transmission.

The sandfly *L. longipalpis* preferentially probes on the cutaneous lesions produced when mice are infected with *Leishmania mexicana mexicana*. These lesions are rich in amastigotes, the parasite stage that infects the vector, and selective feeding on them could facilitate *Leishmania* transmission (Coleman and Edman, 1988). In the same study,

Coleman and Edman (1988) suggested that the number of landings on the lesions was not significantly greater than on an uninfected mouse and that the selection of landing sites was essentially random. The implication is that, once the lesion has been encountered, the sandfly will choose to feed at that site rather than select another site. However, lesions on infected animals tend to occur at those sites that are preferentially probed in any case, i.e. the exposed areas of skin on the ears, nose and paws.

It has been suggested (Davies, 1990; Moloo *et al.*, 2000; Taylor and Hurd, 2001) that the epidemiological consequences of enhanced vector feeding on the parasitized host may be increased biting and thus increased transmission. However, once the parasite has developed within the vector and is at an infective stage, further parasitic manipulations may ensure enhanced host contact.

Landing and probing by a parasitized vector

When the vector is infected with the parasite, the situation changes. Laboratory studies have shown that, in *A. aegypti* mosquitoes, *P. gallinacium* sporozoites invade the salivary gland and reduce apyrase secretion. Apyrase is an enzyme that inhibits the aggregation of platelets and has been found in all haematophagous insects so far examined (Law *et al.*, 1992). Thus, confronted with the host blood-vessel wounds healing more quickly because of reduced amounts of apyrase and the consequent increased difficulty in feeding, the insect spends more time probing (Rossignol *et al.*, 1984, 1986; Ribeiro *et al.*, 1985). There is no reduction in saliva production because of infection and thus, as the malaria parasite is transmitted in mosquito saliva during probing and feeding, transmission is not reduced (Griffiths and Gordon, 1952) and, indeed, is likely to be enhanced. Wekesa *et al.* (1992) also showed an increase in the number of probes and in the likelihood of probing beginning when *Anopheles* mosquitoes in western Kenya were infected with *P. falciparum*. In a similar set of experiments, *A. aegypti* infected with *P. gallinaceum* probe for longer and more often (Rossignol *et al.*, 1986). The effect of increased probing is increased transmission (Kelly and Edman, 1992) and the ability of a smaller number of vectors to maintain infection within a population (Koella *et al.*, 1998a).

Recent studies have shown that parasitic modifications of mosquito behaviour may be more closely related to the developmental stage of the parasite than hitherto seen and that changes in vector behaviour may be specifically timed to maximize sporozoite transmission. *A. stephensi* infected with *P. yoelii nigeriensis* sporozoites (the infective stage) were significantly more persistent in their feeding attempts than uninfected mosquitoes. However, when infected with oocysts (an uninfective, developmental stage), the mosquitoes were significantly less persistent than those that were uninfected and thereby possibly were able to reduce dangerous host contact (Anderson *et al.*, 1999).

Results of work conducted in the field reveal generally similar changes in mosquito behaviour due to parasitic infection. Koella and

Packer (1996) showed that individual *A. punctulatus* infected with *P. falciparum* or *P. vivax* sporozoites in Papua New Guinea fed maximally throughout the night, in comparison with uninfected mosquitoes, which took increasingly large blood meals as the night wore on. They attributed this difference to more tenacious blood-feeding behaviour, i.e. infected mosquitoes fed for longer and several times on the same or separate hosts, in both instances, putting themselves at greater risk from host defensive behaviour. In a further separate study, Koella *et al.* (1998b) showed that *A. gambiae* infected with *P. falciparum* sporozoites engorged more fully than uninfected mosquitoes and were more likely to bite more than one host. These studies are somewhat contradicted by that of Bockarie *et al.* (1996), who showed that, in Papua New Guinea, *A. punctulatus* infected with *P. falciparum* sporozoites tended to bite later in the night than uninfected mosquitoes. Other workers have reported similar changes in feeding behaviour of naturally infected populations in the field. For example, Maxwell *et al.* (1998) showed that *A. gambiae* and *Anopheles funestus* infected with *P. falciparum* sporozoites in Tanzania would bite either earlier or later in the evening than uninfected mosquitoes. Robert and Carnevale (1991) demonstrated similar changes in the behaviour of *A. ambiae* and *A. funestus* in Burkina Faso and Gillies (1957) in that of *A. gambiae* in Tanzania.

Together, these results suggest that the malaria parasite may manipulate the behaviour of the mosquito vector to enhance transmission (Koella, 1999).

Leishmania-infected sandflies, like malaria-infected mosquitoes, probe more often than uninfected flies, i.e. they may probe the same host repeatedly or they may probe several hosts. As a consequence, several hosts can be infected by one fly (Beach *et al.*, 1985). In mature *Leishmania* infections, promastigote forms (flagellated *Leishmania*) secrete a gel-like plug (promastigote secretory gel (PSG)), which occludes and distorts the cardia region of the gut, increasing the volume of the lumen by two to three times and holding open the stomodeal valve (Stierhof *et al.*, 1999). Further, it has been shown that transformation from the promastigote to the metacyclic promastigote form occurs within the PSG plug and infective metacyclic promastigotes (fast-swimming infective forms) are regurgitated from a position in front of the PSG plug into the wound caused by feeding (Rogers *et al.*, 2002). Infected flies experience difficulty feeding, due to the presence of PSG, and feed more frequently and/or for longer, thus increasing the opportunity for transmission. In addition, PSG material is egested during sandfly feeding and is able to enhance metacyclic infectivity and lesion development, whereas little enhancement is observed in the presence of sandfly saliva (Rogers *et al.*, 2002). It would appear that *Leishmania* parasites produce a substance that acts to modulate vector-feeding success and thereby enhances their transmission.

Jenni *et al.* (1980) showed that *G. morsitans morsitans* infected with *Trypanosoma* (*Trypnozoon*) *brucei* probed more frequently and fed more voraciously than uninfected flies. The increased probing may be caused by the trypanosomes attaching directly to mechanoreceptors on the tsetse-fly labrum and physically blocking them, thus impairing their function (Molyneux *et al.*, 1979).

Vector Reproduction

Vector reproductive behaviour includes mating and oviposition. There are few reports of either being affected by infection, and changes in the former are unlikely, because most vectors have mated before they become infected. Reports of changes in oviposition behaviour are rare, because oviposition is often used as a measure of egg production and few studies have assessed the effect of parasites on egg retention as well as oviposition. However, El Sawaf *et al.* (1994) report that *Leishmania major* and *L. infantum* cause a significant reduction in both egg retention and oviposition in the sandfly *Phlebotomus papatasi*.

Parasite-induced fecundity reduction is reported in diverse parasite–vector associations, such as *Leishmania* spp. and sandflies, malaria and mosquitoes and filarial nematodes and mosquitoes and blackflies (Hurd, 1993; Hurd *et al.*, 1995), thus fulfilling one of Poulin's (1995) criteria for adaptive manipulation. Although curtailment of reproduction may be a by-product of manipulation of feeding behaviour, it also occurs when blood-meal size is not affected by infection.

A. stephensi and *A. gambiae* are being used to investigate the mechanisms underlying parasite-induced fecundity reduction caused by *P. yoelii nigeriensis*. Infection results in a significant reduction in egg production and egg hatch rate during an initial gonotrophic cycle after feeding on an infected host and also during subsequent gonotrophic cycles, when oocysts are present in the mosquito midgut and when sporozoites are in the salivary glands (Hogg and Hurd, 1995; Jahan and Hurd, 1997; Ahmed *et al.*, 1999). Fecundity reduction also occurs in wild-caught *A. gambiae* infected with *P. falciparum* (Hogg and Hurd, 1997). There is considerable physiological evidence to suggest that, in these associations, fecundity reduction is not a by-product of nutrient depletion (reviewed by Hurd, 2001). In the fat body of infected females, the abundance of messenger RNA (mRNA) of the yolk protein precursor, vitellogenin, is decreased by infection (Ahmed *et al.*, 2001) and, in the ovaries, cells of the follicular epithelium undergo apoptosis, thus precipitating the resorption of developing follicles (Hopwood *et al.*, 2001). These complex responses to infection may be host or parasite adaptations (Hurd, 2001).

If vector fecundity reduction is a parasite adaptive strategy, it appears to have evolved several times among different taxa (Hurd, 1993). This suggests that fitness benefits must accrue to the parasite from reducing vector egg production, which, as discussed above, is likely to result in enhanced vector survivorship over and above that of a vector that sustains a parasite infection and matures a full batch of eggs (Hurd, 2001).

Longevity

A number of studies have been carried out to investigate the effect of parasitic infection on the longevity of the vector. Decreased longevity of infected vectors would reduce the proportion of infected vectors and thus reduce parasite transmission. The results suggest that, when the infection is atypical, the infection is pathological and the lifespan of the parasite is significantly reduced. On the other hand, when the parasite–vector association is natural, there is no apparent effect on longevity. The effect on longevity and daily survival rates for wild-caught *A. gambiae* and *A. funestus* naturally infected with *P. falciparum* was investigated by Chege and Beier (1990), who showed that *Plasmodium* infections did not reduce vector survival. Laboratory studies on the same vector and parasites either show similar results to the field results (e.g. Freier and Friedman, 1987) or the parasite reduces the longevity of the vector (e.g. Klein *et al.*, 1986). These conflicting results may be due to the intensity of the infection in wild mosquitoes, which may not be sufficiently intense to decrease survival (Chege and Beier, 1990). Bates (1949) reported that there was no evidence that, while *Plasmodium* or viral infection did not have any 'unfavourable' effect on mosquito longevity, nematodes, particularly filariae and mermithids, cause very heavy mortality.

Conclusions

During the course of the preparation of this chapter, it has become apparent that there is an acute lack of recent studies on the effect of parasites on vector behaviour. Of those that exist, malaria/mosquito interactions dominate the field. This may result from a commonly held misconception that parasites induce few changes in their vectors. Both old and very recent studies refute this concept. Studies may be technically difficult and time-consuming and definitely require verification in field conditions, but the rewards that they would bring in terms of a greater understanding of vector transmission and vector control will make them worth the effort.

References

Ahmed, A.M., Taylor, P. Maingon, R. and Hurd, H. (1999) The effect of *Plasmodium yoelii nigeriensis* on the reproductive fitness of *Anopheles gambiae. Invertebrate Reproductive Development* 36, 217–222.

Ahmed, A.M., Maingon, R., Romans, P. and Hurd, H. (2001) Effects of malaria infection on vitellogenesis in *Anopheles gambiae* during two gonotropic cycles. *Insect Molecular Biology* 10, 347–356.

Anderson, R.A. and Roitberg, B.D. (1999) Modelling trade-offs between mortality and fitness associated with persistent blood feeding by mosquitoes. *Ecological Letters* 2, 98–105.

Anderson, R.A., Koella, J.C. and Hurd, H.H. (1999) The effect of *Plasmodium yoelii nigeriensis* infection on the feeding persistence of *Anopheles stephensi* Liston throughout the sporogonic cycle. *Proceedings of the Royal Society of London, Series B* 266, 1729–1733.

Bates, M. (1949) *The Natural History of Mosquitoes*. Macmillan, London, 379 pp.

Baylis, M. and Mbwabi, A.L. (1995) Feeding behavior of tsetse flies (*Glossina pallidipes* Austen) on *Trypanosoma*-infected oxen in Kenya. *Parasitology* 110, 297–305.

Beach, R., Kiilu, G. and Leeuwenburg, J. (1985) Modification of sand fly biting behavior by *Leishmania* leads to increased parasite transmission. *American Journal of Tropical Medicine and Hygiene* 34, 278–282.

Berry, W.J., Rowley, W.A. and Christensen, B.M. (1986) Influence of developing *Brugia pahangi* on spontaneous flight activity of *Aedes aegypti* (Diptera: Culicidae). *Journal of Medical Entomology* 23, 441–445.

Berry, W.J., Rowley, W.A. and Christensen, B.M. (1987) Influence of developing *Dirofilaria immitis* on the spontaneous activity of *Aedes aegypti* (Diptera: Culicidae). *Journal of Medical Entomology* 24, 699–701.

Berry, W.J., Rowley, W.A. and Christensen, B.M. (1988) Spontaneous flight activity of *Aedes trivittatus* infected with *Dirofilaria immitis*. *Journal of Parasitology* 74, 970–974.

Bockarie, M.J., Alexander, N., Bockarie, F., Ibam, E., Barbish, G. and Alpers, M. (1996) The late biting habit of parous *Anopheles* mosquitoes and pre-bedtime exposure of humans to infective female mosquitoes. *Transactions of the Royal Society of Tropical Medicine and Hygiene* 90, 23–25.

Braks, M.A.H., Anderson, R.A. and Knols, B.G.J. (1999) Info-chemicals in mosquito host selection: human skin micro flora and *Plasmodium* parasites. *Parasitology Today* 4, 409–413.

Burkot, T.R. (1988) Non-random host selection by anopheline mosquitoes. *Parasitology Today* 4(6), 156–162.

Burkot, T.R., Narara, A., Paru, R., Graves, P.M. and Garner, P. (1989) Human host selection by anophelines: no evidence for preferential selection of malaria or microfilariae-infected individuals in a hyperendemic area. *Parasitology* 98, 337–342.

Chege, G.M.M. and Beier, J.C. (1990) Effect of *Plasmodium falciparum* on the survival of naturally infected Afrotropical *Anopheles* (Diptera: Culicidae). *Journal of Medical Entomology* 27, 454–458.

Clements, A.N. (1963) *The Physiology of Mosquitoes*. Pergamon Press, Oxford, 393 pp.

Clements, A.N. (1992) *The Biology of Mosquitoes*, Vol. 1. Chapman & Hall, London.

Coleman, R.E. and Edman, J.D. (1988) Feeding-site selection of *Lutzomyia longipalpis* (Diptera: Psychodidae) on mice infected with *Leishmania mexicana amazonensis*. *Journal of Medical Entomology* 25, 229–233.

Coleman, R.E., Edman, J.D. and Semprevivo, L.H. (1988) Interactions between malaria (*Plasmodium yoelii*) and Leishmaniasis (*Leishmania mexicana amazonensis*): effect of concomitant infection on host activity, host body temperature and vector engorgement success. *Journal of Medical Entomology* 25, 467–471.

Daniel, T.L. and Kingsolver, J.G. (1983) Feeding strategy and the mechanics of blood sucking insects. *Journal of Theoretical Biology* 105, 661–672.

Davies, C.R. (1990) Interrupted feeding of blood-sucking insects: causes and effects. *Parasitology Today* 6, 19–22.

Davies, S., Spanel, P. and Smith, D. (1997) Quantitative analysis of ammonia on the breath of patients in end-stage renal failure. *Kidney International* 52, 223–228.

Davis, E.E. (1984) Development of lactic acid-receptor sensitivity and host seeking behaviour in newly emerged female *Aedes aegypti* mosquitoes. *Journal of Insect Physiology* 30, 211–215.

Davis, E.E., Haggart, D.A. and Bowden, M.F. (1987) Receptors mediating host-seeking behaviour in mosquitoes and their regulation by endogenous hormones. *Insect Science and its Applications* 8, 637–641.

Dawkins, R. (1982) *The Extended Phenotype*. Oxford University Press, Oxford, 307 pp.

Dawkins, R. (1990) Parasites, desiderata lists and the paradox of the organism. *Parasitology* 100, S63-S75.

Day, J.F. and Edman, J.D. (1983) Malaria renders mice susceptible to mosquito feeding when gametocytes are most infective. *Journal of Parasitology* 69, 163–170.

Day, J.F. and Edman, J.D. (1984a) Mosquito engorgement on normally defensive hosts depends on host activity patterns. *Journal of Medical Entomology* 21, 732–740.

Day, J.F. and Edman, J.D. (1984b) The importance of disease-induced changes in mammalian body temperature to mosquito blood-feeding. *Comparative Physiology and Biochemistry* 77, 447–452.

Day, J.F., Ebert, K.M. and Edman, J.D. (1983) Feeding patterns of mosquitoes (Diptera: Culicidae) simultaneously exposed to malarious and healthy mice, including a method for separating blood meals from conspecific hosts. *Journal of Medical Entomology* 20, 120–127.

Detinova, T.S. (1962) *Age Grouping Methods in Diptera of Medical Importance*. WHO Monograph Series No. 47, WHO, Geneva.

Dye, C. (1990) Epidemiological significance of vector–parasite interactions. *Parasitology* 101, 409–415.

Edman, J.D. (1989) Are mosquitoes gourmet or gourmand? *Journal of the American Mosquito Control Association* 5, 487–497.

El Sawaf, B.M., El Sattar, S.A., Shehata, M.G., Lane, P.R. and Morsy, T.A. (1994) Reduced longevity and fecundity in *Leishmania*-infected sand flies. *American Journal of Tropical Medicine and Hygiene* 51, 767–770.

Ferdig, M.T., Beerntsen, B.T., Spray, F.J., Li, J. and Christensen, B.M. (1993) Reproductive costs associated with resistance in a mosquito–filarial worm system. *American Journal of Tropical Medicine and Hygiene* 49, 756–762.

Freier, J.E. and Friedman, S. (1976) Effect of host infection with *Plasmodium gallinaceum* on the reproductive capacity of *Aedes aegypti*. *Journal of Invertebrate Pathology* 28, 161–166.

Freier, J.E. and Friedman, S. (1987) Effect of *Plasmodium gallinaceum* infection on the mortality and body weight of *Aedes aegypti* (Diptera: Culicidae). *Journal of Medical Entomology* 24, 6–10.

Garrett-Jones, C. (1964) Prognosis for the interruption of malaria transmission through assessment of the mosquito's vectorial capacity. *Nature* 204, 1173–1175.

Garrett-Jones, C. and Shidrawi, G.R. (1969) Malaria vectorial capacity of a population of *Anopheles gambiae*. *Bulletin of the World Health Organization* 40, 531–545.

Gillett, J.D. and Connor, J. (1976) Host temperature and the transmission of arboviruses by mosquitoes. *Mosquito News* 36, 472–477.

Gillies, M.T. (1957) Age group and the biting cycle in *Anopheles gambiae*. *Bulletin of Entomological Research* 48, 553–559.

Griffiths, R.B. and Gordon, R.M. (1952) An apparatus which enables the process of feeding by mosquitoes to be observed in the tissues of a live rodent: together with an account of the ejection of saliva and its significance in malaria. *Annals of Tropical Medicine and Parasitology* 46, 311–319.

Hogg, J.C. and Hurd, H. (1995) *Plasmodium yoelii nigeriensis*: the effect of high and low intensity of infection upon the egg production and blood meal size of *Anopheles stephensi* during three gonotrophic cycles. *Parasitology* 111, 555–562.

Hogg, J.C. and Hurd, H. (1997) The effect of natural *Plasmodium falciparum* infection on the fecundity and mortality of *Anopheles gambiae s.l.* in north east Tanzania. *Parasitology* 114, 325–331.

Hopwood, J.A., Ahmed, A.M., Polwart, A., Williams, G.T. and Hurd, H. (2001) Malaria-induced apoptosis in mosquito ovaries: a mechanism to control vector egg production. *Journal of Experimental Biology* 204, 2773–2780.

Horton, D.R. and Moore, J. (1993) Behavioral effects of parasites and pathogens in insect hosts. In: Beckage, N.E., Thompson, S.N. and Federici, B.A. (eds) *Parasites and Pathogens of Insects*, Vol. 1, *Parasites*. Academic Press, San Diego, pp. 107–124.

Hurd, H. (1990) Physiological and behavioural interactions between parasites and invertebrate hosts. *Advances in Parasitology* 29, 271–317.

Hurd, H. (1993) Reproductive disturbances induced by parasites and pathogens of insects. In: Beckage, N.E., Thompson, S.N. and Federici, B.A. (eds) *Parasites and Pathogens of Insects*, Vol. 1, *Parasites*. Academic Press, San Diego, pp. 87–105.

Hurd, H. (1998) Parasite manipulation of insect reproduction: who benefits? *Parasitology* 116, S13–S22.

Hurd, H. (2001) Host fecundity reduction: a strategy for damage limitation? *Trends in Parasitology* 17, 363–368.

Hurd, H., Hogg, J.C. and Renshaw, M. (1995) Interactions between blood-feeding, fecundity and infection in mosquitoes. *Parasitology Today* 11, 411–416.

Jahan, N. and Hurd, H. (1997) The effects of infection with *Plasmodium yoelii nigeriensis* on the reproductive fitness of *Anopheles stephensi*. *Annals of Tropical Medicine and Parasitology* 91, 365–369.

Jenni, L., Molyneux, D.H., Livesey, J.L. and Galun, R. (1980) Feeding behaviour of tsetse flies infected with salivarian trypanosomes. *Nature* 283, 383–385.

Kelly, R. and Edman, J.D. (1992) Multiple transmission of *Plasmodium gallinaceum* (Eucoccida: Plasmodiidae) during serial probing by *Aedes aegypti* on several hosts. *Journal of Medical Entomology* 29, 329–331.

Kettle, D.S. (1990) *Medical and Veterinary Entomology*. CAB International, Wallingford, UK, 658 pp.

Kingsolver, J.G. (1987) Mosquito host choice and the epidemiology of malaria. *American Naturalist* 130, 811–827.

Klein, T.A., Harrison, B.A., Grove, J.S., Dixon, S.V. and Andre, R.G. (1986) Correlation of the survival rates of *Anopheles dirus* A (Diptera: Culicidae) with different infection densities of *Plasmodium cynomolgi*. *Bulletin of the World Health Organization* 64, 901–907.

Klowden, M.J. (1987) Distention-mediated egg maturation in the mosquito *Aedes aegypti*. *Journal of Insect Physiology* 33, 83–87.

Klowden, M.J. (1995) Blood, sex and the mosquito. *Bioscience* 45, 326–331.

Klowden, M.J., Davis, E.E., and Bowden, M.F. (1987) Role of the fat body in the regulation of host seeking behaviour in the mosquito, *Aedes aegypti*. *Journal of Insect Physiology* 33, 643–646.

Koella, J.C. (1999) An evolutionary view of the interactions between anopheline mosquitoes and malaria parasites. *Microbes and Infection* 1, 303–308.

Koella, J.C. and Packer, M.J. (1996) Malaria parasites enhance blood-feeding of their naturally infected vector *Anopheles punctulatus*. *Parasitology* 113, 105–109.

Koella, J.C., Agnew, P. and Michalakis, Y. (1998a) Coevolutionary interactions between host life histories and parasite life cycles. *Parasitology* 116, S47–S55.

Koella, J.C., Sørensen, F.L. and Anderson, R.A. (1998b) The malaria parasite, *Plasmodium falciparum*, increases the frequency of multiple feeding of its mosquito vector, *Anopheles gambiae*. *Proceedings of the Royal Society of London, Series B* 265, 763–768.

Laarman, J.J. (1958) The host-seeking behaviour of anopheline mosquitoes. *Tropical and Geographical Medicine* 10, 293–305.

Law, J.H., Ribeiro, J.M. and Wells, M.A. (1992) Biochemical insights derived from diversity in insects. *Annual Review of Biochemistry* 61, 87–111.

Leak, S.G.A. (1999) *Tsetse Biology and Ecology: Their Role in the Epidemiology and Control of Trypanosomosis*. CAB International, Wallingford, UK, 568 pp.

Lehane, M.J. (1991) *The Biology of Blood-sucking Insects*. Harper Collins Academic, London, 288 pp.

Lines, J.D., Wilkes, T.J. and Lyimo, E.O. (1991) Human malaria infectiousness measured by age-specific sporozoite rates in *Anopheles gambiae* in Tanzania. *Parasitology* 102, 167–177.

Luckhart, S., Vodovotz, Y., Cui, L. and Rosenberg, R. (1998) The mosquito *Anopheles stephensi* limits malaria parasite development with inducible synthesis of nitric oxide. *Proceedings of the National Academy of Sciences USA* 95, 5700–5705.

Lyimo, E.O. and Koella, J.C. (1992) Relationship between body size of adult *Anopheles gambiae s.l.* and infection with the malaria parasite *Plasmodium falciparum*. *Parasitology* 104, 233–237.

Macdonald, G. (1957) *The Epidemiology and Control of Malaria*. Oxford University Press, London.

Mahon, R. and Gibbs, A. (1982) Arbovirus-infected hens attract more mosquitoes. In: Mackenzie, J.S. (ed.) *Viral Diseases in South-East Asia and the Western Pacific*. Academic Press, Sydney, pp. 502–505.

Maxwell, C.A., Wakibara, J., Tho, S. and Curtis, C.F. (1998) Malaria-infective biting at different hours of the night. *Medical and Veterinary Entomology* 12, 325–327.

Minchella, D.J. (1985) Host life-history variation in response to parasitism. *Parasitology* 90, 205–216.

Moloo, S.K., Sabwa, C.L. and Baylis, M. (2000) Feeding behaviour of *Glossina pallidipes* and *G. morsitans centralis* on Boran cattle infected with

Trypanosoma congolense or *T. vivax* under laboratory conditions. *Medical and Veterinary Entomology* 14, 290–299.

Molyneux, D.H., Lavin, D.R. and Elce, B. (1979) A possible relationship between salivarian trypanosomes and *Glossina* labrum mechanoreceptors. *Annals of Tropical Medicine and Parasitology* 73, 288–290.

Moore, J. (1993) Parasites and the behaviour of biting flies. *Journal of Parasitology* 79, 1–16.

Moore, J. and Gotelli, N.J. (1990) A phylogenetic perspective on the evolution of altered host behaviours: a critical look at the manipulation hypothesis. In: Barnard, C.J. and Behnke, J.M. (eds) *Parasitism and Host Behaviour.* Taylor and Francis, London, pp. 193–230.

O'Shea, B., Rebollar-Téllez, E., Ward, R.D., Hamilton, J.G.C., El Naiem, D. and Polwart, A. (2002) Enhanced sandfly attraction to *Leishmania* infected hosts. *Transactions of the Royal Society of Tropical Medicine and Hygiene* 96, 1–2.

Paige, C.J. and Craig, G.B., Jr (1975) Variation in filarial susceptibility among East African populations of *Aedes aegypti. Journal of Medical Entomology* 12, 485–493.

Penn, D. and Potts, W.K. (1998) Chemical signals and parasite mediated sexual selection. *Trends in Ecology and Evolution* 13, 391–396.

Poulin, R. (1994) The evolution of parasite manipulation of host behaviour: a theoretical analysis. *Parasitology* 109, S109–S118.

Poulin, R. (1995) 'Adaptive' changes in the behaviour of parasitized animals: a critical review. *International Journal for Parasitology* 25, 1371–1383.

Poulin R. (1998) Evolution and phylogeny of behavioural manipulation of insects by parasites. *Parasitology* 116, S3–S9.

Price, P.W. (1980) *Evolutionary Biology of Parasites.* Princeton University Press, Princeton, New Jersey, 237 pp.

Randolph, S.E., Williams, B.G., Rogers, D.J. and Connor, H. (1992) Modelling the effect of feeding-related mortality on the feeding strategy of tsetse (Diptera: Glossinidae). *Medical and Veterinary Entomology* 6, 231–240.

Rebollar-Téllez, E.A. (1999) Kairomone-mediated behaviour of members of the *Lutzomyia longipalpis* complex (Diptera: Psychodidae). PhD thesis, Keele University, Keele, UK.

Ribeiro, J.M.C., Rossignol, P.A. and Spielman, A. (1985) Salivary gland apyrase determines probing time in anopheline mosquitoes? *Journal of Insect Physiology* 31, 689–692.

Richman, A.M., Dimopoulos, D.S. and Kafatos, F.C. (1997) *Plasmodium* activates the innate immune response of *Anopheles gambiae* mosquitoes. *EMBO Journal* 16, 6114–6119.

Robert, V. and Carnevale, P. (1991) Influence of deltamethrin treatment of bednets on malaria transmission in the Kou valley, Burkina Faso. *Bulletin of the World Health Organization* 69, 735–740.

Rogers, D.J. and Packer, M.J. (1993) Vector-borne diseases, models, and global change. *Lancet* 342, 1282–1284.

Rogers, M.E., Chance, M.L. and Bates, P.A. (2002) The role of promastigote secretory gel in the origin and transmission of the infective stage of *Leishmania mexicana* by the sandfly *Lutzomyia longipalpis. Parasitology* 124, 495–507.

Rossignol, P.A., Ribeiro, J.M.C. and Spielman, A. (1984) Increased intradermal probing time in sporozoite-infected mosquitoes. *American Journal of Tropical Medicine and Hygiene* 33, 17–20.

Rossignol, P.A., Ribeiro, J.M.C., Jungery, M., Turell, M.J., Spielman, A. and Bailey, C.L. (1985) Enhanced mosquito blood-finding success on parasitemic hosts: evidence for vector–parasite mutualism. *Proceedings of the National Academy of Sciences USA* 82, 7725–7727.

Rossignol, P.A., Ribeiro, J.M.C. and Spielman, A. (1986) Increased biting rate and reduced fertility in sporozoite-infected mosquitoes. *American Journal of Tropical Medicine and Hygiene* 35, 277–279.

Rowland, M.W. and Boersma, E. (1988) Changes in the spontaneous flight activity of the mosquito *Anopheles stephensi* by parasitization with the rodent malaria *Plasmodium yoelli*. *Parasitology* 97, 221–227.

Rowland, M.W. and Lindsay, S.W. (1986) The circadian flight activity of *Aedes aegypti* parasitised with the filarial nematode *Brugia pahangi*. *Physiological Entomology* 11, 325–334.

Ryan, L. (1984) The effect of trypanosome infections on a natural population of *Glossina longipalpis* Wiedemann (Diptera: Glossinidae) in Ivory Coast. *Acta Tropica* 41, 355–359.

Schwartz, A. and Koella, J.C. (2001) Trade-offs, conflicts of interest and manipulation in *Plasmodium*–mosquito interactions. *Trends in Parasitology* 17, 189–194.

Shieh, J.N. and Rossignol, P.A. (1992) Opposite influences of host anaemia on blood feeding rate and fecundity of mosquitoes. *Parasitology* 105, 159–163.

Smith, D., Spanel, P., Thompson, J.M., Rajan, B., Cocker, J. and Rolfe, P. (1998) The selected ion flow tube method for workplace analyses of trace gases in air and breath: its scope, validation and applications. *Applied Occupational Environmental Hygiene* 13, 817–823.

Sokolova, M.I. (1994) A redescription of the morphology of mosquito (Diptera: Culicidae) ovarioles during vitellogenesis. *Bulletin of the Society of Vector Ecology* 19, 53–68.

Stierhof, Y.-D., Bates, P.A., Jacobson, R.L., Rogers, M.E., Schlein, Y., Handman, E. and Ilg, T. (1999) Filamentous proteophosphoglycan secreted by *Leishmania* promastigotes forms gel-like three-dimensional networks that obstruct the digestive tract of infected sandfly vectors. *European Journal of Cell Biology* 78, 675–689.

Sutcliffe, J.F. (1987) Distance orientation of biting flies to their hosts. *Insect Science and its Application* 8, 611–616.

Takken, W. (1991) The role of olfaction in host-seeking of mosquitoes: a review. *Insect Science and its Application* 12, 287–295.

Takken, W., van Loon, J.J.A. and Adam, W. (2001) Inhibition of host-seeking response and olfactory responsiveness in *Anopheles gambiae* following blood feeding. *Journal of Insect Physiology* 47, 303–310.

Taylor, P. (2001) The effects of malaria infection on the blood feeding behaviour of anopheline mosquitoes. PhD thesis, Keele University, Keele, UK.

Taylor, P. and Hurd, H.H. (2001) The influence of host haematocrit on the blood feeding success of *Anopheles stephensi*: implications for enhanced malaria transmission. *Parasitology* 122, 491–496.

Townson, H. (1970) The effect of infection with *Brugia pahangi* on the flight of *Aedes aegypti*. *Annals of Tropical Medicine and Parasitology* 64, 411–420.

Turell, M.J., Rossignol, P.A. and Spielman, A. (1984) Enhanced arboviral transmission by mosquitoes that concurrently ingested microfilaria. *Science* 225, 1039–1041.

Vale, G.A. (1980) Flight as a factor in the host-finding behaviour of tsetse flies (Diptera: Glossinidae). *Bulletin of Entomological Research* 70, 299–307.

Waladde, S.M. and Rice, M.J. (1982) The sensory basis of tick feeding behavior. In: Obenchain, F.D. and Galun, R. (eds) *The Physiology of Ticks.* Pergamon Press, Oxford, pp. 71–118.

Wekesa, J.W., Copeland, R.S. and Mwangi, R.W. (1992) Effect of *Plasmodium falciparum* on blood-feeding behaviour of naturally infected *Anopheles* mosquitoes in western Kenya. *American Journal of Tropical Medicine and Hygiene* 47, 484–488.

Parasite Virulence 14

Jos J. Schall

Department of Biology, University of Vermont, Burlington, VT 05405, USA

The Problem

Some parasites exact a terrible price from their hosts, causing severe pathology and reducing the host's fitness, whereas other parasites are essentially benign. Several kinds of comparisons highlight this observation. Least interesting are comparisons of parasites with very different life histories or types of host tissues invaded (compare human immuno-deficiency virus (HIV) and rhinovirus infection in humans). In other cases, the same parasite causes great harm in one species of host, but is tolerable to another, such as the rabies virus, which kills canid hosts but can reside in mustelid populations as non-lethal infections (Kaplan, 1985). Again, this may result from different types of tissues invaded by the parasite. Most intriguing are examples of very different levels of pathology caused to the same host species by closely related parasite species or even different genetic strains within a parasite species.

Examples of this last situation are abundant. Strains of *Trypanosoma brucei*, the causative agent of African sleeping sickness in humans, differ in the severity of the pathology they cause, so much so that they were long classified as different species, based on symptomatology (Toft and Aeschlimann, 1991). Malaria has the reputation as the malignant 'million-murdering death', but only *Plasmodium falciparum* kills a significant number of victims outright and, within each species, the morbidity and mortality associated with infection vary geographically (Arnot, 1998). The rabies virus, so notorious for its lethality for humans and extremely high mortality for dogs, has evolved an African strain that produces non-fatal oulou fato in dogs (Kaplan, 1985). *Entamoeba histolytica* and *Giardia lamblia*, widespread and important intestinal parasites of humans, vary in their pathology by genotype, which led to different species names for polyphyletic clusters of strains (Mehlotra, 1998;

Thompson, 2000). These examples suggest that virulence may be a 'life-history trait' of the parasite and is part of the adaptive picture of parasite–host associations. If this is so, what drives the evolution of virulence?

The dominant view among medical parasitologists for generations held that selection favours a reduction in virulence because the parasite's home site is ephemeral (hosts eventually die) and a parasite should not exacerbate this situation by reducing the host's lifespan. Recently established associations may show poor adaptation on the part of the parasite, but older associations will reveal accommodation of the parasite to its host and low cost of infection (Burnet and White, 1972). Ball (1943) long ago demonstrated the theoretical and empirical weakness of this reasoning. He found that data comparing the likely ages of different parasite–host systems with their virulence did not support the 'association age' hypothesis. Despite the cogent argument of Ball, one of the century's most eminent parasitologists, and the rejection of the classical prudent-parasite image by evolutionary biologists, it is still current in the medical literature (review in Ewald, 1994).

A growing list of hypotheses have supplanted the venerable prudent-parasite view (reviews in Ewald, 1994, 1995; Groisman and Ochman, 1994; Read, 1994; Bull, 1995; Frank, 1996; Poulin, 1998; Ebert, 1999). Coalescing these many hypotheses into a general theory on the evolution of virulence remains an elusive challenge. A general theory must respect the great systematic and ecological diversity of parasites and incorporate such factors as:

- The phylogenetic history of both parasites and hosts.
- Kind of tissue invaded by the parasite.
- Mode of transmission (consumption of one host by another vs. use of a vector, for example).
- Ease of transmission (including aspects of the biology of both parasite and host).
- Number of host species exploited by a parasite.
- Population structure of the parasite–host system.

Can any single theory gather together the details of the biology of parasites as different as viruses, tapeworms and ticks? Perhaps the diversity of 'parasites' in the broadest sense cannot be included within the rubric of a general theory. Ewald (1995) argues persuasively that such a view is pessimistic and unproductive, because a single theory could cast light on a broad range of associations, from viruses to large predators.

What is Virulence?

Defining virulence is problematic (Read, 1994; Ebert, 1999; Poulin and Combes, 1999), in part because disciplines in the life sciences differ in their perspective when interpreting the importance of parasitic diseases.

For those with medical or veterinary interests, virulence is any harm done to the host by another organism (usually limiting discussion to small organisms). For medical workers, virulence is an issue of mechanism (such as so-called 'virulence genes or traits' (Poulin and Combes, 1999)) and the practical goal of reducing illness. An interesting example is botulism. Carl Bromwich, a physician practising in Kuujjuaq in the Canadian Arctic, reports that he has extensive experience in treating botulism. Local people hunt sea mammals and prefer to eat the meat only after it has aged. *Clostridium botulinum* probably lives on the mammals as a commensal (a parasite with no costs to the host), but rises to huge population densities when it reproduces on the carcass. The bacteria produce a toxin (perhaps as interference competition with other bacteria) and humans become ill, not from an infection with *C. botulinum*, but from ingesting the toxin. Different strains of *C. botulinum* could vary in the quantity and nature of the toxin produced. Medical personnel would reasonably discuss the variation in virulence of these strains, but the selective events leading to such variation in the biochemistry of the bacterial strains had nothing to do with their effects on humans. This example is transparent, but many other cases of supposed parasite 'virulence' involve accidental contact of limited significance for the evolution of parasite or host.

The public-health community is concerned with illness associated with infectious disease, as well as the ease of transmission of pathogens. Thus, the term virulence is often applied to some measure of rate of transmission; indeed, for many microbiologists and epidemiologists, that may be the primary definition of the term (Lipsitch and Moxon, 1997). Natural selection will certainly favour more ready transmission by parasites, but this meaning of the term is not relevant for the present discussion.

Wildlife ecologists, in contrast, are normally unconcerned with morbidity or mortality induced in individual hosts, but instead ask if parasitism can regulate host population density (Hudson *et al.*, 1998). For conservation ecologists, the possible reduction in population size of endangered species is a concern (Holmes, 1996). Thus, for an ecologist, consequences of parasitism for host lifespan and fecundity and how they influence host population density are the appropriate measures of virulence, and the evolutionary origin of virulent vs. benign parasites is not of concern.

In the discussion presented here, another outlook on virulence is required: virulence is a trait under selection, either directly or indirectly. Natural selection will work asymmetrically on hosts and parasites, so parasite virulence has two meanings, one for each species in the association. Making a distinction between virulence from the perspective of the host vs. the parasite is not just an exercise in term-mongering but allows us to recognize that the final harm done to the host by infection depends on a composite of two selective forces, one acting on the host and the other on the parasite.

Selection and Virulence – from the Host's Perspective

For the host, virulence is any consequence of infection that reduces the host's lifetime reproductive success (fitness). A fitness cost could result from the direct damage done by the parasite (destruction of cells, usurpation of resources), the expenditure of resources in mounting an immune response and collateral damage done to the host by its own immune system. To better grasp how parasites can reduce reproductive success, we can partition fitness into components, such as lifespan, fecundity, number of reproductive episodes, ability to find and court mates and health of offspring. Trade-offs between these components of fitness are a universal challenge faced by organisms (Bell, 1997), so we can imagine that infection may hinder one component of fitness while benefiting another. For example, castration of the host may be beneficial to the parasite if infected hosts partition more resources towards growth and body maintenance, which could provide more resources for the parasite and a longer-lasting host (Boudoin, 1975). Although host lifespan may increase, its overall fitness is reduced to zero. Another example is a parasite that is transmitted via the host's offspring and manipulates the host's reproductive biology. Such a parasite could increase the short-term fecundity of the host while reducing the host's overall lifetime reproduction. Venereal-transmitted parasites could manipulate the host to increase its attractiveness as a mate or expand its period of courting and mating. An expanded mating effort could also reduce the host's reproductive success if other components of fitness are reduced, such as lifespan. All of these examples demonstrate the importance of keeping the focus on those consequences of infection that would apply a selective pressure on the host.

Selection on the host favours adaptations that prevent or eliminate infection (host immunity in the broadest sense) or reduce the fitness costs of infection. But what would the optimal strategy be for the host – an intense defence that would eliminate the infection or a lesser attack that allows the infection to remain at some low level? Some trade-off must exist that balances the costs of the defence (to the host's fitness) against the benefit (also to its fitness). Figure 14.1 suggests how selection may favour some intermediate level of host response. Thus, part of the variation seen in parasite virulence could result from differing solutions to the trade-off between the costs and benefits of mounting antiparasite tactics by the host.

Antiparasite tactics include natural resistance, behavioural mechanisms to avoid infection and the immune system. Unfortunately, actual measures of the costs vs. benefits of antiparasite mechanisms are scarce (Gemmill and Read, 1998). One of the better examples of the costs vs. benefits of host resistance is the elegant study of Yan *et al.* (1997), who examined two genotypes of the mosquito *Aedes aegypti* in relation to the insect's resistance to the malaria parasite *Plasmodium gallinaceum*. Being refractory, or resistant, to the parasite has associated costs to fitness,

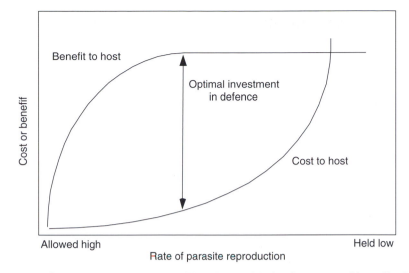

Y-axis: Cost or benefit

Labels within figure: Benefit to host; Optimal investment in defence; Cost to host

X-axis: Allowed high ... Held low

Rate of parasite reproduction

Fig. 14.1. Curves showing one possible relationship for the cost and benefit of differing levels of immune attack against a parasite. The horizontal axis shows the level of the host's antiparasite effort necessary to hold the parasite's rate of replication to low levels, or which will allow the parasite to replicate rapidly. The 'Cost to host' curve shows the cost to the host's fitness from mounting an immune attack. A low-level immune attack, one that allows the parasite reproductive rate to remain high, would have low cost, but a potent immune attack that holds the parasite to a low rate of reproduction would have a high cost. The 'Benefit to host' curve shows the benefits of limiting the rate of parasite reproduction. A fairly low-level immune attack with a marginal reduction in the parasite growth rate would be highly beneficial, so the benefit curve is steep. However, further reduction in parasite growth rate would have a declining additional benefit for the host. The shape of the cost curve shows that the initial small investment in the immune attack would have low cost, but completely curtailing the parasite would have an extreme cost to the host. The optimal solution is an intermediate level of immune response. Parasites with effective methods to evade the host immune system may select for very limited immune attack. The host may well be able to trump the parasite's defences, but the cost would outweigh the benefits.

including smaller body size, production of fewer eggs per clutch and shorter lifespan. Intuition suggests that this genotype remains in nature because it has the benefit of avoiding pathology induced by the parasite. However, Yan *et al.* (1997) could detect no fitness costs of infection. This presents a perplexing question. Why does the genetic polymorphism continue in natural habitats if resistance to infection has a fitness cost but no benefit when the mosquito is infected?

More common are circumstantial stories. One such example concerns an odd anatomical feature found on lizards of at least five families – skin invaginations on the neck, axilla and other body regions that seem particularly prone to infestation by mites and ticks. Arnold (1986) suggested that the pockets function to draw ectoparasites away from areas of the body where they could be more harmful. Immune-system cells are clustered just under the skin of these pockets, perhaps waiting to ambush any pathogens that might be passed by the bites of ectoparasites. Arnold's (1986) hypothesis was supported by Salvador *et al.* (1999), who experimentally plugged the neck pockets of the lacertid lizard *Psammodromus algirus* and observed a movement of ticks to areas of the body where they were likely to hinder normal locomotion. Thus, the pockets may serve as a way of reducing the cost of infestation and may be a lower-cost (to fitness) adaptation than any mechanism that would completely eliminate the ectoparasite burden. Actually measuring this trade-off would be a challenge.

Selection and Virulence — from the Parasite's Perspective

Virulence from the parasite's perspective is any injury to the host that reduces the parasite's fitness – that is, any consequences to the host of the infection that would cut the number of successful transmissions of the parasite to other hosts. Fitness for a parasite genotype is the number of daughter infections spawned into other hosts (May and Anderson, 1983). The following equation:

$$N_d = (N_{to} \times P_t) \, D$$

shows that the number of daughter infections (N_d) is a product of the number of transmission opportunities per unit time (N_{to}), the probability of each of these opportunities being successful (P_t) and the duration of the infection (D). Thus, virulence from the parasite's perspective is any reduction in N_d that results from its own biology within the host. Some examples include an infection-induced degradation of the host environment (the infection then fails, reducing D, or parasite density falls very low, reducing P_t), death of the host (reduced D) or changes in the host that reduce N_{to} (an immobilized host would greatly reduce the transmission opportunities for a directly transmitted parasite!).

Most discussion of parasite virulence assumes that there must be some trade-off between acute, or short-term, transmission success and the duration of the transmission period (typically the host's survival) – that is, N_{to}, P_t and D cannot be maximized simultaneously, because high parasite density increases the probability of transmission but also reduces host survival and thus the duration of the transmission period. There are actually two assumptions being made: (i) parasitaemia is negatively correlated with host survival; and (ii) parasitaemia is positively correlated with the probability of transmission. The final result is that a parasite can maximize either P_t or D but not both.

This argument is intuitively pleasing but is not often explicitly presented (although it underlies most of the hypotheses on virulence that have been proposed – see below), and even more rarely is it actually tested (Mackinnon and Read, 1999a). The assumption that higher parasite loads lead to reduction in the host survival is supported by data on some pathogens (examples include HIV (Mellors *et al.*, 1996) and rodent malaria (Mackinnon and Read, 1999a)), but the literature is replete with exceptions (Messenger *et al.*, 1999). The actual relationship between parasitaemia and pathology may be non-linear, especially if the host's immune system is responsible for some part of the injury 'caused' by the parasite. In such cases, low parasite loads may be as harmful as moderate or even fairly heavy loads. For example, the destruction of liver and other organs seen in human schistosomiasis is in part a result of the host's immune attack on the microscopic eggs of the worm (Warren, 1975) and the anaemia presenting with malaria derives in large part from destruction of uninfected red blood cells by an overactive immune system (Wetherall, 1988). The assumption that higher parasitaemia will increase the ease of movement of the parasite from host to host (increase in P_t) also seems reasonable, but the overall ecology of transmission can be complex and confound the expected simple relationship between parasitaemia and transmission success (Lipsitch and Moxon, 1997). Natural-history studies on directly transmitted parasites must include data on the distribution and survival of the parasite stages once they leave the host. Vector-borne parasites have ecologies that often fly in the face of biologists' intuition. For example, malaria parasites replicate asexually in the vertebrate host's blood and produce gametocytes that are taken up by the biting vector, where they undergo the sexual phase of their life cycle. More rapid asexual replication results in larger numbers of transmissible gametocytes in the blood (Mackinnon and Read, 1999a; Eisen and Schall, 2000). Although higher numbers of gametocytes would seem to favour more efficient transmission, data from experiments on experimental transmission of some malaria parasites most often fail to confirm this relationship (reviewed in Schall, 2000) – that is, infections with a high density of gametocytes in the vertebrate host's blood are not necessarily those with highest transmission success into the vector.

Hypotheses on the Evolution of Virulence

Transmission-opportunity hypothesis

This hypothesis proposes that the ecology of transmission is the central factor driving the evolution of virulence (Gill and Mock, 1985; Ewald, 1994). When the parasite has many opportunities to move from host to host, selection will tilt towards rapidly reproducing parasite genotypes. In contrast, when transmission opportunities are rare, a very rapidly reproducing parasite could kill its host before transmission is possible. The most extreme case would be parasites with seasonal transmission or other periods of 'impossible transmission', when the infected host leaves the area where vectors are present (Gill and Mock, 1985). This parasite should reduce its reproduction and virulence, perhaps rebounding when cues indicate that the transmission period is approaching.

For vector-borne parasites, N_{to} will often be high when vectors are abundant in the environment, but vector abundance may not always be a good predictor of selection for high parasitaemia. Variation in vector competence among sites (such as differences in host-seeking behaviour or the physiological environment presented to the parasite) could obscure any pattern between vector abundance and opportunities for transmission. Also, if hosts behaviourally avoid sites where transmission is likely or if they flick off alighting vectors (Hart, 1994), N_{to} could drop and select for low parasitaemia and consequently low virulence. Thus, host behaviour can apply selective pressure on the parasite to reduce virulence.

Overall, in environments or times with low N_{to}, infections of prudent, slow-growing genotypes would always be prone to invasion by mutation to rapidly reproducing genotypes of parasites, but such infections would fail to yield many (if any) daughter infections. There is thus a trade-off between P_t and D, which is driven by N_{to}.

Mobility hypothesis

Parasites that require their host to be mobile for successful transmission cannot cause disabling morbidity and still enjoy successful transmission. Transmission opportunities would be few for directly transmitted parasites that disable their host (Ewald, 1995), but this may not be the case for vector-borne parasites (indeed, vectors may prefer non-ambulatory hosts). Parasites should always be avirulent for vectors that must remain mobile to allow transmission (Ewald and Schubert, 1989). Parasites using non-living vectors, such as flowing water, should be among the most virulent pathogens (Ewald, 1988). Parasites can also increase N_{to} by developing long-lived, highly durable transmission stages, reducing the need for a mobile host. This is termed the 'curse of the pharaoh', from the folk-tales that ancient corpses may harbour viable and exotic pathogens (Bonhoeffer *et al.*, 1996).

Host-demography hypothesis

Ebert and Mangin (1997) suggest that selection favours a high reproductive rate (and thus high virulence) for parasites exploiting hosts with naturally short lifespans – that is, if D is naturally short (a short-lived host could support an infection only of short duration), selection would favour high parasite replication to increase the probability of transmission.

Immunity hypothesis

The nature of a host's antiparasite tactics and the ability of the parasite to respond should influence virulence. As noted above, behavioural defences that reduce the opportunities for transmission should select for decreased virulence. However, a potent immune attack once transmission is successful could produce an opposite selective force (van Baalen, 1998; Gandon and Michalakis, 2000). An effective immune response would limit the duration of the infection, and selection would favour more rapid replication of the parasite (low D favours high P_t, just as presented for the host-demography hypothesis) (Anita *et al.*, 1994).

Host-specificity hypothesis

Parasites that specialize on a single (or very few) host species should evolve highly specific ways of dealing with the host immune response and are thus more likely to exploit their hosts efficiently and produce severe pathology (Ewald, 1983). Frank and Jeffrey (2001) turn this view on its head. They propose that newly established parasite–host associations (the equivalent of a parasite with a very large range in host species) will be among the most virulent because of the lack of an efficient, specialized host defence.

Transmission-mode hypothesis

Parasites typically move from host to host in the environment (infectious or horizontal transmission). Some parasites, though, are transmitted via the host's offspring (congenital or vertical transmission). In such cases, the fitness of the host translates to fitness of the parasite, and the fitness 'desiderata lists' (Dawkins, 1990) of both species in the association coincide. Thus, vertically transmitted parasites should be less virulent than those using horizontal transmission (Messenger *et al.*, 1999). If the death of the host is required for transmission (one host must eat the other, for example), the desiderata lists diverge completely, and the parasite may actually manipulate its host to increase its chance of being killed (Poulin, 1998).

Small worlds – diminishing-returns hypothesis

Parasites with low dispersal could have a small world of potential hosts (Herre, 1993; Lipsitch *et al.*, 1995; Boots and Sasaki, 1999). Over time, there would be diminishing returns on rapid transmission, because the opportunity for new hosts would decline as a parasite genotype fully exploits available hosts. Thus, parasites with a small world of potential hosts must reduce their virulence.

Clonal-diversity hypothesis

For parasites that replicate within the host (malaria parasites and viruses are examples), the presence of multiple genotypes, or clones, may lead to competition for resources or simply to be the clone most likely to be transmitted. When infections typically consist of many clones, this would select for high parasite replication and higher virulence (van Baalen and Sabelis, 1995; Frank, 1996). Even if each clone is prudent and replicates slowly, the sum of the clone densities would result in higher virulence. High clonal diversity may also lead to some proto-cooperation by the parasites to elude the host immune system more efficiently, thus resulting in a higher rate of parasite replication and higher virulence.

A Successful General Theory of Virulence?

The review of hypotheses presented here must be incomplete (the literature on parasite virulence is large and growing rapidly), but at least presents a flavour of the discussion. Note that all of the reviewed hypotheses centre on how selection works on the parasite. Virulence from the host perspective has received too little attention (but see an exception below). Can these hypotheses be merged to produce a general theory of parasite virulence? A successful general theory in science must explain a broad array of observations and suggest numerous predictions for future testing. The review suggests that no simple selective process drives the evolution of virulence. We could easily produce numerous thought experiments to design parasites with high (or low) virulence that came to that state via very different evolutionary trajectories. Frank (1996) concludes that 'the models [on virulence] cannot be applied without careful consideration of the biology of particular host–parasite interactions'. Any general theory can take a very broad sweep, ignore the annoying complexity of natural history and still yield useful insights. The notion of 'desiderata lists' in Dawkins (1990), for example, shows that, when the fitness of the parasite depends on the fitness of the host, virulence should evaporate and the parasite may evolve towards a mutualisitic relationship. Such a perspective has heuristic value (that was Dawkins's stated objective), but may not be of much use for medical or experimental parasitologists. Ewald, in testing the mobility hypothesis (Ewald, 1983, 1988, 1994; Ewald and Schubert, 1989), used large among-species comparisons – another broad sweep through parasites with quite different biologies.

Untidy results that emerge from broad tests of the theory do not mean that the theory has failed, but point out which species are likely to be particularly interesting for future study. For example, Jaenike (1996) found that nematode parasites of *Drosophila* are less likely to fit a model of parasite virulence when they exploit multiple host species – and thus no 'optimal' virulence is possible for any particular species of host. Ebert and Mangin (1997) conducted a selection experiment with a micro-sporidian parasite of *Daphnia* to test the host-demography hypothesis. They found that the selective regime that should have favoured low virulence (longer lifespan of *Daphnia*) actually resulted in higher virulence, because multiple infections became more common in the longer-lived infections, resulting in within-cell competition among the parasites. Thus, the specific natural history of the parasite 'can lead to wrong predictions' (Ebert and Mangin, 1997). A last example is a selection experiment for high and low virulence of a malaria parasite of mice (Mackinnon and Read, 1999b). Over time, both the selected lines increased in virulence; again, the details of the natural history of this malaria parasite most probably confounded the expectations of simple models of virulence.

The theory becomes most interesting when we recognize that parasite and host coevolve, so we must consider how selection acts from the parasite and host perspective, but simultaneously. The model of van Baalen (1998) uses this tactic. The model takes into account the trade-off between the cost of mounting an immune response by the host and the benefit of eliminating/reducing the parasite, as well as the prevalence of the parasite and the parasite's ability to coevolve with the host. The results are intriguing. Coevolution of parasites and hosts can lead to two stable situations, one with avirulent, common parasites and low host investment in the immune response, and one with rare, but virulent para-sites and high host defence costs. These two genotypes of parasite could well exist in a mixed strategy, which would account for the presence of high- and low-virulence genotypes in metapopulations of *P. falciparum* (Gupta *et al.*, 1994), as well as other parasites.

A Case-study in Parasite Virulence: Lizard Malaria

Systematic and ecological diversity

The malaria parasites, genus *Plasmodium*, are taxonomically and ecologically diverse; ±170 described species exploit reptiles, birds and mammals as their vertebrate hosts (Schall, 1996). Of these, approximately 70 species of *Plasmodium* infect lizards on all the warm continents (except Europe) and are found in a wide range of habitats (wet tropical forest to dry, deciduous, temperate savannah). The great ecological and systematic diversity of lizard malaria parasites makes them a good model system for among-species tests of the theory of virulence. This has been one of my goals over the past 23 years – to compare the costs of infection for lizard–*Plasmodium* associations from distinct ecological situations.

Here I report a summary of the results for six *Plasmodium* species, and then examine the data to evaluate the hypotheses presented earlier.

Six malaria parasites of lizards

Plasmodium mexicanum infects the western fence lizard, *Sceloporus occidentalis*, in the western USA and Mexico (Ayala, 1970). Its life cycle is the only one known in detail for lizard malaria parasites, but is typical for *Plasmodium*. Repeated cycles of asexual replication occur in red blood cells. With each reproduction, the mother cell (schizont) releases about 14 daughter parasites (merozoites); the cell is destroyed and the merozoites enter new erythrocytes to begin a new reproductive cycle. After an initial period of asexual growth, some parasite cells develop into sex cells, or gametocytes, which cease replication in the blood. The vectors, two species of sandfly (*Lutzomyia vexator* and *Lutzomyia stewarti*), take up the parasite cells when they consume a blood meal from an infected lizard (Fialho and Schall, 1995). Only the gametocytes survive, and some undergo sexual reproduction followed by asexual replication to produce cells that travel to the insect's salivary glands. The parasites are passed back to a lizard during the next blood feeding by the vector. They then travel to the liver and other organs to undergo a cycle of asexual replication before entering the blood cells. Transmission success into the vector is only weakly related to the density of gametocytes in the lizard's blood (Fig. 14.2). Among lizard hosts of *P. mexicanum*, there is substantial variation in life-history traits, such as rate of asexual replication and final parasite levels (Eisen, 2000), with genetic variation explaining part of this variation in life histories among infections (Eisen and Schall, 2000).

My students and I have studied *P. mexicanum* at the Hopland Field Station in Mendocino County, California, since 1978. Although prevalence of *P. mexicanum* varies among years and sites (Schall and Marghoob, 1995), typically about 25% of lizards are infected. The environment at Hopland is strongly seasonal, with wet, cool winters and hot, dry summers. Transmission is thus seasonal. The lizards suffer substantial winter mortality during their inactive brumation period. The parasite density in the blood drops during the winter months to very low levels, but rebounds again the next spring (Bromwich and Schall, 1986; Eisen, 2000).

Two parasite species were studied in the rainbow lizard, *Agama agama*, in Sierra Leone, West Africa: *Plasmodium agamae* and *Plasmodium giganteum* (Schall, 1990; Schall and Bromwich, 1994). Both parasites were common in the lizards surveyed at 22 sites in several habitat types, including savannah, riparian forest and urban zones; typically, 25–75% of lizards were infected. The two species have strikingly different life histories. *P. agamae* is a small parasite, producing only eight merozoites during asexual replication in the blood, whereas *P. giganteum* is a true giant, filling the host cell and producing > 100

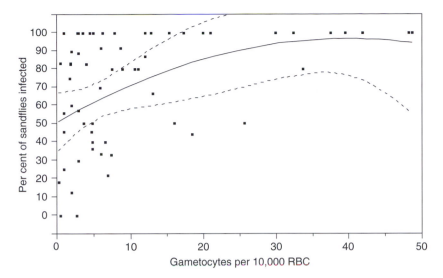

Fig. 14.2. Relationship between density of parasite gametocytes and transmission efficiency for *Plasmodium mexicanum*, a parasite of fence lizards in California, into its vector, the sandfly *Lutzomyia vexator*. A curve fitted to the data is shown, with 95% confidence intervals. Transmission efficiency increases with density of gametocytes, but appears to level off at about 20 gametocytes per 10,000 erythrocytes. Why do any infections increase above this threshold level when higher parasitaemia could increase the cost to the host? If multiple clones are present, each clone could rise to the threshold number of gametocytes. Thus, multiple clone infections may be more virulent. (Data from Schall, 2000.)

merozoites. There are more mixed infections (both parasites within a single lizard) than expected by chance. *P. giganteum* uses primarily immature red blood cells and *P. agamae* the mature cells. Immature cells are rare in uninfected lizards, but increase rapidly once the lizard becomes infected with either species of *Plasmodium*. Thus, *P. giganteum* may have difficulty in becoming established in a lizard unless *P. agamae* is already present (Schall and Bromwich, 1994). The use of a rare habitat (immature red blood cells) by *P. giganteum* may also keep its growth rate and parasitaemia to lower levels than expected, based on the large reproductive output of each giant schizont.

The habitat in Sierra Leone, although tropical, is also strongly seasonal because of pronounced wet and dry seasons. Transmission of the parasites may also be seasonal (Schall, 1990). Rainbow lizards are large, long-lived animals.

Three species of malaria parasite were studied in the Caribbean islands: *Plasmodium floridense*, *Plasmodium azurophilum* and an unidentified species, referred to here as *Plasmodium 'red'*. These three species infect *Anolis* lizards throughout the eastern Caribbean islands (Staats and Schall, 1996a,b; Perkins, 2001). *P. floridense* is found on the northern islands in the eastern Caribbean; it is a small parasite, producing very few merozoites. The name '*P. azurophilum*' was originally erected for a parasite infecting both erythrocytes and two classes of white blood cells of the anoles (Telford, 1975). Recently, this parasite has been revealed to be two species, each with a wide distribution in the eastern Caribbean islands (Perkins, 2000, 2001). I retain here the name *P. azurophilum* for the species that infects only white blood cells and use *P.* '*red*' for the species that infects erythrocytes. These two species are morphologically indistinguishable (Perkins, 2000) and are giant parasites, each schizont producing about 65 merozoites. Phylogenetic evidence indicates that *P. azurophilum* originated from *P.* '*red*' on St Kitts or a nearby island (Perkins, 2001).

The number of white cells in the blood limits the density of *P. azurophilum*. Although infection is associated with an increase in white blood cells (Ayala and Hertz, 1981; Schall, 1992), the number of the cell type needed by the parasite is never near the abundance of erythrocytes in the lizard's circulation. The derived life history of *P. azurophilum* limits its population density in the host and perhaps its ability to be transmitted by the biting vectors. The use of white blood cells by *P. azurophilum* must also lead to quite different pathology for the lizard host from that of the more typical *Plasmodium* infecting erythrocytes.

The vectors for the Caribbean lizard malaria species are unknown. Prevalence is about 30% for *P.* '*red*' and 10% for the other two species on Puerto Rico (Schall *et al.*, 2000), and on Saba, Netherlands Antilles, about half the lizards are infected by at least one species of *Plasmodium* (Staats and Schall, 1996a). No difference in prevalence was noted among seasons for either the Puerto Rico or Saba sites (where long-term studies have been under way for the past decade), so transmission may be year-round. However, periodic hurricanes and droughts strike the islands and could result in unpredictable periods of reduced transmission.

In summary, six species of lizard malaria parasite will be discussed here: *P. mexicanum* in temperate, seasonal California; *P. agamae* and *P. giganteum* in a seasonal, tropical region of West Africa and *P. floridense*, *P. azurophilum* and *P.* '*red*' on the tropical, aseasonal Caribbean islands, which are regularly disturbed by severe weather. A seventh species, *Plasmodium chiricahuae* of temperate and seasonal high elevations in Arizona, was studied by Foufopoulos (1999), so some comparative data are also presented for that species. *P. chiricahuae* is particularly interesting because it is the closest sister taxon to *P. mexicanum* (Perkins and Schall, 2002).

Measuring the virulence of malaria infections

Comparisons of malaria-infected vs. non-infected lizards have revealed many health consequences of infection, including effects on haematology, physiology, behaviour and reproduction. Figure 14.3 summarizes the data for the *P. mexicanum*–fence-lizard association in California. Simple

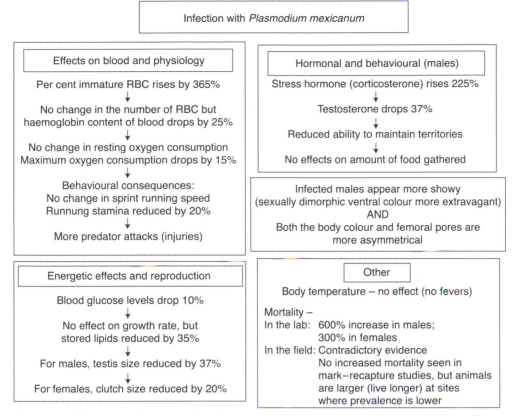

Fig. 14.3. Summary of costs suffered by western fence lizards (*Sceloporus occidentalis*) when infected with the malaria parasite *Plasmodium mexicanum*.

comparisons of infected and non-infected lizards may be misleading for two reasons. First, such comparisons may obscure actual cause-and-effect relationships, i.e. perhaps sickly lizards may simply be more prone to infection with parasites. Foufopoulos (1999) has manipulated infections of *P. chiricahuae* in mountain spiny lizards and observed changes in health status similar to some of those shown in Fig. 14.3, which argues that malaria infection is the cause of the pathologies to be described here. Secondly, some kinds of pathology may be dependent on parasitaemia, and density-dependent virulence would be missed by simple comparisons of infected vs. non-infected hosts. Our studies have found that some effects of parasitism are more severe with increasing parasitaemia (change in social behaviour appears related to parasitaemia (Schall and Dearing, 1987)), but other effects appear to be more or less independent of parasitaemia (effects on blood haemoglobin levels (Schall, 1982)). Some other measures are difficult to relate to parasitaemia – clutch size, for example, is determined by long-term events and parasitaemia is determined from a single blood smear (Schall, 1983).

The goal of the research was to take as many measures of pathology as possible, with the expectation that the total picture will reveal some measure of virulence from both the parasite and host perspectives.

Measuring host mortality

The most difficult data to obtain on virulence may be the most interesting: reduction in host lifespan. Several methods were used to detect any increase in mortality induced by *Plasmodium* infection in lizards. Naturally infected fence lizards suffer two to six times higher mortality in laboratory cages and, for all parasite species, infected animals are more prone to attack by predators (Schall, 1996). Prevalence (percentage infected) rises with the lizard's age for every system studied, but prevalence typically levels off or even drops for the oldest lizards (Fig. 14.4). Long-term mark–recapture studies of infected fence lizards (Bromwich and Schall, 1986; Eisen, 2000) and laboratory-held anoles and rainbow lizards show that infections are seldom eliminated. Thus, the dip in prevalence suggests that mortality increases for older lizards when they are infected. Among sites at Hopland and Saba, there is a negative relationship between maximum body size and the prevalence of malaria infection (Fig. 14.5). Thus, lizards tend to be smaller in areas where prevalence of the parasites is highest. Lizards typically grow throughout

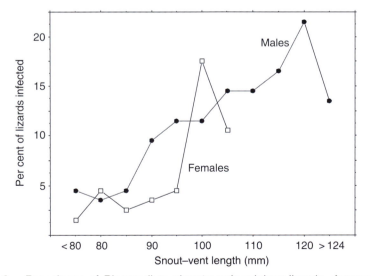

Fig. 14.4. Prevalence of *Plasmodium giganteum* in rainbow lizards, *Agama agama*, in Sierra Leone, West Africa. The pattern shown here is typical for lizard malaria parasites. Prevalence increases with size (= age), as expected if lizards seldom lose infection. Males are more often infected than females. Prevalence drops for the oldest animals, suggesting that the parasite increases mortality for these animals.

their life and reduce growth rate as they age. Two studies (*P. mexicanum* in fence lizards and *P. agamae* and *P. giganteum* in rainbow lizards) show no reduction in growth rate for infected lizards. Therefore, Fig. 14.5 suggests that infected lizards simply have an abbreviated lifespan and thus never grow to their largest possible size.

In summary, infected fence lizards suffer higher mortality in captivity, and infected lizards are more prone to attack by predators and other aggressive animals. There is a drop in prevalence for older lizards, and lizards are generally smaller (younger) at sites with a higher prevalence of malaria infection. These observation suggest that lizard malaria can cause an increase in host mortality. However, the most direct measure of mortality comes from mark–recapture studies on fence lizards in California, and these reveal no indication of an increase in mortality associated with infection (Bromwich and Schall, 1986; Eisen, 2001). That is, the duration for which a lizard was known to be alive did not differ for lizards infected or not infected with *P. mexicanum*. The second study (Eisen, 2001) was most striking because it followed marked lizards over several warm seasons and is the most detailed such study ever done for a malaria parasite of non-humans. No similar studies have been done for the Caribbean or African systems, but infected anoles and rainbow lizards

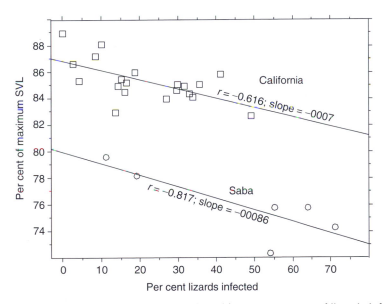

Fig. 14.5. Body size (SVL, snout-to-vent length) vs. percentage of lizards infected with malaria parasites among sites for two study locations, California (fence lizards, *Sceloporus occidentalis*, infected with *Plasmodium mexicanum*) and Saba island (the endemic anole, *Anolis sabanus*, infected with *P. floridense*, *P. azurophilum* and *P. 'red'*). For each location, lizards grow larger at sites where malaria is less common. Malaria does not reduce body growth, so the two relationships suggest that lizards on the average live longer where they are less likely to be infected with the parasites.

brought into the laboratory suffer no increase in mortality compared with non-infected animals.

Measuring reduction in host reproduction

Malaria infection can disrupt the reproduction of both male and female lizards. The data are most complete for fence lizards infected with *P. mexicanum*; these data are highlighted here, but some comparative data for other species are also presented. Figure 14.6 shows results for fence lizards and rainbow lizards and reveals a significant reduction in fecundity for infected females amounting to one to two eggs for fence lizards infected with *P. mexicanum* and about four to five for rainbow lizards infected with both *P. agamae* and *P. giganteum*. This averages out to approximately 20% and 60% reduction in fecundity, respectively. No measure of an effect on female reproductive output for *Anolis* was possible. Anoles produce one egg per clutch, so only longitudinal studies of individual lizards will reveal any reduction in the number of reproductive periods per lifetime.

The origin of this substantial reduction in fitness may derive from the ability of the lizard to assimilate and store resources. Fence lizards store

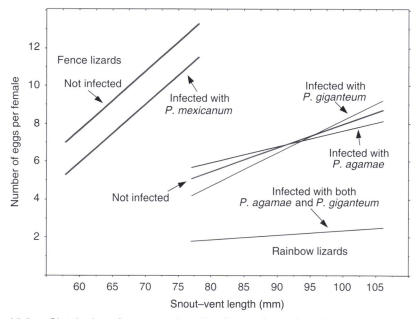

Fig. 14.6. Clutch size of eggs produced by female fence lizards and rainbow lizards, comparing females infected with a malaria parasite with those not infected. Regression lines are fitted to the data. Clutch size increases with body size for both lizard species. Clutch size is significantly reduced for fence lizards infected with *Plasmodium mexicanum*. For rainbow lizards, clutch size is reduced for females infected with both *P. agamae* and *P. giganteum*.

fat during the end of the warm season and then recycle this source of energy into eggs the next year. Infected fence lizards store less fat and the energetic loss is equal to one to two eggs, the deficit in observed fitness (Schall, 1982, 1983). Similar data are available only for the tropical rainbow lizard, which stores very little fat, so no effect can be detected for malaria infection (Schall, 1990). The loss in fat stored by fence lizards is not a result of infected lizards having reduced foraging success. The mass of faeces produced within 24 h after capture, an indication of foraging success, is not lower for infected fence lizards (Eisen and Schall, 1997). A similar result emerges for the anole of Saba island infected with all three Caribbean *Plasmodium* species (Schall and Staats, 2002). Thus, the cost to fitness of malaria infection for fence lizards appears to result from the parasite consuming resources normally used to produce offspring.

Infected male fence lizards also suffer a fitness cost. They are less active socially (Schall and Sarni, 1987), are less able to hold a territory against non-infected conspecific males (Schall and Houle, 1992) and fare poorly in male–male interactions (Schall and Dearing, 1987). Infected males also produce less testosterone and higher levels of corticosterone, a 'stress' hormone (Dunlap and Schall, 1995), and have smaller testes (Schall, 1983). The sexually dimorphic ventral colour of infected male lizards is altered, which may allow females to determine infection status (Ressel and Schall, 1989). Male Saban *Anolis* appear to be less harmed by malaria infection, because there is no effect on male–male interactions or on body colour (Schall and Staats, 2002).

Other consequences of infection for lizard hosts

Infection with *Plasmodium* initiates a cascade of effects on lizard hosts, beginning with haematological changes, which drive physiological and behavioural alterations. The data are most complete for fence lizards and *P. mexicanum*, so these results are presented first (reviewed in detail in Schall, 1996). When red blood cells are consumed by the parasite, the number of immature erythrocytes increases in the blood. These cells house less haemoglobin than mature cells, so blood haemoglobin levels drop, sometimes by as much as 20%. As a consequence, the ability of the blood to deliver oxygen to tissues declines (maximal oxygen consumption is reduced). This translates into effects on locomotive performance. Sprint running in lizards is funded by anaerobic respiration and is not affected by infection, but aerobically sustained stamina running (measured as the distance the lizard can run) is reduced.

Immature red blood cells increase in the circulation and blood haemoglobin declines for infections of *P. floridense* and *P. 'red'* in Caribbean anoles (Schall, 1992; Schall and Staats, 2002) and for infections of *P. agamae* and *P. giganteum* in African rainbow lizards, and running stamina is reduced for malarious rainbow lizards (Schall, 1990). Thus, similar physiological and behavioural changes are apparent for these parasite–host systems. Blood haemoglobin levels do not drop for infections of *P. azurophilum*, as might be expected, because this parasite infects only white blood cells. However, white blood cells infected with *P. azurophilum* produce less acid phosphatase, an important enzyme in the functioning of these immune-system cells (Schall, 1992). This suggests that anoles infected with *P. azurophilum* may have reduced resistance to infection with other parasites if their immune system is compromised. Unfortunately, no data on this issue are available (indeed, no data on the impact of malaria infection on resistance to other parasites exist for any lizard host).

Comparisons among *Plasmodium*–lizard systems

Clearly, malaria parasites can harm their lizard hosts (Fig. 14.3). The consequences of infection are broad, including changes in hormone levels, haematology, physiology, running stamina, social and courtship behaviour, colour, fecundity and perhaps survival. But the consequences of infection differ among parasite and host species. *P. mexicanum* is particularly virulent for fence lizards, but the Caribbean lizard malaria parasites seem rather benign overall. For example, in a survey of the consequences of infection for the Saba island lizard, *Anolis sabanus*, few indications of harm were noted (Schall and Staats, 2002), and infected *Anolis gundlachi* on Puerto Rico do not suffer a reduction in body condition (mass vs. length) (Schall and Pearson, 2000). Thus, the species discussed here, with some additional data on *P. chiricahuae*, can be used to compare levels of virulence for malaria parasites and to test the hypotheses presented above on the evolution of virulence.

Table 14.1 compares the various measures of virulence for the six species of lizard malaria parasite. *P. mexicanum* is clearly the most virulent species, affecting every aspect of the physiology, behaviour and reproduction of infected fence lizards. Data on the consequences for mortality, however, are equivocal. *P. agamae* and *P. giganteum* also appear virulent for rainbow lizards, although the most severe harm depends on mixed infection of the two parasite species. The three Caribbean species also have effects on their hosts' haematology, reducing haemoglobin (*P. floridense* and *P. 'red'*) or altering the physiology of white blood cells (*P. azurophilum*), but studies on the costs of infection for the *Anolis* hosts detected no other consequences of infection. Thus, these three species may almost be benign.

Cross-species comparisons can be confounded if the phylogenetic relationships among those species are not understood (Harvey and Pagel, 1991). Two species of *Plasmodium* could have similar effects on their hosts simply because they are close sister taxa: that is, virulence may be a conservative trait, not subject to rapid alteration by differences in ecological conditions. To eliminate this potential source of error, a portion of the overall phylogeny for malaria parasites of Perkins and Schall (2002) is

Table 14.1. Inventory of known costs to lizard hosts of infection by six species of malaria parasite (*Plasmodium* species). The various consequences of infection and the parasite–host systems are described in the text. 'Age × prevalence' is the relationship between body size (an indication of lizard age) and percentage of lizards infected at a site.

	P. mexicanum	*P. agamae*	*P. giganteum*	*P. floridense*	*P. azurophilum*	*P. 'red'*
Lizard host	*Sceloporus*	*Agama*	*Agama*	*Anolis*	*Anolis*	*Anolis*
Parasitaemia per 10,000 RBC						
Modal	< 50–500	85	840	87		22
High	2500	119	2317	1130		2180
Mortality						
In laboratory	+300–600%	Nil	Nil	Nil	Nil	Nil
Injuries	+10–18%	No effect	No effect	No effect	No effect	No effect
Age × prevalence	Negative			Negative	Negative	Negative
Survival in field	No effect					
Behaviour						
Foraging success	No effect			No effect	No effect	No effect
Male–male status	Reduced			No effect	No effect	No effect
Percentage time social	Reduced					
Body colour						
Showy trait	Altered			No effect	No effect	No effect
Symmetry	Reduced			No effect	No effect	No effect
Haematology						
% Immature RBC	+365%	+750%	+908%	+142%	No effect	+148–268%
RBC density	No effect	No effect	No effect			
Haemoglobin in blood	−25%	−22%		No effect	No effect	No effect
Acid phosphatase in WBC				No effect	−45%	No effect
Number WBC				No effect	Increased	No effect
Physiology						
Blood glucose	−10%					
Blood testosterone	−36%					
Blood corticosterone	+225%					
Resting oxygen use	No effect	No effect	No effect			
Maximal oxygen use	−38%	−11%	−27%			
Spring speed	No effect	No effect	No effect			
Stamina	−20%	−15%				
Body temperature	No effect	No effect	No effect	No effect	No effect	No effect
Body condition				No effect	No effect	No effect
Reproduction						
Clutch size	−20%	No effect	No effect			
		Mixed infection −60 to −75%				
Testis mass	−37%	No effect	No effect			
Fat stored	−20 to 45%	No effect	No effect			

RBC, red blood cells; WBC, white blood cells.

presented in Fig. 14.7, with the species of interest indicated. Examination of this tree reveals no pattern for the virulent vs. avirulent lizard malaria species. The most virulent, *P. mexicanum*, is a sister species of *P. chiricahuae*. Recent studies by Foufopoulos (1999) on mountain spiny lizards (*Sceloporus jarrovi*) infected with *P. chiricahuae* in Arizona reveals no effects of infection on male spiny lizards, but a reduction in body condition and clutch size of females. *P. floridense*, an avirulent species, is most closely related to the *P. chiricahuae* + *P. mexicanum* pair. *P. agamae* and *P. giganteum* are sister taxa, as are *P. 'red'* and *P. azurophilum*. These two pairs of species are more closely related than each is to *P. floridense* or *P. mexicanum*, and yet *P. 'red'* and *P. azurophilum* are low-virulent species and *P. agamae* and *P. giganteum* are more virulent. Overall, the virulent parasites do not cluster, and neither do the avirulent species.

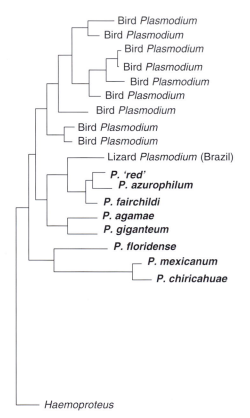

Fig. 14.7. Phylogeny for lizard and bird malaria parasites (extracted from a larger phylogeny recovered from cytochrome b sequences for 52 parasite taxa) (Perkins and Schall, 2002)). The outgroup for this tree are species of *Haemoproteus*, another group of malaria parasites, from lizards and birds. Species of *Plasmodium* isolated from bird hosts are indicated as 'Bird *Plasmodium*' on the tree. Other taxa on the tree are *Plasmodium* isolated from lizards. Species discussed in this review are indicated in bold.

The lack of any phylogenetic influence on virulence suggests that ecological reasons underlie the variation in virulence of the lizard malaria parasites. Therefore, I now revisit the hypotheses presented previously to determine if the present ecology of the lizard malaria parasites differs in ways that would predict variation in virulence.

Evaluation of hypotheses

Host perspective
No data are available on the cost to lizards of mounting an immune response to malaria infection. This is not unusual, as data on this issue are rare (Gemmill and Read, 1998). I have followed both captive and free-ranging animals infected with the *Plasmodium* species discussed here, as well as lizards infected with various haemogregarines, microfilarial worms and a trypanosome, and have rarely noted elimination of the parasite. The only exception was a virus that forms huge assembly pools within erythrocytes of *A. agama*. These infections were acute and were eliminated within a few weeks. This anecdotal evidence suggests that lizards do not mount a particularly effective immune response to protozoan blood parasites. This is surprising, because there is a substantial fitness cost (reduction in clutch size) associated with infection by *P. mexicanum* and mixed infections of the two African species.

Transmission-opportunity hypothesis
Comparisons among the lizard malaria systems do not give support for this hypothesis. *P. mexicanum* in fence lizards appears to be the most virulent of the seven species surveyed here and yet transmission is strongly seasonal. Overall prevalence of *P. mexicanum* at the California site is somewhat lower than for the tropical systems, suggesting that transmission is less intense. For the two tropical systems, vector biting activity may be more uniform in the Caribbean, where there is no clear dry season with reduced transmission. However, the three Caribbean lizard malaria species appear to be least harmful to their host.

Mobility hypothesis and transmission-mode hypothesis
These hypotheses should be irrelevant for comparisons among lizard malaria, as all are transmitted by blood-feeding Diptera.

Host-demography hypothesis
Mark–recapture programmes at all sites reveal that anoles, fence lizards and rainbow lizards can all live for several years. However, rainbow lizards are much larger animals and would be expected to have longer lifespans, and fence lizards suffer heavy winter mortality in California. This suggests that the virulence of malaria in fence lizards should be the

most severe and in rainbow lizards the least harmful. This is not the trend observed.

Immunity hypothesis

As noted above, lizards may mount a weak immune response to blood parasites. This should select for avirulent strains of parasite. *P. mexicanum* is the most virulent of the parasites studied, and there is no evidence that the immune response of fence lizards to *P. mexicanum* is stronger than that produced by the other lizards.

Host-specificity hypothesis

Surveys of common lizard species at each of the study sites revealed that each of the parasite species is highly host-specific. For example, at the Puerto Rico study area, five species of *Anolis* are common and yet only one, *A. gundlachi*, is a host for the three Caribbean lizard malaria parasites (Schall and Vogt, 1993). Thus, there is no variation in degree of host specificity for any of the parasites, so this hypothesis cannot explain the observed variation in virulence.

Small worlds – diminishing-returns hypothesis

Although data on vector behaviour for any lizard malaria parasite are scant (indeed, the vector is known with any certainty only for *P. mexicanum*), some intriguing data suggest that dispersal of the parasite is limited. First, high-prevalence vs. low-prevalence sites can be very local at the California and Saba study sites (Schall and Marghoob, 1995; Staats and Schall, 1996b). Sites only a few hundred metres apart can differ substantially in the proportion of lizards infected there, and this pattern can remain for many years. Secondly, 'hot spots' for malaria prevalence at the California site can be very small and local – just 100 m^2 (Eisen and Wright, 2001). If this indicates that dispersal of the parasite is low, this would suggest that selection should favour reduction in virulence. Again, the data do not support this prediction, because *P. mexicanum* in California is the most virulent parasite and the three species on Saba are the least harmful to their host.

Clonal-diversity hypothesis

No direct measures of clonal diversity of lizard malaria infections are available. Highly variable surface proteins known for *P. falciparum* and *Plasmodium vivax*, or variable microsatellite loci for *P. falciparum* allow such measures for human malaria, but similar variable loci have not been identified for any lizard malaria species. However, an indirect measure of clonal diversity is possible by determining the sex ratio of gametocytes in infections (Read *et al.*, 1992; Schall, 2000). Sex-ratio theory predicts that low clonal diversity, and thus low inbreeding of gametes within the vector, will select for a strongly female-biased gametocyte sex ratio. In contrast, high clonal diversity will lead to a 50 : 50 sex ratio. Data on gametocyte sex ratio are available for only three of the lizard malaria parasites: *P. mexicanum*, *P. agamae* and *P. giganteum*. Of these, *P. mexicanum* has the lowest proportion of female gametocytes, and the proportion is much lower than those for human or bird malaria parasites that have been described (Read *et al.*, 1992). For *P. mexicanum*, the sex ratio is correlated with two potential measures of virulence: infection growth rate and final parasitaemia (Schall, 2000). Perhaps the transmission biology of *P. mexicanum* leads to high genetic diversity within infections and competition among those clones for resources and transmission, and hence the high virulence observed for this parasite. Other studies suggest that clonal diversity drives the virulence of infections within a species of malaria parasite (Taylor *et al.*, 1998; Pickering *et al.*, 2000), so variation in the way infections are established (leading to high vs. low clonal diversity) could drive differences in virulence among species.

Prospects

The good news is that the venerable 'association-time' hypothesis, which dominated parasitology for generations, has now been supplanted by a growing literature of sophisticated verbal and mathematical models on the evolution of virulence. The less satisfying news is that tests of the theories are relatively rare and, more often than not, cast doubt on the most discussed of the models. None the less, manipulative and comparative studies reveal that tests of the theory on parasite virulence provide some of the most intriguing findings in all of ecology and evolutionary biology (Ewald, 1983; Bull and Molineaux, 1991; Herre, 1993; Jaenike, 1996; Ebert and Mangin, 1997; Taylor *et al.*, 1998; Mackinnon and Read, 1999b; Messenger *et al.*, 1999). Elegant laboratory systems that allow careful manipulation of relevant factors and yet retain a close resemblance to natural parasite–host associations are particularly productive and desirable (Ebert and Mangin, 1997; Mackinnon and Read, 1999b; Messenger *et al.*, 1999). Also needed are careful tests of the assumptions that underlie each of the hypotheses, such as the relationship between transmission efficiency and cost to the host (Mackinnon and Read, 1999a).

This review ends with a plea. We need more data. Data on the actual costs of parasitism – costs for natural parasite–host systems – are notoriously scant (Dobson and Hudson, 1995). Obtaining such data is laborious and time-consuming and not particularly helpful to those wishing to increase their academic fitness. The relevant measures of virulence may not be obvious and they certainly differ among species (insects vs. vertebrates). Most of our data on parasite virulence come from human medicine (the best-known comparative studies centre on human pathogens (for example, Ewald, 1983, 1988, 1994)), simply because those are the available data. Our perspective on parasite virulence may well be biased by this scarcity of comparative data. Thus, we need not just more data on costs of infection, but data on a greater variety of taxa, both of parasites and hosts. Some taxa are simply more charismatic and likely to draw the attention of researchers (*Plasmodium* among parasites and especially lizards among hosts!), but a broader view will certainly lead to exciting and unexpected findings on the evolution of parasite virulence.

Acknowledgements

I thank those who shared their views and offered lively debate on parasite virulence. Their insights shaped this review: Lori Stevens, Chris Staats, Becky Eisen, Paul Ewald, Andy Read, Susan Perkins, Rob Fialho, Sarah Osgood, Mike Sukhdeo and Doug Gill. The research on lizard malaria was funded by numerous grants from the National Science Foundation (NSF), the National Institutes of Health (NIH), the National Geographic Society and the University of Vermont.

References

Anita, R., Levin, B.R. and May, R.M. (1994) Within-host population dynamics and the evolution and maintenance of microparasite virulence. *American Naturalist* 144, 457–472.

Arnold, E.N. (1986) Mite pockets of lizards, a possible means of reducing damage by ectoparasites. *Biological Journal of the Linnean Society* 29, 1–21.

Arnot, D. (1998) Clone multiplicity of *Plasmodium falciparum* infections in individuals exposed to variable levels of disease transmission. *Transactions of the Royal Society of Tropical Medicine and Hygiene* 92, 580–585.

Ayala, S.C. (1970) Lizard malaria in California; description of a strain of *Plasmodium mexicanum*, and biogeography of lizard malaria in North America. *Journal of Parasitology* 56, 417–425.

Ayala, S.C. and Hertz, P.E. (1981) Malaria infection in *Anolis* lizards on Martinique, Lesser Antilles. *Revista do Instituto de Medicine Tropicale São Paulo* 23, 12–17.

Ball, G.H. (1943) Parasitism and evolution. *American Naturalist* 77, 345–364.

Bell, G. (1997) *Selection: the Mechanism of Evolution.* Chapman & Hall, New York, 699 pp.

Bonhoeffer, S., Lenski, R. and Ebert, D. (1996) The curse of the pharaoh: the evolution of virulence in pathogens with long living propagules. *Proceedings of the Royal Society of London B* 263, 715–721.

Boots, M. and Sasaki, A. (1999) 'Small worlds' and the evolution of virulence: infection occurs locally and at a distance. *Proceedings of the Royal Society of London B* 266, 1933–1938.

Boudoin, M. (1975) Host castration as a parasite strategy. *Evolution* 29, 335–352.

Bromwich, C.R. and Schall, J.J. (1986) Infection dynamics of *Plasmodium mexicanum*, a malarial parasite of lizards. *Ecology* 67, 1227–1235.

Bull, J.J. (1995) Perspective: virulence. *Evolution* 48, 1423–1437.

Bull, J.J. and Molineaux, I.J. (1991) Selection of benevolence in a parasite–host system. *Evolution* 45, 875–882.

Burnet, F.M. and White, D.O. (1972) *Natural History of Infectious Disease*, 4th edn. Cambridge University Press, Cambridge, UK, 278 pp.

Dawkins, R. (1990) Parasites, desiderata lists and the paradox of the organism. *Parasitology* 27, S63–S73.

Dobson, A.P. and Hudson, P.J. (1995) Microparasites: observed patterns in wild animal populations. In: Grenfell, B.T. and Dobson, A.P. (eds) *Ecology of Infectious Diseases in Natural Populations*. Cambridge University Press, Cambridge, UK, pp. 52–89.

Dunlap, K.D. and Schall, J.J. (1995) Hormonal alterations and reproductive inhibition in male fence lizards (*Sceloporus occidentalis*) infected with the malarial parasite *Plasmodium mexicanum*. *Physiological Zoology* 68, 608–621.

Ebert, D. (1999) The evolution and expression of parasite virulence. In: Stearns, S. (ed.) *Evolution in Health and Disease*. Oxford University Press, Oxford, pp. 161–172.

Ebert, D. and Mangin, K.L. (1997) The influence of host demography on the evolution of virulence of a microsporidian gut parasite. *Evolution* 51, 1828–1837.

Eisen, R.J. (2000) Variation in life-history traits of *Plasmodium mexicanum*, a malaria parasite infecting western fence lizards: a longitudinal study. *Canadian Journal of Zoology* 78, 1230–1237.

Eisen, R.J. (2001) Absence of measurable malaria-induced mortality in western fence lizards (*Sceloporus occidentalis*) in nature: a four year study of annual and overwintering mortality. *Oecologia* 127, 586–589.

Eisen, R.J. and Schall, J.J. (1997) Comparing foraging success in submissive malaria-infected and territorial noninfected fence lizards (*Sceloporus occidentalis*). *Journal of Herpetology* 31, 147–149.

Eisen, R.J. and Schall, J.J. (2000) Life history of a malaria parasite (*Plasmodium mexicanum*): assessment of independent traits and origin of variation. *Proceedings of the Royal Society of London B* 267, 793–799.

Eisen, R.J. and Wright, N.M. (2001) Landscape features associated with infection by a malaria parasite (*Plasmodium mexicanum*) and the importance of multiple scale studies. *Parasitology* 122, 507–513.

Ewald, P.W. (1983) Host–parasite relations, vectors, and the evolution of disease severity. *Annual Reviews of Ecology and Systematics* 14, 465–485.

Ewald, P.W. (1988) Cultural vectors, virulence, and the emergence of evolutionary epidemiology. *Oxford Surveys in Evolutionary Biology* 5, 215–245.

Ewald, P.W. (1994) *Evolution of Infectious Disease*. Oxford University Press, New York, 298 pp.

Ewald, P.W. (1995) The evolution of virulence: a unifying link between parasitology and ecology. *Journal of Parasitology* 81, 659–669.

Ewald, P.W. and Schubert, J. (1989) Vertical and vector-borne transmission of insect endocytobionts and the evolution of benignity. In: Schwemmler, W. and Gassner, G. (eds) *Insect Endocytobiosis: Morphology, Physiology, Genetics, Evolution*. CRC Press, Boca Raton, Florida, pp. 21–35.

Fialho, R.F. and Schall, J.J. (1995) Thermal ecology of a malarial parasite and its insect vector: consequences for the parasite's transmission success. *Journal of Animal Ecology* 64, 553–562.

Foufopoulos, J. (1999) Host–parasite interactions in the mountain spiny lizard *Sceloporus jarrovi* (Trombiculid mites, *Plasmodium chiricahuae*). PhD dissertation, University of Wisconsin, Madison.

Frank, S.A. (1996) Models of parasite virulence. *Quarterly Review of Biology* 71, 37–78.

Frank, S.A. and Jeffrey, J.S. (2001) The probability of severe disease in zoonotic and commensal infections. *Proceedings of the Royal Society of London B* 268, 53–60.

Gandon, S. and Michalakis, Y. (2000) Evolution of parasite virulence against qualitative or quantitative host resistance. *Proceedings of the Royal Society of London B* 267, 985–990.

Gemmill, A.W. and Read, A.F. (1998) Counting the costs of disease resistance. *Trends in Ecology and Evolution* 13, 8–9.

Gill, D.E. and Mock, B.A. (1985) Ecological and evolutionary dynamics of parasites: the case of *Trypanosoma diemyctyli* in the red spotted newt *Notophthalmus viridescens*. In: Rollinson, D. and Anderson, R.M. (eds) *Ecology and Genetics of Host–Parasite Interaction*. Academic Press, London, pp. 157–183.

Groisman, E.A. and Ochman, H. (1994) How to become a pathogen. *Trends in Microbiology* 2, 289–294.

Gupta, S., Hill, A.V.S., Kwiatkowski, D., Greenwood, A.M., Greenwood, B.M. and Day, K.P. (1994) Parasite virulence and disease patterns in *Plasmodium falciparum* malaria. *Proceedings of the National Academy of Sciences USA* 91, 3715–3719.

Hart, B.L. (1994) Behavioural defense against parasites: interaction with parasite invasiveness. *Parasitology* 109, S139–S151.

Harvey, P.H. and Pagel, M.D. (1991) *The Comparative Method in Evolutionary Biology*. Oxford University Press, Oxford, 239 pp.

Herre, E.A. (1993) Population structure and the evolution of virulence in nematode parasites of fig wasps. *Science* 259, 1442–1445.

Holmes, J.C. (1996) Parasites as threats to biodiversity in shrinking ecosystems. *Biodiversity and Conservation* 5, 975–983.

Hudson, P.J., Dobson, A.P. and Newborn, D. (1998) Prevention of population cycles by parasite removal. *Science* 282, 2256–2258.

Jaenike, J. (1996) Suboptimal virulence of an insect-parasitic nematode. *Evolution* 50, 2241–2247.

Kaplan, C. (1985) Rabies: a world-wide disease. In: Bacon, P.J. (ed.) *Population Dynamics of Rabies in Wildlife*. Academic Press, London, pp. 1–21.

Lipsitch, M. and Moxon, E.R. (1997) Virulence and transmissibility of pathogens: what is the relationship? *Trends in Microbiology* 5, 31–36.

Lipsitch, M., Herre, E.A. and Nowak, M.A. (1995) Host population structure and the evolution of virulence: a 'law of diminishing returns.' *Evolution* 49, 743–748.

Mackinnon, M.J. and Read, E.R. (1999a) Genetic relationships between parasite virulence and transmission in the rodent malaria *Plasmodium chabaudi*. *Evolution* 53, 689–703.

Mackinnon, M.J. and Read, A.F. (1999b) Selection for high and low virulence in the malaria parasite *Plasmodium chabaudi*. *Proceedings of the Royal Society of London B* 266, 741–748.

May, R.M. and Anderson, R.M. (1983) Parasite–host coevolution. In: Futuyma, D.J. and Slatkin, M. (eds) *Coevolution*. Sinauer, Sunderland, Massachusetts, pp. 186–206.

Mehlotra, R.K. (1998) Differentiation of pathogenic and nonpathogenic Entamoeba: has the question been answered? *Indian Journal of Gastro-enterology* 17, 58–60.

Mellors, J.W., Rinaldo, C.R., Gupta, P., White, R.M., Todd, J.A. and Kingsley, L.A. (1996) Prognosis in HIV-1 infection predicted by the quantity of virus in plasma. *Science* 272, 1167–1170.

Messenger, S.L., Molineux, I.J. and Bull, J.J. (1999) Virulence evolution in a virus that obeys a trade-off. *Proceedings of the Royal Society of London B* 266, 397–404.

Perkins, S.L. (2000) Species concepts and malaria parasites: detecting a cryptic species of *Plasmodium*. *Proceedings of the Royal Society of London B* 267, 2345–2350.

Perkins, S.L. (2001) Phylogeography of Caribbean lizard malaria: tracing the history of vector-borne parasites. *Journal of Evolutionary Biology* 14, 34–45.

Perkins, S.L. and Schall, J.J. (2002) A molecular phylogeny of malaria parasites recovered from cytochrome b gene sequences. *Journal of Parasitology* (in press).

Pickering, J., Read, A.F., Guerrero, S. and West, S.A. (2000) Sex ratio and virulence in two species of lizard malaria parasite. *Evolutionary Ecology Research* 2, 171–184.

Poulin, R. (1998) *Evolutionary Ecology of Parasites*. Chapman & Hall, London, 212 pp.

Poulin, R. and Combes, C. (1999) The concept of virulence: interpretations and implications. *Parasitology Today* 15, 474–475.

Read, A.F. (1994) The evolution of virulence. *Trends in Microbiology* 2, 73–76.

Read, A.F., Narara, A., Nee, S., Keymer, A.E. and Day, K. (1992) Gametocyte sex ratios as indirect measures of outcrossing rates in malaria. *Parasitology* 104, 387–395.

Ressel, S. and Schall, J.J. (1989) Parasites and showy males: malarial infection and color variation in fence lizards. *Oecologia* 78, 158–164.

Salvador, A., Veiga, J.P. and Givantos, E. (1999) Do skin pockets of lizards reduce the deleterious effects of ectoparasites? An experimental study with *Psammodromus algirus*. *Herpetologica* 55, 1–7.

Schall, J.J. (1982) Lizard malaria: parasite–host ecology. In: Huey, R.B., Schoener, T.W. and Pianka, E.R. (eds) *Lizard Ecology: Studies on a Model Organism*. Harvard University Press, Cambridge, Massachusetts, pp. 84–100.

Schall, J.J. (1983) Lizard malaria: cost to vertebrate host's reproductive success. *Parasitology* 87, 1–6.

Schall, J.J. (1990) Virulence of lizard malaria: the evolutionary ecology of an ancient parasite-host association. *Parasitology* 100, S35-S52.

Schall, J.J. (1992) Parasite-mediated competition in *Anolis* lizards. *Oecologia* 92, 58–64.

Schall, J.J. (1996) Malarial parasites of lizards: diversity and ecology. *Advances in Parasitology* 37, 255–333.

Schall, J.J. (2000) Transmission success of the malaria parasite *Plasmodium mexicanum* into its vector: role of gametocyte density and sex ratio. *Parasitology* 121, 575–580.

Schall, J.J. and Bromwich, C.R. (1994) Interspecific interactions tested: two species of malarial parasite in a west African lizard. *Oecologia* 97, 326–332.

Schall, J.J. and Dearing, M.D. (1987) Malarial parasitism and male competition for mates in the western fence lizard, *Sceloporus occidentalis*. *Oecologia* 73, 389–392.

Schall, J.J. and Houle, P.R. (1992) Malarial parasitism and home range and social status of male western fence lizards, *Sceloporus occidentalis*. *Journal of Herpetology* 26, 74–76.

Schall, J.J. and Marghoob, A.B. (1995) Prevalence of a malarial parasite over time and space: *Plasmodium mexicanum* in its vertebrate host, the western fence lizard, *Sceloporus occidentalis*. *Journal of Animal Ecology* 64, 177–185.

Schall, J.J. and Pearson, A.R. (2000) Body condition of a Puerto Rican anole, *Anolis gundlachi*: effect of a malaria parasite and weather variation. *Journal of Herpetology* 34, 489–491.

Schall, J.J. and Sarni, G.A. (1987) Malarial parasitism and the behavior of the lizard, *Sceloporus occidentalis*. *Copeia* 1987, 84–93.

Schall, J.J. and Staats, C.M. (2002) The virulence of lizard malaria: three species of *Plasmodium* infecting *Anolis sabanus*, the endemic Caribbean anole of Saba, Netherlands Antilles. *Copeia* 2002, 39–43.

Schall, J.J. and Vogt, S. (1993) Distribution of malaria in *Anolis* lizards of the Luquillo Forest, Puerto Rico: implications for host community ecology. *Biotropica* 25, 229–235.

Schall, J.J., Pearson, A.R. and Perkins, S.L. (2000) Prevalence of malaria parasites (*Plasmodium floridense* and *Plasmodium azurophilum*) infecting a Puerto Rican lizard (*Anolis gundlachi*): a nine year study. *Journal of Parasitology* 86, 511–515.

Staats, C.M. and Schall, J.M. (1996a) Distribution and abundance of two malarial parasites of the endemic *Anolis* lizard of Saba island, Netherlands Antilles. *Journal of Parasitology* 82, 409–413.

Staats, C.M. and Schall, J.J. (1996b) Malarial parasites (*Plasmodium*) of *Anolis* lizards: biogeography in the Lesser Antilles. *Biotropica* 28, 388–393.

Taylor, L.H., Mackinnon, M.J. and Read, A.F. (1998) Virulence of mixed-clone and single-clone infections of the rodent malaria *Plasmodium chabaudi*. *Evolution* 52, 583–591.

Telford, S.R., Jr (1975) Saurian malaria in the Caribbean: *Plasmodium azurophilum* sp. nov., a malarial parasite with schizony and gametogony in both red and white cells. *International Journal for Parasitology* 5, 383–394.

Thompson, R.C.A. (2000) Giardiasis as a re-emerging infectious disease and its zoonotic potential. *International Journal for Parasitology* 30, 1259–1267.

Toft, C.A. and Aeschlimann, A. (1991) Introduction: coexistence or conflict? In: Toft, C.A. and Aeschlimann, A. (eds) *Parasite–Host Associations, Coexistence or Conflict*. Oxford University Press, New York, pp. 1–12.

van Baalen, M. (1998) Coevolution of recovery ability and virulence. *Proceedings of the Royal Society of London B* 265, 317–325.

van Baalen, M. and Sabelis, M.W. (1995) The dynamics of multiple infection and the evolution of virulence. *American Naturalist* 146, 881–910.

Warren, K.S. (1975) Hepatosplenic schistosomiasis mansoni: an immunologic disease. *Bulletin of the New York Academy of Medicine* 51, 545–550.

Wetherall, D.J. (1988) The anemia of malaria. In: Wernsdorfer, W.H. and McGregor, I. (eds) *Malaria: Principles and Practices of Malariology*. Churchill Livingstone, Edinburgh, pp. 735–751.

Yan, G., Severson, D.W. and Christensen, B.M. (1997) Costs and benefits of mosquito refractoriness to malaria parasites: implications for genetic variability of mosquitoes and genetic control of malaria. *Evolution* 51, 441–450.

The Behavioural Ecology of **15** Social Parasitism in Ants

Robin J. Stuart

*Citrus Research and Education Center, Institute of Food and Agricultural
Sciences, University of Florida, 700 Experiment Station Road,
Lake Alfred, FL 33850, USA*

Introduction

I shall next bring forward a scene still more astonishing, which at first,
perhaps you will be disposed to regard as a mere illusion of a lively
imagination. What will you say when I tell you that certain ants are affirmed
to sally forth from their nests on predatory expeditions, for the singular
purpose of procuring slaves to employ in their domestic business . . . .

(Kirby and Spence, 1859, p. 328)

it is possible that pupae originally stored as food might become developed;
and the ants thus unintentionally reared would then follow their proper
instincts, and do what work they could. If their presence proved useful to
the species which had seized them – if it were more advantageous to this
species to capture workers than to procreate them – the habit of collecting
pupae originally for food might by natural selection be strengthened and
rendered permanent for the very different purpose of raising slaves . . . .

(Darwin, 1859, pp. 341–342)

Social parasitism refers to the coexistence of two species of animals in
which one is parasitic on the society of the other (Wilson, 1971, 1975a).
In social parasitism, the parasite or parasite society exploits the labour
performed by a host society, often through social interactions. In this form
of parasitism, it is the altruistic or selfless behaviour of the hosts that
would normally be directed towards other members of their society, i.e.
their relatives, that is diverted by the parasite to its own advantage and to
the detriment of the hosts. Brood parasitism in birds is a familiar example
of social parasitism (Rothstein and Robinson, 1998), but it is among the
social insects in general and ants in particular where we find the greatest
abundance and diversity of socially parasitic relationships. Indeed, over
200 cases of interspecific symbiosis have been documented among the

8800 described ant species (Hölldobler and Wilson, 1990), and they encompass almost every conceivable mode of commensalism and parasitism. Commensalism is a relationship in which one species benefits and the other is unaffected, whereas, in parasitism, one species (the parasite) benefits and the other (the host) suffers negative consequences. Mutualism, in which two species cooperate for the benefit of both, has yet to be clearly documented among ant species, with all known cases of symbiosis that have been closely examined being at least somewhat unilateral and, in most cases, decidedly parasitic. Hölldobler and Wilson (1990) provide an extensive list of the ant species involved in socially parasitic relationships.

Why are social insects so prone to the formation of socially parasitic relationships? Social-insect colonies, with their nests, territories and trails and the various food sources that they defend, control or maintain, constitute valuable resources that can be exploited by any organism that can successfully penetrate and either elude or be accepted by colony members. Indeed, to paraphrase Wilson (1971), the lines of communication among the members of social-insect colonies are decidedly tenuous and, through the course of evolution, have been repeatedly opened, tapped, pried apart or rerouted through a variety of mechanisms to encompass alien individuals from a broad range of taxonomic groups. In some cases, these relationships between species involve little more than banditry, as stealthy, aggressive, well-armoured or fast-moving interlopers appropriate colony resources and evade colony defences. However, on occasion, they have evolved into long-lasting and intimate associations and the interlopers have become highly specialized obligatory social parasites, with various behavioural, physiological and morphological adaptations that facilitate their penetration and integration into the societies of their hosts. Although social parasitism among social insects has been recognized and studied for almost 200 years, we are only beginning to understand the various causes, constraints and consequences of parasitic evolution in these societies, and there is much that remains unexplained.

The study of social parasitism among social insects has a long and illustrious history. The dramatic slave raids of the legendary amazon-ants (see below) were first described by Huber (1810) and were commented upon by Darwin in *The Origin of Species* (1859). Indeed, Darwin (1859) was the first to suggest a plausible scenario for the evolution of slavery in ants (see above quotation). Subsequently, a vast bestiary of socially parasitic social insects, with a variety of life-history strategies, has been documented (see Wilson, 1971; Hölldobler and Wilson, 1990). Many of these species are highly evolved, obligatory parasites and exhibit an astounding array of adaptations that rival the raids of the amazons. For students of behavioural ecology and evolution, the socially parasitic social insects provide a rich menagerie for detailed study and experimentation. Indeed, since socially parasitic social insects are often exceedingly rare, many more species with diverse and perhaps bizarre life-history strategies probably still await discovery.

In this chapter, I review social parasitism among ant species and emphasize the behavioural aspects of these relationships. I examine the major life-history strategies of socially parasitic ants, explore some of the behavioural mechanisms exhibited by these species and discuss some especially salient aspects of their evolution, focusing on their behavioural evolution. The major emphasis will be on the relatively well-studied and most highly specialized species, particularly the obligatory slave-making, inquiline and xenobiotic ants. However, other free-living and facultatively parasitic species will also be discussed, especially when their characteristics provide insight into the evolution of the more specialized forms. For more extensive reviews, see Wilson (1971, 1975a) and Hölldobler and Wilson (1990).

Life-history Strategies

Traditionally, the life histories of socially parasitic ants are considered to be of two fundamental types: those that involve 'compound nests' and those that involve 'mixed colonies'. In compound nests, the parasite species and the host species maintain their broods separately, each tending their own in separate areas. In mixed colonies, the brood is kept communally, with the parasite's brood mixed in with that of the host (Wheeler, 1910; Wilson, 1971; Hölldobler and Wilson, 1990).

The distinction between compound nests and mixed colonies provides a useful framework for ordering and distinguishing various kinds of associations among ant species and also underlines an important variable in socially parasitic relationships, namely, the degree of intimacy obtained between parasite and host. Compound nests occur between pairs of species that tend to be only distantly related taxonomically. Relationships involving compound nests range from somewhat casual and fairly unspecialized or facultative associations that are not necessarily parasitic to highly evolved and obligatory forms of social parasitism. In contrast, mixed colonies often reflect a very close taxonomic relationship between the species and tend to be characteristic of some form of either facultative or obligatory social parasitism. Thus, the distinction between compound nests and mixed colonies often entails important and fundamental differences in the origin and evolution of parasitic relationships.

Generally, the degree of integration between closely related species in mixed colonies far exceeds that in even the most highly evolved cases involving distantly related species in compound nests. Indeed, close taxonomic relatedness appears to be an important prerequisite for the formation of fully integrated mixed colonies and, therefore, for the evolution of some of the more advanced forms of social parasitism. This pattern of close taxonomic relationship between social parasites and their hosts, often referred to as 'Emery's rule' (Hölldobler and Wilson, 1990; Bourke and Franks, 1995), contrasts sharply with the much more distant taxonomic relationship between most other kinds of parasites and their hosts, as exemplified by the other chapters in this volume.

The commingling and communal care of brood in mixed colonies provides important additional opportunities for parasitic adaptation and for the further elaboration and specialization of parasitic relationships that are not possible in compound nests. In compound nests, the parasite species must retain the ability to maintain brood, the chambers in which it is contained and the queens that produce it. However, in mixed colonies, where the parasite colony's brood and queens are fully integrated into the host colony and cared for by the host workers, the domestic abilities of the parasite workers are no longer required; and these abilities, and often the workers themselves, can be discarded. An evolutionary progression towards the loss of domestic abilities in workers of obligatory slave-making (or dulotic) species and towards the elimination of the worker caste in inquilines is a hallmark of the highly specialized social parasites that exist in mixed colonies (Wilson, 1971, 1975b; Dobrazańska, 1978; Buschinger, 1986; Stuart and Alloway, 1985; Mori and Le Moli, 1988; Hölldobler and Wilson, 1990).

Compound nests

The various kinds of compound nests that have been described constitute an array of different types of relationships, generally among distantly related ant species. They encompass what might be fairly casual, perhaps accidental, facultative relationships but also include some that are clearly highly specialized obligatory social parasitism. The more rudimentary associations might be quite common in nature and might constitute important evolutionary precursors to the more specialized forms.

Plesiobiosis

The most rudimentary and 'least intimate' association involving compound nests is referred to as plesiobiosis and involves a spatially close nesting association between species with little, if any, direct communication between the societies. Two or more nests of different species under the same stone but otherwise quite separate are a common example. Plesiobiosis tends to occur among species that are not closely related taxonomically, since closely related species rarely tolerate the close proximity of one another's nests. None the less, such associations are not always completely amiable and, if the nest chambers of such colonies are broken open, then fighting, brood pilfering and predation are likely to ensue. The benefits from plesiobiotic associations might involve the creation of a suitable microenvironment for one species by the other or inadvertent protection from certain predators or competitors (see Wheeler, 1910; Wilson, 1971).

Cleptobiosis

In cleptobiosis, one species builds its nests in close proximity to another and feeds from its refuse piles or steals food from returning foragers of the other species. Foragers of a *Crematogaster* species in India have been observed to 'lie in wait' for returning foragers of a *Holcomyrmex* species along the other species' trails, to aggressively interact with the returning foragers and to appropriate the food that they were carrying (cited in Wheeler, 1910). This cleptobiotic behaviour is apparently an extremely common and important foraging strategy for this species. In the southern USA, workers of *Conomyrma* (= *Dorymyrmex*) *pyramica* devote a considerable portion of their foraging activities to collecting dead insects discarded into kitchen middens by *Pogonomyrmex* harvester-ant colonies (Wilson, 1971).

Lestobiosis

Small species of ants in particular taxonomic groups, most notably some of the 'thief ants' of the *Solenopsis* subgenus *Diplorhoptrum*, exhibit what is referred to as lestobiosis. These ants nest in the walls of the nests of larger species of ants and termites, enter the nest chambers of the larger species, steal food and prey on the brood (Wheeler, 1910; Wilson, 1971). As remarked by Lubbock (1883: p. 79), 'It is as if we had small dwarfs, about eighteen inches to two feet long, harbouring in the walls of our houses, and every now and then carrying off some of our children into their horrid dens.' The sharing of the nest is considered parasitic, whereas consumption of the host brood is predatory (Hölldobler and Wilson, 1990).

Parabiosis

Parabiosis refers to an association in which two or more species share a common nest and sometimes use the same foraging trails, while keeping their broods separate. Forel (1898) coined this term to refer to the nesting association of *Crematogaster parabiotica* and *Dolichoderus debilis* in Colombia, and it also appears to apply to the nesting associations for a variety of ant species nesting in epiphytes in Mexico (Wheeler, 1910) and in arboreal ant gardens in French Guiana (Orivel *et al.*, 1997). Although there is potential in these relationships for true mutualism, it appears that one species often dominates and exploits the other (Hölldobler and Wilson, 1990).

Xenobiosis

Xenobiosis is a much more highly evolved and specialized relationship than those described above and is an obligatory form of social parasitism. The typically diminutive parasite species lives in the walls or chambers of the host species' nest, moves freely among its hosts and obtains food from them, usually by actively soliciting regurgitation (i.e. trophallaxis). None the less, the two species are not closely related taxonomically and the parasite tends its own brood and queen(s) (Hölldobler and Wilson, 1990).

Perhaps the best-known case of xenobiosis is that of the shampoo ant, originally described by Wheeler (1903, 1910), in which *Formicoxenus provancheri* (= *Leptothorax emersoni*) parasitizes colonies of *Myrmica incompleta* (= *Myrmica brevinoda*). Ants in the tribe Formicoxeni (= Leptothoracini) tend to be small and inconspicuous and often nest in tiny cavities in the soil or in hollow twigs, stems and nuts. They are common elements in the north temperate ant fauna, and nest densities can be very high in local areas. Interestingly, they also appear to be especially prone to the formation of socially parasitic relationships (Alloway, 1980, 1997; Alloway *et al.*, 1982; Stuart and Alloway, 1982, 1983; Buschinger, 1986; Hölldobler and Wilson, 1990). *M. incompleta* forms nests in soil, often in clumps of moss or under logs or stones, especially in damp areas; and *F. provancheri* nests in the soil, next to *M. incompleta* nests, joined to them by short galleries. The *Formicoxenus* move freely through the nests of their hosts but their galleries are too small to permit entrance by the *Myrmica*. The *Formicoxenus* appear to rely almost exclusively on liquid food obtained from their hosts. The *Formicoxenus* workers solicit regurgitation from the *Myrmica* workers and occasionally mount them and 'lick' them in an excited manner, moving their softer mouth-parts back and forth over the host ant's body, the shampoo behaviour of Wheeler's original description. Other *Formicoxenus* species have similar associations with various *Myrmica*, *Formica* and *Manica* species (Francoeur *et al.*, 1985).

Wheeler (1925) reported an additional case of xenobiosis, in which the small myrmicine ant *Megalomyrmex symmetochus* parasitizes the fungus-growing ant, *Sericomyrmex amabilis*. The parasite and the host both maintain their broods within the fungus gardens of the host, but apparently tend them quite separately; and the parasite feeds on the fungus and occasionally licks the bodies of its host.

Mixed colonies

Mixed colonies typically represent some form of facultative or obligatory social parasitism and three major forms are recognized: temporary social parasitism, slavery (or dulosis) and inquilinism. These forms share certain common elements and might have evolved from one to another in certain phylogenetic lines.

Temporary social parasitism

In temporary social parasitism, a newly mated queen of the parasite species manages to secure adoption into a colony of the host species by some mechanism, often involving aggression, stealth, some form of apparent 'conciliation' or a combination of these tactics. Subsequently, usurpation of the host colony is completed, as the host queen or queens are killed either by the parasite or by her own workers. The parasite queen then assumes the egg-laying chores of the colony, with the host workers tending her and rearing her brood. As the parasite's brood ecloses, the colony's worker force becomes a mixture of the two species. Later, as the host workers gradually die off, the colony becomes a pure colony of the parasite species, with no indication of its parasitic origin. In some temporary social parasites, adoption can be either intra- or interspecific, whereas, in others, it is strictly interspecific and obligatory. Temporary social parasitism has been documented in various ant genera, including *Formica*, *Lasius*, *Acanthomyops* and *Bothriomyrmex* (Hölldobler and Wilson, 1990).

Slavery

Slavery in ants, often referred to as dulosis, is a form of social parasitism in which established colonies of the parasite species obtain functional workers ('slaves') from other established colonies (Hölldobler, 1976; Buschinger *et al.*, 1980; Stuart and Alloway, 1982, 1983, 1985; Stuart, 1984; Hölldobler and Wilson, 1990; Topoff, 1997). Typically, slave makers conduct what are referred to as slave raids, during which they attack other colonies, kill or drive away the adults and appropriate the brood. This captured brood is reared in the slave maker's nest, and the workers that eclose from it become functional members of the slave-maker colony. In some cases, adult host-species workers are taken and enslaved, a phenomenon referred to as eudulosis. Slave makers vary in their degree of specialization; and their parasitic relationships may be either intra- or interspecific and either facultative or obligatory. Obligatory slave makers are often highly specialized social parasites, conduct dramatic, well-organized slave raids and display a broad range of morphological, behavioural and physiological adaptations for their parasitic lifestyle, especially in terms of fighting and recruiting nest mates during raids.

Obligatory slave makers also tend to exhibit various degrees of ineptitude in their ability to forage, to care for their brood and to perform various other domestic tasks that are routinely performed by their slaves (Wilson, 1975b; Dobrazańska, 1978; Stuart and Alloway, 1985; Mori and Le Moli, 1988; Hölldobler and Wilson, 1990). Moreover, all known obligatory slave makers are unable to found colonies independently and must rely on some form of parasitic colony foundation. Typically, this involves colony usurpation, similar to that seen in temporary social parasites, with a newly mated slave-maker queen aggressively invading a small host-species colony or colony fragment, capturing its brood and subsequently being adopted by the eclosing host workers. In some cases, the parasite is adopted by adult workers present in the host colony at the time of usurpation (e.g. Stuart, 1984; Topoff, 1997).

Perhaps the most specialized and well-known slave makers are the amazon-ants of the genus *Polyergus* (see Hölldobler and Wilson, 1990; Topoff, 1997; and references therein). Five *Polyergus* species are recognized: *P. rufescens* in Europe and North America, *P. breviceps* and *P. lucidus* in North America, *P. nigerrimus* in Russia and *P. samurai* in Japan and eastern Siberia. These slave makers parasitize ants in the genus

Formica, are large and shiny red or black, have extraordinary toothless, sickle-shaped mandibles and conduct dramatic, well-organized slave raids, usually during warm, sunny afternoons in July and August. Typically, a slave-maker worker discovers a nest of the host species and returns to its nest laying a chemical trail. Soon, slave makers pour from their nest, and proceed to the target nest in a fast-moving (3 cm s^{-1}), compact column, which can contain thousands of slave-maker workers. Upon arriving at the target nest, sometimes as much as 150 m away, the slave makers force an immediate entrance, seize brood and return home. Resisting host-species workers are attacked and killed, the slave makers piercing their heads with their specialized mandibles. The slave makers do not forage for food, feed their brood or queen or build or maintain their nests, and are entirely dependent on their slaves for the performance of these tasks. Young *Polyergus* queens display similar aggressive behaviours during colony foundation, as they attack small host-species colonies, kill the host queen(s) and are subsequently adopted by the host workers. Obligatory slavery has been discovered in three tribes, belonging to two subfamilies of ants, and has apparently evolved repeatedly in various phylogenetic lines, which include the genera *Polyergus*, *Rossomyrmex*, *Leptothorax*, *Harpagoxenus*, *Protomognathus*, *Chalepoxenus*, *Epimyrma* and *Strongylognathus* (Alloway, 1980, 1997; Buschinger, 1986; Hölldobler and Wilson, 1990).

Inquilinism

Inquilines are obligatory social parasites and spend their entire life cycle within colonies of their host species, except when they leave to mate or disperse to found new colonies (Hölldobler and Wilson, 1990; Bourke and Franks, 1991). Typically, inquilines produce few, if any, workers, and the workers that are produced contribute little to the economy of the colony. Often, inquiline queens and host queens coexist within the same colony and live in close association with one another. Indeed, in certain extreme cases, the parasites have evolved to the point of being essentially ectoparasitic and ride about on the bodies of their hosts, especially on the host queens (see below). When the host queen(s) is present, she continues to lay eggs, which are reared to produce new workers, but the parasite queen(s) usually produces only reproductives. In some cases, the host queen(s) is killed either by the parasites or by the host queen's own workers, in a manner similar to that seen during colony usurpation by the queens of temporary social parasites or obligatory slave makers. Without a resident host queen to lay eggs for the production of new workers, such parasite colonies can exist only for the lifetime of the current host workers, perhaps only a few years. None the less, even this limited life expectancy can be sufficient for the production of large numbers of new parasites. Interestingly, since many inquilines completely lack a worker caste and, therefore, no longer exhibit any division of labour between reproductive and non-reproductive conspecifics, they no longer satisfy the definition of being highly social (i.e. eusocial) species and therein differ from all other ants (see Hölldobler and Wilson, 1990). Some authors have suggested that inquilines that tolerate the host queen are fundamentally different and might often have a very different evolutionary origin from those that do not tolerate the host queen (Buschinger, 1986, 1990; Bourke and Franks, 1991, 1995).

Perhaps the most extreme and highly specialized social parasite yet described is the inquiline *Teleutomyrmex schneideri* (see Buschinger, 1986, 1987; Hölldobler and Wilson, 1990; and references therein). *Teleutomyrmex*, literally 'final ant', is a workerless inquiline parasite of two *Tetramorium* species, *T. caespitum* and *T. impurum*, and occurs in small, isolated populations in the Swiss Alps, French Alps, French Pyrenees and Spanish Sierra Nevada. The queens are small compared with their hosts, generally inactive and spend most of their time riding around on the backs of their hosts, especially the host queen. The ventral body surfaces of the parasites are strikingly concave and well adapted to this ectoparasitic lifestyle, and they have unusually large tarsal claws and a strong tendency to grasp objects. Large numbers of unicellular glands under the cuticle appear to be the source of a powerful attractant for the host workers, which frequently lick the parasites. Older queens are highly physogastric, their abdomens swollen with fat body and ovarioles. Several physogastric parasite queens often coexist in the same host nest, and each can lay an egg every 30 s. The *Tetramorium* queen continues to lay eggs and produce workers and the colony's brood is a mixture of the two species. The bodies of *T. schneideri* females appear to have undergone extensive morphological degeneration. The integument is thin and shiny, with little pigment or sculpturing. The mandibles, sting and poison apparatus are reduced. There is degeneration in the exocrine system, with reduced labial and postpharyngeal glands. The maxillary and metapleural glands are completely absent. There is even degeneration in the nervous system, with reduced and fused ganglia. The males are also morphologically degenerate and appear almost pupoid, with a thin, greyish cuticle, broad petiole and postpetiole and soft abdomen, deflected down at the tip. As in many inquilines, mating takes place in the nest, and newly mated females either lose their wings and join the egg layers in their maternal nest or fly off to found new parasite colonies elsewhere.

Inquilinism has apparently evolved repeatedly and independently in various ant subfamilies, but the factors responsible for its evolution remain a matter of debate (see Buschinger, 1986, 1990; Hölldobler and Wilson, 1990; Bourke and Franks, 1991, 1995; and references therein). None the less, comparative studies suggest that, once a species has evolved an inquiline life history, it quickly becomes completely dependent on its host and tends to acquire some or all of an array of characteristics that have been referred to as the 'inquiline syndrome' (Wilson, 1971; Hölldobler and Wilson, 1990). These characteristics include:

1. Loss of the worker caste.
2. Winglessness of queens and males (i.e. ergatomorphism).

3. Mating within the nest.
4. Limited dispersal of mated queens.
5. Coexistence of multiple queens in the same nest (i.e. polygyny).
6. Rare, patchy populations.
7. Reduced body size, mouth-parts, antennal segmentation and propodeal spines.
8. Reduced thickness, pigmentation and sculpturing of the exoskeleton.
9. Reduced nervous and exocrine systems.
10. Broadening of petiole and postpetiole.
11. Enhanced attractiveness to hosts, probably chemically mediated and indicated by frequent licking and grooming by the hosts.

Behavioural Mechanisms

The behavioural mechanisms associated with the adaptations of various organisms for a parasitic existence can be summarized using Doutt's (1964) four sequential steps of host selection: host habitat finding, host finding, host acceptance and host suitability. This framework can be applied to socially parasitic social insects, but most studies of the behavioural ecology of these insects have focused on the interactions that occur between parasite and host once the parasite has located the host nest and is attempting to establish a parasitic association. At this point, the first three steps in Doutt's sequence are already completed and only host suitability remains to be resolved. None the less, this final step is of considerable interest, since it involves direct interactions between the parasite and the host and is the point at which the parasite must overcome host defences if it is to be successful.

Host-habitat finding

Locating and colonizing host populations is undoubtedly an important aspect of the biology of actively dispersing social parasites, but very little research has addressed this issue directly. In general, socially parasitic social insects tend to be relatively rare, compared with their hosts, and many are known from only a limited number of records and from very localized and sometimes very patchy populations (Wilson, 1963, 1971; Hölldobler and Wilson, 1990). Thus, the known range of many social parasites appears to be a small subset of the range of their host species. Moreover, in many social parasites, mating appears to occur within the nest and dispersal is sometimes very limited. Indeed, some social parasites have evolved to become wingless and have thereby lost the ability for long-range dispersal (Wilson, 1963, 1971; Buschinger and Heinze, 1992). None the less, in those species that do retain wings and actively disperse from the vicinity of the parental nest, the ability to locate suitable habitat containing host colonies is clearly essential to survival.

When the host species is common and suitable habitat is widespread, then host-habitat finding is unlikely to be a problem or to require any special adaptations. Indeed, such appears to be the case for most social parasites in the temperate region, which include most known social parasites (Hölldobler and Wilson, 1990). Clearly, socially parasitic relationships will only evolve among species with overlapping habitat requirements, and the mechanisms for habitat finding that served the ancestral species will probably continue to serve the evolving parasite and might further evolve to more closely match the host species.

Host finding

Socially parasitic social insects are generally highly specific to one or a few host species (see Hölldobler and Wilson, 1990; Heinze *et al.*, 1992), and finding colonies of these species is crucial to the formation of new parasite colonies. For socially parasitic relationships to evolve, the nesting requirements of the two species must be sufficiently similar to bring them into close contact and, as the relationship evolves, the parasite is likely to respond much more strongly to the various cues that facilitate the appropriate close contact and the ability to locate suitable host nests. The level of host specificity observed certainly indicates that finding host nests of the appropriate species is important, but little research has been conducted on the cues and mechanisms involved.

For many social parasites, the task of locating a host colony falls exclusively to the newly mated parasite female. However, for slave makers, the workers must also be able to locate host nests if they are to conduct effective slave raids. In most cases, parasitic colony foundation is probably an evolutionary derivative of secondary polygyny, a phenomenon in which at least some of the young newly mated queens from a colony are adopted back into their parental nest after mating and become active reproductives, together with the other queens in that nest (Buschinger, 1970, 1986; Wilson, 1971; Alloway, 1980, 1997; Stuart *et al.*, 1993). Primary polygyny refers to colonies that are initially founded by multiple, inseminated, fully reproductive queens. Secondary polygyny occurs in a sizeable minority of ant species and is characteristic of many free-living species in the particular phylogenetic groups that have given rise to most social parasites. Independent colony foundation appears to be extremely hazardous for young queens, with few incipient colonies surviving long enough to produce a new generation of reproductives (Pollock and Rissing, 1989). Achieving adoption back into the parental nest, a conspecific nest or an allospecific nest would protect young queens from these hazards and could provide an enormous boost to a queen's lifetime reproductive success. However, young queens are not always adopted back into their parental nests, and adoptions into alien conspecific and allospecific nests can be extremely rare and hazardous as well (Pollock and Rissing, 1989; Hölldobler and Wilson, 1990). None the less, adoptions into alien conspecific nests are apparently possible and are quite common in some species (Alloway, 1980, 1997; Stuart *et al.*, 1993; Bourke and Franks, 1995). The nests of closely related species might provide cues that are very similar to conspecific cues, and the response of young queens to these cues and their attempts to gain acceptance probably form an important and perhaps general basis for the evolution of interspecific socially parasitic relationships. Indeed, the similarity of both nest cues and the cues emanating from the young queens themselves among closely related sympatric species is likely to facilitate the adoption of young queens into alien nests and probably explains, at least in part, why socially parasitic relationships tend to evolve among closely related species.

As mentioned above, the slave-raiding behaviour of slave-maker workers also requires effective host finding. The evolutionary basis for slave raiding in various groups of ants appears to be the aggressive territorial behaviour of free-living species (Wilson, 1975b; Alloway, 1980; Stuart and Alloway, 1982, 1983; Hölldobler and Wilson, 1990; see below), and an effective aggressive response to appropriate competitors was probably characteristic of the free-living ancestral species that gave rise to these parasites. Again, the precise cues and mechanisms involved are unknown, but the high level of host specificity achieved indicates that such cues and mechanisms undoubtedly exist, and evolving parasites would be under considerable selection pressure to further develop and refine their responses to these host-specific cues.

Host acceptance

The ability of socially parasitic social insects to recognize and accept their hosts and to respond to them appropriately seems strongly analogous to their ability to recognize, accept and respond appropriately to their nest mates, an ability that social insects possess in abundance (Hölldobler and Michener, 1980; Stuart, 1987, 1988a,b, 1992; Hölldobler and Wilson, 1990). Typically, social insects are highly aggressive in defending their nests, territories, trails and food sources against intruding members of their own or other species. Nest-mate recognition in these contexts generally involves colony-specific chemical cues (or odours), located on the surface of the body and learned by colony members. Generally, this learning first occurs in the early adult stage, shortly after the adult emerges from the pupa, and, therefore, might constitute imprinting. However, the learning of new nest-mate recognition cues can apparently continue throughout adult life under appropriate circumstances and, in the absence of a critical period, would not be considered imprinting (Stuart, 1988a,b, 1992). Imprinting has been implicated in the learning of species-specific brood-recognition cues in some species but not in others (Jaisson and Fresneau, 1978; Le Moli and Mori, 1982; Alloway, 1997). For social parasites, learning the species-specific chemical cues of the hosts that were present in the parasite colony where they eclosed could be an important aspect of host specificity (Topoff, 1997). Moreover, the learning of colony-specific nest-mate-recognition cues by newly eclosed workers helps to explain how new slaves can be continually incorporated into slave-maker colonies, whereas the acceptance of these new slaves by older adult colony members clearly indicates that learning of nest-mate-recognition cues is not confined exclusively to the early adult stage, at least in these species (Stuart, 1988a,b). Thus, paradoxically, the learning of nest-mate recognition cues forms the basis for much of colony defence in social insects, but it is also one of the primary factors that appears to make them vulnerable to exploitation by social parasites.

Host suitability

Once a young queen of a socially parasitic ant species finds and accepts a host colony, it must penetrate the host society, overcome the host's defences, gain acceptance and reproduce; and the parasite's brood and eclosing offspring must also be accepted and nurtured by the host colony. Much of the literature on socially parasitic social insects is devoted to detailed descriptions of the broad range of diverse mechanisms used by social parasites to penetrate host colonies. As referred to above, social insect colonies are typically highly aggressive in the defence of their nests, with well-developed nest-mate-recognition systems, which enable them to discriminate between legitimate nest mates and various kinds of intruders (Hölldobler and Michener, 1980; Stuart, 1988b). To be successful, socially parasitic social insects must have mechanisms to breech these defences, and various tactics have apparently evolved in this context.

Obligatory slave makers are highly aggressive in their attacks on host nests and exhibit various fighting and recruitment adaptations for conducting slave raids, which enable them to defeat their hosts in direct frontal assaults (see Buschinger, 1986; Hölldobler and Wilson, 1990; and references therein). Workers of obligatory slave-making ants are often quite literally highly evolved fighting machines, with an array of adaptations for successful slave raiding. In many species, the workers have specialized, clipper-shaped or sickle-shaped toothless mandibles, which they use for dismembering their opponents or piercing their exoskeletons during slave raids. In some cases, the workers have enlarged poison glands and the propensity to sting their opponents to death, whereas others have apparent chemical weapons, so-called 'propaganda substances', which they use to confuse and disperse their opponents or induce them to attack one another during raids (e.g. Regnier and Wilson, 1971; Allies *et al.*, 1986). Moreover, all known slave makers actively recruit nest mates for coordinated assaults on target nests, often using chemical trails. Once a slave-maker colony launches a raid on a target nest, the outcome is seldom in doubt.

Colony founding by slave-maker queens is also typically aggressive, with newly mated queens often appearing to use the same lethal weapons and fighting tactics during their solitary assaults on host-species nests as do their workers during slave raids. However, in some cases, less aggressive tactics come into play, and chemical propaganda or appeasement substances are sometimes used (Stuart, 1984; Buschinger, 1986, 1989; Topoff, 1997).

Queens of various *Formica* species in Europe that are facultative temporary social parasites illustrate a range of tactics for dealing with colony defences (see Hölldobler and Wilson, 1990, and references therein). Young queens of *Formica rufa* plunge directly into host nests and, although many are killed, enough survive to make this ant extremely common and widespread. Queens of *Formica exsecta* exhibit a more cautious approach and initially 'stalk' host nests and then either enter by stealth or are carried in by host workers, which display relatively little hostility to the parasite. Queens of *Formica pressilabris* reportedly lie down and 'play dead', assuming a pupal position, and are then picked up by host workers and carried into the nest without any apparent hostility.

For certain *Lasius* species, temporary parasitism is apparently obligatory (see Hölldobler and Wilson, 1990, and references therein). Colony-founding *Lasius umbratis* queens first attack and kill a host worker and carry it around for a time, before attempting to enter the host nest. Queens of *Lasius reginae* eliminate host queens by rolling them over and 'throttling' them, biting the ventral neck region until they are dead. Queens of the temporary social parasites *Bothriomyrmex decapitans* and *Bothyriomyrmex regicidus* exhibit similar behaviours (see Hölldobler and Wilson, 1990, and references therein). The parasites are initially attacked and dragged into the host nest by host workers but ultimately find their way on to the back of the host queen, where they slowly and meticulously cut off her head. Often the parasite queens appear extremely attractive to host workers, probably because of chemical attractants or appeasement substances, and in some cases it is the host workers that kill their own queens, once the parasite has become accepted into the colony. Thus, social parasites appear to use a variety of behavioural and chemical means to penetrate host nests, eliminate host queens and secure adoption by host workers.

Evolution

The evolution of social parasitism in social insects has been a topic of speculation and debate since the time of Darwin, and numerous factors are likely to have contributed to the evolution of this phenomenon and its various manifestations in different phylogenetic lines. Hölldobler and Wilson (1990) summarized the possible evolutionary relationships among free-living and socially parasitic species (see Fig. 15.1) and most of the links indicated are well supported by comparative data. However, some

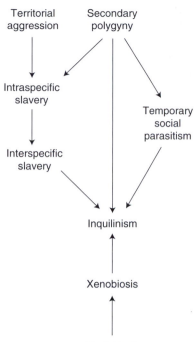

Fig. 15.1. Possible evolutionary pathways for social parasitism in ants (redrawn with modifications from Hölldobler and Wilson, 1990).

relationships and origins remain unresolved. For example, it is unclear whether xenobiosis or temporary social parasitism ever lead to inquilinism.

Social parasitism in ants and other social insects appears to be largely a temperate-zone phenomenon and to be concentrated in particular phylogenetic lines. Various authors have discussed factors that might be responsible for these patterns (see Alloway, 1980, 1997; Alloway *et al.*, 1982; Stuart and Alloway, 1982, 1983; Buschinger, 1986, 1990; Hölldobler and Wilson, 1990; Stuart, 1990, 1993; Topoff, 1997; and references therein). Because most known socially parasitic social insects occur in the temperate regions around the world, it has been suggested that cooler temperatures might facilitate the acceptance of alien individuals into colonies. As already discussed, the adoption of young mated queens back into their parental nests (i.e. secondary polygyny) would seem to be an important precursor to parasitic colony foundation. It has also been suggested that the occurrence of closely related sympatric species in large dense populations, which, to some extent, compete with each other, would promote the kinds of interactions necessary for the evolution of social parasitism; and these factors also tend to be characteristic of temperate regions. The tendency of colonies to fragment into small, often queenless, nest units, a trait often associated with secondary polygyny and colony reproduction by 'budding', could also be a factor, since it might facilitate the evolution of parasitic colony foundation by ensuring large numbers of easily invaded nests. Moreover, as mentioned above, the ability to learn relatively diverse nest-mate recognition cues might also be necessary to facilitate the integration of alien individuals into colonies. Taken together, these factors might explain why social parasitism appears to be predominantly a temperate-zone phenomenon and why certain taxonomic groups among the ants with certain arrays of traits might be much more prone to the formation of socially parasitic relationships among closely related species than are other groups.

The various considerations discussed above could facilitate cross-species adoption and the evolution of the specialized forms of parasitic colony foundation that characterize temporary social parasites, inquilines and obligatory slave makers. The evolution of the various behavioural and chemical tactics used in colony usurpation, as described above, are some-what enigmatic but are probably based on the various mechanisms that exist within species in the context of queen dominance, reproductive competition and queen-number regulation in polygynous species (Heinze, 1990, 1992; Buschinger and Heinze, 1992). However, a complete explanation for the evolution of obligatory slavery in ants also requires an explanation for the evolution of slave raiding, a behaviour that appears to be exceedingly complex and highly specialized. In this respect, three hypotheses have been advanced: the predation hypothesis, the territorial hypothesis and the transport hypothesis (Buschinger, 1986; Hölldobler and Wilson, 1990).

Darwin (1859) first advanced the predation hypothesis and suggested that the ancestral species to slave makers might have been predatory species that attacked other colonies to obtain food. In this case, the first slaves would be prey items that were not consumed, managed to eclose in the predator's nest and were accepted into the colony. The territorial hypothesis suggests that aggressive territorial interactions among colonies of the same or closely related species were the evolutionary precursors to slave raiding (Wilson, 1975b; Alloway, 1980; Stuart and Alloway, 1982, 1983). Territorial interactions often lead to opportunistic brood pilfering and predation; and, again, captured brood that is not consumed might survive to be the first slaves. The transport hypothesis suggests that slave raiding evolved from the regular transport of brood among the multiple nests of a single colony (i.e. a polydomous colony). According to this scenario, attempts to transport brood between more distant and increasingly less familiar nests of a single polydomous colony might give rise to aggression and ultimately lead to the evolution of slave raiding (Buschinger, 1970, 1986).

In comparing the three hypotheses presented above, the territorial hypothesis appears to provide the most complete and reasonable explanation for how slave raiding in ants might have evolved from the known behaviour patterns of free-living ancestral species. Indeed, territorial behaviour, as observed in free-living species encompasses important aspects of the other two hypotheses and, to some extent, might be considered a synthesis of the various ideas that have been presented. The predation hypothesis is unlikely to be a complete explanation, since the ants in the various phylogenetic lines in which slave makers occur are not strongly predatory and do not attack other ant colonies as a major component of their foraging strategies (but see Topoff, 1991, for a possible exception). None the less, opportunistic brood predation as a result of territorial interactions is relatively common in these groups (Wilson, 1975b; Alloway, 1980; Stuart and Alloway, 1982, 1983). The transport hypothesis provides a ready explanation for the transport of pilfered brood and the recruitment of nest mates at the conclusion of a slave raid but does so in a non-aggressive intracolonial context, which is quite different from and not easily translated into the aggressive intercolonial context of a slave raid. Indeed, if fragments of a polydomous colony became unfamiliar and aggressive towards one another (as they might well do (see Stuart, 1987)), then the behaviours expressed in this context would probably be aggressive territorial behaviours, and the hypotheses merge. Moreover, point-by-point comparisons of the slave raids of slave makers and the territorial interactions of related free-living species indicate that territorial behaviour includes all of the behavioural elements seen during slave raids: scouting, fighting, recruitment of nest mates prior to the assault on the opposing nest and thereafter, the invasion of the target nest and the transport of the remaining brood back to the victorious colony's nest (Stuart and Alloway, 1982, 1983). The refining of these behaviours by increasing the distance and frequency of raids and increasing the acceptance of captured brood, coupled with an increased tendency for parasitic colony foundation, would put such species well on the way to being obligatory slave makers. Moreover, the combination of secondary polygyny and territorial behaviour is common in the various ancestral groups that gave rise to the slave makers; and facultative interspecific slavery as a consequence of territorial interactions has been reported in these groups (Wilson, 1975b; Alloway, 1980, 1997; Hölldobler and Wilson, 1990). Thus, secondary polygyny and territorial behaviour appear to have been key elements in the evolution of slavery in ants.

The evolution of socially parasitic relationships between closely related sympatric species is the typical model presented for the evolution of the relationships that we see in mixed colonies, with two, formerly independent, free-living species evolving to form a host–parasite dyad (Hölldobler and Wilson, 1990). However, some authors have also suggested that sympatric speciation, in which a species essentially generates its own parasitic sibling species, is also possible under certain circumstances and might better explain the evolution of many inquiline species (see Buschinger, 1986, 1990; Bourke and Franks, 1991, 1995; and references therein).

Perhaps the best example of apparent evolution from one form of social parasitism to another is seen in the myrmicine genus *Epimyrma*, where slavery might have evolved into a form of inquilinism (Buschinger,

1989; Hölldobler and Wilson, 1990). Members of the genus *Epimyrma* are parasites of various *Leptothorax* species. Four *Epimyrma* species, *E. algeriana*, *E. bernardi*, *E. ravouxi* and *E. stumperi*, conduct well-organized slave raids that involve group recruitment, using processions and effective sting fighting. *Epimyrma kraussei* produces few workers and their raids are rare and ineffectual, whereas *Epimyrma corsica* and *Epimyrma adlerzi* are workerless inquilines. A corresponding progression from mating flights to within-nest mating is also observed in these species. Colony foundation by *Epimyrma* queens follows the typical pattern seen in many temporary social parasites and obligatory slave makers, and involves the forceful usurpation of a host colony. However, the method of invasion varies greatly among species. *E. kraussei* queens approach host colonies in an aggressive manner. Once the parasite penetrates the host nest, she kills the host queen and is adopted by the host workers. The queens of *E. ravouxi* use a more 'conciliatory' approach initially, grooming and stroking host workers, but, once inside the nest, the parasite mounts the host queen from behind and kills her by seizing her around the neck with her sabre-shaped mandibles. *E. stumperi* queens crouch down, freeze and seem to feign death during their initial encounters with host workers, but subsequently begin to mount workers from behind and to groom them, perhaps acquiring chemical recognition cues in the process. However, once inside the nest, the parasite queen systematically eliminates the host queens by mounting them, rolling them over and grasping their necks in her mandibles until they succumb.

A similar evolutionary progression from dulosis to inquilinism might also have occurred in the myrmicine genus *Strongylognathus*. However, in *Strongylognathus testaceus*, the only known inquiline, the host queen is apparently tolerated (see Hölldobler and Wilson, 1990, and references therein).

Conclusions

Studies of the behavioural ecology of social parasitism among ant species have revealed an array of complex relationships, led to the discovery of diverse mechanisms by which these relationships are initiated and maintained and provided insight into how these relationships initially evolve and become further elaborated. Continuing research is likely to reveal many additional socially parasitic species, new variations in life histories and novel behavioural and chemical strategies of social exploitation. Moreover, our understanding of the evolution of social parasitism in ants is far from complete, and much remains to be determined regarding the broader ecological causes, constraints and consequences of this phenomenon. Some of the current issues, debates and outstanding questions regarding the evolution of social parasitism in ants include the following. What pre-existing species characteristics facilitate, promote or are necessary for the evolution of various forms of social parasitism? Under what ecological circumstances does a species evolve a particular socially parasitic life history? Do species generate their own social parasites through sympatric speciation? If so, under what circumstances does this occur, and to what extent does this explain the evolution of various forms of social parasitism? Do different forms of social parasitism evolve into one another and, if so, under what circumstances? To what extent do social parasites influence the evolution of their host species? Do social parasites switch host species and, if so, under what circumstances? Is the apparent extreme rarity of social parasites real or illusory? Social parasitism in ants has fascinated biologists for over a century, but there is considerable potential for further research, new discoveries and a much more thorough understanding of this amazing phenomenon.

References

Allies, A.B., Bourke, A.F.G. and Franks, N.R. (1986) Propaganda substances in the cuckoo ant *Leptothorax kutteri* and the slave-maker *Harpagoxenus sublaevis*. *Journal of Chemical Ecology* 12, 1285–1293.

Alloway, T.M. (1980) The origins of slavery in leptothoracine ants (Hymenoptera: Formicidae). *American Naturalist* 115, 247–261.

Alloway, T.M. (1997) The role of workers and queens in the colony-member recognition systems of ants: are there any differences that predispose some kinds of ants to social parasitism? In: Greenberg, G. and Tobach, E. (eds) *Evolution of Social Behavior and Integrative Levels*. Erlbaum, Hillsdale, New Jersey, pp. 193–219.

Alloway, T.M., Buschinger, A., Talbot, M., Stuart, R. and Thomas, C. (1982) Polygyny and polydomy in three North American species of the ant genus *Leptothorax* Mayr (Hymenoptera: Formicidae). *Psyche* 89, 249–274.

Bourke, A.F.G. and Franks, N.R. (1991) Alternative adaptations, sympatric speciation and the evolution of parasitic, inquiline ants. *Biological Journal of the Linnean Society* 43, 157–178.

Bourke, A.F.G. and Franks, N.R. (1995) *Social Evolution in Ants*. Princeton University Press, Princeton, New Jersey, xiii + 529 pp.

Buschinger, A. (1970) Neue Vorstellungen zur Evolution des Sozialparasitimus und der Dulosis bei Ameisen (Hym., Formicidae). *Biologisches Zentralblatt* 89, 273–299.

Buschinger, A. (1986) Evolution of social parasitism in ants. *Trends in Ecology and Evolution* 1, 155–160.

Buschinger, A. (1987) *Teleutomyrmex schneideri* Kutter 1950 and other parasitic ants found in the Pyrenees. *Spixiana* 10, 81–83.

Buschinger, A. (1989) Evolution, speciation, and inbreeding in the parasitic ant genus *Epimyrma* (Hymenoptera, Formicidae). *Journal of Evolutionary Biology* 2, 265–283.

Buschinger, A. (1990) Sympatric speciation and radiative evolution of socially parasitic ants – heretic hypotheses and their factual background. *Zeitschrift für Zoologische Systematik und Evolutions-forschung* 28, 241–260.

Buschinger, A. and Heinze, J. (1992) Polymorphism of female reproductives in ants. In: Billen, J. (ed.) *Biology and Evolution of Social Insects*. Leuven University Press, Leuven, Belgium, pp. 11–23.

Buschinger, A., Ehrhardt, W. and Winter, U. (1980) The organization of slave raids in dulotic ants – a comparative study (Hymenoptera; Formicidae). *Zeitschrift für Tierpsychologie* 53, 245–264.

Darwin, C.R. (1859) *On the Origin of Species by Means of Natural Selection, or the Preservation of Favoured Races in the Struggle for Life* (Facsimile of 1st edn, 1964). Harvard University Press, Cambridge, Massachusetts, xxvii + 502 pp.

Dobrazańska, J. (1978) Evaluation of functional degeneration of the amazon-ant *Polyergus rufescens* Latr. under an influence of socially parasitic way of life. *Acta Neurobiologiae Experimentalis* 38, 133–138.

Doutt, R.L. (1964) Biological characteristics of entomophagous adults. In: DeBach, P. (ed.) *Biological Control of Insect Pests and Weeds*. Reinhold, New York, pp. 145–167.

Forel, A. (1898) La parabiose chez les fourmis. *Bulletin de la Société Vaudoise des Sciences Naturelles* 34, 380–384.

Francoeur, A., Loiselle, R. and Buschinger, A. (1985) Biosystématique de la tribu Leptothoracini (Formicidae, Hymenoptera), 1: Le genre *Formicoxenus* dans la région holarctique. *Naturaliste Canadien* 112, 343–403.

Heinze, J. (1990) Dominance behavior among ant females. *Naturwissenschaften* 77, 41–43.

Heinze, J. (1992) Reproductive competition in colonies of the ant *Leptothorax gredleri*. *Ethology* 90, 265–278.

Heinze, J., Stuart, R.J., Alloway, T.M. and Buschinger, A. (1992) Host specificity in the slave-making ant, *Harpagoxenus canadensis*. *Canadian Journal of Zoology* 70, 167–170.

Hölldobler, B. (1976) Tournaments and slavery in a desert ant. *Science* 192, 912–914.

Hölldobler, B. and Michener, C.D. (1980) Mechanisms of identification and discrimination in social Hymenoptera. In: Markl, H. (ed.) *Evolution of Social Behavior: Hypotheses and Empirical Tests*. Verlag Chemie, Weinheim.

Hölldobler, B. and Wilson, E.O. (1990) *The Ants*. Belknap/Harvard, Cambridge, Massachusetts, xii + 732 pp.

Huber, P. (1810) *Recherches sur les moeurs des fourmis indigènes*. Paschoud, Paris, xvi + 328 pp.

Jaisson, P. and Fresneau, D. (1978) The sensitivity and responsiveness of ants to their cocoons in relation to age and methods of measurement. *Animal Behaviour* 26, 1064–1071.

Kirby, W. and Spence, W. (1859) *An Introduction to Entomology; or, Elements of the Natural History of Insects: Comprising an Account of Their Meta-morphosis, Food, Strategems, Habitations, Societies, Motions, Noises, Hybernation, Instinct, etc. etc.*, 7th edn. Longman, Brown, Green, Longmans and Roberts, London.

Le Moli, F. and Mori, A. (1982) The effect of early learning on recognition, acceptance and care of cocoons in the ant *Formica rufa* L. *Bolletino di Zoologia* 49, 93–97.

Lubbock, J. (1883) *Ants, Bees, and Wasps: a Record of Observations on the Habits of the Social Hymenoptera*. Appleton, New York, xix + 448 pp.

Mori, A. and Le Moli, F. (1988) Behavioural plasticity and domestic degeneration in facultative and obligatory slave-making ant species (Hymenoptera: Formicidae). *Monitore Zoologico Italiano* 22, 271–285.

Orivel, J., Errard, C. and Dejean, A. (1997) Ant gardens: interspecific recognition in parabiotic ant species. *Behavioral Ecology and Sociobiology* 40, 87–93.

Pollock, G.B. and Rissing, S.W. (1989) Intraspecific brood raiding, territoriality, and slavery in ants. *American Naturalist* 133, 61–70.

Regnier, F.E. and Wilson, E.O. (1971) Chemical communication and 'propaganda' in slave-maker ants. *Science* 172, 267–269.

Rothstein, S.I. and Robinson, S.K. (1998) *Parasitic Birds and Their Hosts*. Oxford University Press, New York, xii + 444 pp.

Stuart, R.J. (1984) Experiments on colony foundation in the slave-making ant *Harpagoxenus canadensis* M.R. Smith (Hymenoptera; Formicidae). *Canadian Journal of Zoology* 62, 1995–2001.

Stuart, R.J. (1987) Transient nestmate recognition cues contribute to a multicolonial population structure in the ant *Leptothorax curvispinosus*. *Behavioral Ecology and Sociobiology* 21, 229–235.

Stuart, R.J. (1988a) Collective cues as a basis for nestmate recognition in polygynous leptothoracine ants. *Proceedings of the National Academy of Sciences USA* 85, 4572–4575.

Stuart, R.J. (1988b) Development and evolution in the nestmate recognition systems of social insects. In: Greenberg, G. and Tobach, E. (eds) *Evolution of Social Behavior and Integrative Levels*. Erlbaum, Hillsdale, New Jersey, pp. 177–195.

Stuart, R.J. (1990) Aggression, competition, and asymmetries in the evolution of ant slavery. In: Veeresh, G.K., Malik, B. and Viraktamath, C.A. (eds) *Social Insects and the Environment*. Oxford and IBH, New Delhi, pp. 153–154.

Stuart, R.J. (1992) Nestmate recognition and the ontogeny of acceptability in the ant *Leptothorax curvispinosus*. *Behavioral Ecology and Sociobiology* 30, 403–408.

Stuart, R.J. (1993) Differences in aggression among sympatric facultatively polygynous *Leptothorax* ant species. *Animal Behaviour* 46, 809–812.

Stuart, R.J. and Alloway, T.M. (1982) Territoriality and the origin of slave raiding in leptothoracine ants. *Science* 215, 1262–1263.

Stuart, R.J. and Alloway, T.M. (1983) The slave-making ant, *Harpagoxenus canadensis* M.R. Smith, and its host species, *Leptothorax muscorum* (Nylander): slave raiding and territoriality. *Behaviour* 85, 58–90.

Stuart, R.J. and Alloway, T.M. (1985) Behavioural evolution and domestic degeneration in obligatory slave-making ants (Hymenoptera; Formicidae; Leptothoracini). *Animal Behavior* 33, 1080–1088.

Stuart, R.J., Gresham-Bissett, L. and Alloway, T.M. (1993) Queen adoption in the facultatively polygynous and polydomous ant, *Leptothorax curvispinosus*. *Behavioral Ecology* 4, 276–281.

Topoff, H. (1991) *Formica wheeleri*: Darwin's predatory slave-making ant? *Psyche* 98, 309–317.

Topoff, H. (1997) Adaptations for social parasitism in the slave-making ant genus *Polyergus*. In: Greenberg, G. and Tobach, E. (eds) *Evolution of Social Behavior and Integrative Levels*. Erlbaum, Hillsdale, New Jersey, pp. 178–192.

Wheeler, W.M. (1903) Ethological observations on an American ant (*Leptothorax emersoni* Wheeler). *Journal für Psychologie und Neurologie* 2, 31–47.

Wheeler, W.M. (1910) *Ants: Their Structure, Development and Behaviour*. Columbia University Press, New York, xxv + 663 pp.

Wheeler, W.M. (1925) A new guest-ant and other new Formicidae from Barro Colorado Island, Panama. *Biological Bulletin of the Marine Biological Laboratory, Woods Hole* 49, 150–181.

Wilson, E.O. (1963) Social modifications related to rareness in ant species. *Evolution* 17, 249–253.

Wilson, E.O. (1971) *The Insect Societies*. Harvard University Press, Cambridge, Massachusetts, x + 548 pp.

Wilson, E.O. (1975a) *Sociobiology, the New Synthesis*. Harvard University Press, Cambridge, Massachusetts, ix + 697 pp.

Wilson, E.O. (1975b) *Leptothorax duloticus* and the beginnings of slavery in ants. *Evolution* 29, 108–119.

Parasite Behavioural Ecology 16 in a Field of Diverse Perspectives

E.E. Lewis,[1] J.F. Campbell[2] and M.V.K. Sukhdeo[3]

[1]Department of Entomology, Virginia Polytechnic Institute and State University, Blacksburg, VA 24061, USA; [2]Grain Marketing and Production Research Center, USDA-ARS, 1515 College Avenue, Manhattan, KS 66502, USA; [3]Department of Ecology, Evolution and Natural Resources, Rutgers University, 14 College Farm Road, New Brunswick, NJ 08901, USA

Theoretical approaches to behavioural ecology have generally provided the frameworks within which we ask questions about animal behaviour. These questions are usually addressed in terms of causation, development, function and evolution (Tinbergen, 1963). However, traditional studies of animal behaviour, including most of the early studies of parasite behaviour, emphasized descriptive approaches that focused on proximate causation. Parasite behavioural ecology continues to have a strong emphasis on questions related to proximate causation, but, like behavioural ecology in general, is moving towards a more comprehensive approach that takes into account the linkage between proximate and ultimate questions about behaviour (Krebs and Davies, 1997). Our goal in this book has been to emphasize this linkage with a series of chapters that addresses different aspects of parasite–host interactions, from the parasite's point of view.

Parasites infect essentially all multicellular organisms and, as an ecological guild, represent a large proportion of the diversity of life. Despite their ubiquitous distribution and importance in shaping ecosystems, parasites have received a level of attention by biologists that is disproportionately small relative to their significance. This may be because parasites live in a world that is very different from the world that humans normally perceive (Price, 1980; Sukhdeo and Sukhdeo, 2002). Parasites tend to be small organisms with life histories that are not very pleasant to human sensibilities. Indeed, most of the research on parasites has been undertaken from the perspective of the impact parasites have on their hosts, especially in relation to human and veterinary pathogenesis. The importance of parasites has also been underappreciated, because of their incredible diversity of associations, and the separation of research into distinct disciplines that rarely intercommunicate. Thus, conventional parasitology has been focused on viral, bacterial, protozoal, helminth and arthropod parasites of vertebrate animals, but the parasites of most invertebrate animals and plants have largely been excluded. Clearly, an integration and synthesis of disciplinary perspectives and experiences are requisite for continued progress in the understanding of parasite behaviour, but the means to achieve this goal are uncertain. Nevertheless, a good first step is to acquaint ourselves with the ideas of our colleagues outside our specific disciplines.

The definition of parasites remains thorny and controversial (Cheng, 1991). The group is not monophyletic, making it difficult to say just what makes a parasite a parasite (Brooks and McLennan, 1993), and the discrete limits set to delineate parasitism are seldom biologically meaningful (Askew, 1971; Kennedy, 1975). We have adopted the broad and inclusive definition proposed by Toft (1991). Parasitism is a two-trophic level interaction where the parasite is the receiver of some benefit at a cost to the host donor and the parasite exhibits some degree of symbiosis. Symbiosis literally means 'living together' and refers to a physically intimate association (Cheng, 1991). The host could be a single organism, a host in the traditional sense or a defined group of organisms (Stuart, Chapter 15). In terms of behaviour, parasites share a common need to find new hosts, to infect their hosts and to find the appropriate locations within their hosts.

Host Finding

Searching for resources is a critical aspect in every organism's biology, because of the close link between foraging success and fitness. Parasites are adapted to exploiting the free-living environments associated with their hosts, but hosts are patchy and ephemeral resources. In addition, adaptations to exploit a living host can be negatively correlated with a parasite's ability to persist in the external environment. Thus, a central problem faced by all parasites is how to bridge the gap between hosts. Parasites use one of four basic strategies to bridge the gap between hosts: passive transmission using infective stages; active transmission using infective stages; active transmission by female oviposition; or transmission facilitated by another organism (host manipulation, intermediate hosts or vectors).

Species with passive transmission between hosts typically use a strategy of contaminating the hosts' environment with large numbers of persistent infective stages to maximize the probability that they will be encountered and consumed by hosts (Viney, Chapter 6). In active transmission, infective stages generally exhibit behaviours that affect their distribution and probability of encountering a host (Combes *et al.*, Chapter 1; Campbell and Lewis, Chapter 2). In parasitoid wasps (Vet *et al.*, Chapter

3) and seed-parasitic beetles (Messina, Chapter 4), adult females search for hosts in a way analogous to predator foraging. In addition, there are some strategies that are unique to parasites. Some parasites avoid the need to survive in the external environment by manipulating their host to increase the probability that it is consumed by an upstream host (Combes *et al.*, Chapter 1; Poulin, Chapter 12), and some parasites use vectors to facilitate transmission between hosts (Hamilton and Hurd, Chapter 13).

There may be significant ecological constraints on the infective stages of parasites. Parasites are typically small relative to their hosts, they have limited sensory and movement capabilities, and the environment in which they search for a new host is large and complex. Small body size alone has profound effects on how an organism perceives its environment and how it can respond evolutionarily (Wehner, 1997). For example, humidity is critical for parasitic nematodes, because their free-living stages are so small that the thickness of water films on surfaces constrains their behaviour (Campbell and Lewis, Chapter 2; Robinson, Chapter 5). In addition, these infective stages have limited sensory systems to detect the cues emanating from a host and limited behavioural repertoires with which to respond. Nevertheless, even in the more primitive groups, relatively simple sensory systems and limited behaviour repertoires are combined in ways that produce elegant and sophisticated host-search strategies (Combes *et al.*, Chapter 1; Campbell and Lewis, Chapter 2). In parasitoid wasps and seed-parasitic beetles, where the behavioural sophistication and neurosensory systems of the parasite are similar to or greater than those of their hosts, some of the most complex mechanisms of host finding have evolved (Vet *et al.*, Chapter 3; Messina, Chapter 4).

The conceptual model of host finding originally developed by Salt (1935) and Laing (1937) proposes a hierarchical process of host-habitat location, host location, host acceptance and host suitability. However, the distinction between the steps in this hierarchical process is not always clear and the sequence is not necessarily always the same. The process is more dynamic in nature, and experimental studies show that the response to stimuli depends more on the information content of the signals than the sequence in which they are presented (Vet *et al.*, Chapter 3). The behavioural flexibility of infective stages that are actively transmitted is more constrained than that of parasitoid insects. Entomopathogenic nematode species that use different foraging strategies differ in the rigidity of their responses to hierarchical sequences of cues (Campbell and Lewis, Chapter 2). Parasitoid wasps are able to learn while foraging and adjust their behaviour in ways that increase the probability of host encounter (Vet *et al.*, Chapter 3), but, even in these species, learning is likely to be less important for parasite infective stages that search for a single host.

Another conceptual model that has been useful for understanding parasite host-search behaviour is the ambusher/cruiser continuum. Foraging strategies can be divided into two broad categories, cruise (widely foraging) and ambush (sit and wait), which represent end-points on a continuum of strategies (Pianka, 1966; Schoener, 1971; Huey and Pianka, 1981). Ambush foraging, in particular, is likely to be an important search strategy for many parasite species, due to their limited ability to capture hosts via pursuit (Campbell and Lewis, Chapter 2). A system of host-habitat location followed by ambush foraging with short-range behavioural mechanisms to facilitate host attachment may be a common strategy for active parasite infective stages (e.g. Combes *et al.*, Chapter 1; Campbell and Lewis, Chapter 2). However, parasites that search primarily for sedentary hosts (e.g. plant parasites) will have to use a more active search strategy, and most parasitoid wasps also appear to use active host search. For parasites with multiple hosts, different foraging strategies may be used during different phases of the life cycle. Combes *et al.* (Chapter 1) describe how differences in host activity and environment at different phases in a multihost life cycle can explain differences in the search behaviours of trematode cercariae and miracidia.

Optimal-foraging models assume that natural selection favours behavioural strategies that result in the highest reproductive success (MacArthur and Pianka, 1966). By weighing the costs and benefits of different strategies, predictions can be made about which strategy or strategies will maximize a particular currency, usually a surrogate of fitness (Stephens and Krebs, 1986; Krebs and Davies, 1997). These theoretical models have great heuristic value and are excellent tools for addressing several questions on parasite behaviour, including host manipulation (Poulin, Chapter 12). However, there can be problems with the unrealistic assumptions of some optimality models applied to some parasite situations, e.g. in the application of optimal-foraging models to parasite infective stages that seek a single host. These models often use currencies (e.g. rate of gain) that are appropriate for parasites searching for more than one host, such as female parasitoids and seed-parasitic beetles. In addition, there has been a tendency to focus on active cruise foragers. Optimal decisions for ambush foragers, when the assumption of complete knowledge is relaxed, become complex and difficult to solve (e.g. Nishimura, 1994; Beachly *et al.*, 1995). Empirically, many ambush foragers appear to use relatively simple decision rules or rules of thumb (Janetos and Cole, 1981; Janetos, 1982; Kareiva *et al.*, 1989), such as the patch-leaving rules of ambush-foraging entomopathogenic nematode infective stages (Campbell and Lewis, Chapter 2), and these may represent the best solutions in environments where predictions about host encounter are difficult to make.

Host Acceptance and Infection

Habitat assessment is fundamental to many studies of behavioural ecology (Krebs and Davies, 1997). Studies of cost/benefit relationships have shown that most animals make decisions about which habitats they will exploit, based upon some measure of habitat quality. For example, predatory mites remain within a patch of prey only as long as the quality of the patch remains high, as measured by the number of prey encountered per unit time. The cost of an incorrect foraging decision, such as leaving a patch of prey before it is depleted, is difficult to determine, but is potentially much greater for parasites than for most free-living animals. Host acceptance and infection usually result in a physiological commitment on the part of the infecting parasite – often a transition from an infective stage, suited to the outside environment, to a parasitic stage, which is adapted to life inside a host (Viney, Chapter 6). For parasitoid wasps that lay eggs in a host, the cost of laying an egg in an inappropriate host is the loss of that potential offspring (Vet *et al.*, Chapter 3; Strand, Chapter 7). For parasites that infect hosts themselves, the cost of poor host choice is even more striking; a plant-parasitic nematode that infects a plant that will not support its development and reproduction will have no representative in the next generation (Robinson, Chapter 5). On the other hand, encounters with hosts are probably rare events in an infective-stage parasite's life, so rejecting a suitable host may have an outcome similar to that of infecting a poor host: no reproduction. Thus, this scenario suggests that natural selection will act strongly on the abilities of parasites to choose appropriate hosts.

Infection by a mobile infective stage occurs when the parasite penetrates into the interior of the host, but where host finding ends and host recognition begins is a grey area. For most parasites that actively infect the host, the decision of whether or not to infect takes place outside the host, and this limits the information available to the parasite. For example, in entomopathogenic nematodes, the infective juvenile is not likely to be able to enter a host and then leave to find another if the first proves unsuitable, and the infective juvenile must decide on the basis of cues available on the cuticle of the insect (Lewis *et al.*, 1996). From these cues, the parasite must gauge host suitability, host identity and whether or not the host is infected by other parasites (Messina, Chapter 4).

Passive infection, accomplished by either an infective stage or an egg, occurs when the infective stage is consumed by the host. In the tapeworm *Hymenolepis diminuta*, the cysticercoid stage is embedded in the muscle tissue of a beetle intermediate host, which is ingested as prey by the definitive host, a vertebrate. The tapeworm begins development when stimulated by conditions in the definitive host's gut. Just how 'passive' this mode of infection is is sometimes debatable. The infection can alter the behaviour of the intermediate host to make it more easily caught or more apparent to predatory hosts (Combes *et al.*, Chapter 1; Lafferty, Chapter 8). Even in environmentally resistant infective eggs that initiate development inside the host, development can be altered by conditions that are experienced by the infective stage outside the host. Nematodes in the genus *Strongyloides* develop in the host's stomach, but the route to the stomach is influenced by temperature and moisture conditions encountered by the infective stages outside the host (Viney, Chapter 6). Parasites that are passively carried by a vector rely on the vector to make the decision to feed on a particular host. However, the parasite may also influence the behaviour of the vector to increase the chances of transmission (Hamilton and Hurd, Chapter 13).

Parasite Behaviours within the Host

The proximate behaviours of parasites within their hosts remain one of the least understood, but most intriguing, aspects of their biology. For example, consider the tortuous migrations through the host that many parasites make to get to specific tissues and organs. The adaptive value of these behaviours often appears to be related to some aspect of the parasite's transmission. Thus, *Schistosoma mansoni*, the human blood fluke, migrates to the gut because of life-cycle requirements to shed its infective stages in the host's faecal stream. *Dicrocoelium dendriticum*, the sheep fluke, migrates to the brain of its intermediate host to alter its behaviour and increase its chances of being eaten by the parasite's definitive host. Numerous examples of this strategy are presented in this book (Lafferty, Chapter 8; Poulin, Chapter 12; Hamilton and Hurd, Chapter 13). However, it is not at all clear what mechanisms of parasite locomotion and navigation are used in these journeys through the host. In addition, there are several other equally mysterious parasite behaviours that occur within the host. For example, it is not obvious how parasitoid larvae might recognize and then actively seek and destroy competitors in their insect hosts, eggs or seed (Messina, Chapter 4; Strand, Chapter 7).

In almost all cases, the host is treated as a black box, and parasite behaviours are usually defined and interpreted only within an ultimate context. However, the lack of proximate details illustrates the challenges to understanding the behaviour of parasites in their hosts. The major obstacles relate to the technical difficulty of observing parasites *in situ*. Almost all of our understanding of parasite proximate behaviours comes from observations made after the host has been opened and the parasites removed. These actions create dramatic changes in the parasites' environment, which render classical behavioural studies virtually impossible. Additionally, parasite responses may be state-dependent, and their proximate strategies will change with changing conditions in the host. For example, tapeworms alter their reproductive strategies in response to food deprivation in the host (Sukhdeo and Bansemir, 1996), and blood flukes reduce their blood feeding and egg production when in hosts with immune systems weakened by bacterial or viral infections (Davies *et al.*, 2001).

An implicit assumption of most investigators is that parasite behaviours within the host are not significantly different from parasite behaviours outside the host. As several chapters in this book demonstrate, parasite behaviours in the free-living environment are exquisitely linked in ecological time and space to the behaviours of their hosts. Many of these behaviours are similar to the behaviours of free-living organisms foraging for patchy resources. However, while it is sometimes reasonable to assume that parasite behaviours within the host are similar to parasite behaviours outside the host, this may not always be the case. Consider that orientation responses play a critical part in all host-finding strategies by free-living infective stages. For orientation to occur, there is an explicit requirement for a gradient (usually chemical), and the forager must recognize the signal and signal strength and respond appropriately to this gradient (Fraenkel and Gunn, 1940). Orientation responses are assumed to occur in parasites in the host, when, for example, they migrate to the brain of their host or when parasitoid larvae home in on competitors.

The problem is that the chemical gradients required for parasite orientation over long distances have never been demonstrated within hosts. In coelomate organisms, turbulence from the circulation of blood or lymph precludes the formation of gradients. In addition, gradient formation requires an open-ended system or a sink, where the signal can diffuse *ad infinitum*, but hosts are closed systems. Furthermore, in more than half a century of studies on parasite migrations within their hosts, orientation behaviour has never been observed, despite intensive effort by numerous scientists working on dozens of different host–parasite models (Arai, 1980; Kemp and Devine, 1982; Holmes and Price, 1985; Sukhdeo and Mettrick, 1987; Sukhdeo, 1997).

Orientation behaviours have also never been demonstrated in ecto-parasite migrations on their hosts. The monogenean *Entobdella soleae* initially attaches to the dorsal side of its fish host, the sole *Solea solea*, then migrates over the surface of the fish to the head. Water currents generated by gill movements or gradients in mucus quality were thought to provide the cues for orientation during this migration (Kearn, 1984; Whittington *et al.*, 2000), but this could not be demonstrated (Kearn, 1998). In fact, even though these parasites recognize sole-specific cues to attach to the fish, the worms do not show orientation responses to any host product (Kearn, 1967, 1998). Similarly, in terrestrial environments, chemical gradients on the surface of the host were thought to provide orientation cues for ectoparasite migrations. For example, after ticks attach to their host, they often migrate over the surface of the body to the head, neck or perianal areas of the host (Hart *et al.*, 1990; Hart, 1994). Laboratory experiments demonstrate that ticks are capable of orientation and that they recognize components of host breath, including carbon dioxide, hydrogen sulphide (Steullet and Guerin, 1992a,b) and a variety of phenols, alcohols and aldehydes (Yunker *et al.*, 1992). Nevertheless, investigators have not been able to show that orientation responses contribute in any way to the directed migrations of ticks on their hosts (Hart, 1994). As with monogenean ectoparasites on the surface of the fish, tactile cues related to the topography and structure of skin and fur may provide more reliable signals for directional movement than orientation to chemical gradients (Sukhdeo and Sukhdeo, 2002).

A profitable direction for future studies might be founded on the concept that all organisms live in species-specific perceptual worlds. Von

Uexküll (1934) argued that we can only appreciate how and why an animal does what it does if we view it as the subject, rather than the object, of the processes which influence its behaviour. He used the term 'Umwelt' to refer to the world around an animal as the animal sees it, and not as it appears to us. For example, the visual space of an animal is a function of its sensory equipment. Thus, even though dragonflies and frogs live in the same environment, the visual space of the dragonfly, with its compound eyes, will differ from that of the frog. Each will recognize and respond only to those signals in the environment that are important to its own biology. The frog might respond with a feeding strike aimed at a flying dragonfly in the same way as it will to a black spot moving across its visual field (Ingle, 1983). Animals also perceive time differently. For example, the human ear does not discriminate between 18 vibrations s^{-1} but perceives it as a single sound. Eighteen taps on the skin are felt as a single pressure and, as in the movies, images must be presented at 18 frames s^{-1} for us to perceive them as continuous. This temporal moment of $\frac{1}{18}$ s in humans, applies to all sensory modalities. On the other hand, Japanese fighting fish will only recognize and attack images that are presented at greater than 30 frames s^{-1}. At the other extreme, snails have temporal moments of $\frac{1}{4}$ s (von Uexküll, 1934).

Clearly, animals have unique spatial and temporal perceptions, and we may only get close to understanding how and why parasites behave the way they do when we get better at understanding how they perceive their worlds. In addition, the culture and constraints of our specific academic disciplines also have a significant impact on the way we, the researchers, perceive our parasites and the types of questions we ask. Thus, we shall also have to learn to appreciate how our colleagues perceive and think about the behavioural strategies of their parasites.

References

Arai, H.P. (1980) Migratory activity and related phenomena in *Hymenolepis diminuta*. In: Arai, H.P. (ed.) *Biology of the Tapeworm* Hymenolepis diminuta. Academic Press, New York, pp. 615–632.

Askew, R.R. (1971) *Parasitic Insects*. Elsevier, New York, 316 pp.

Beachly, W.M., Stephens, D.W. and Toyer, K.B. (1995) On the economics of sit-and-wait foraging: site selection and assessment. *Behavioral Ecology* 6, 258–268.

Brooks, D.R. and McLennan, D.A. (1993) *Parascript: Parasites and the Language of Evolution*. Smithsonian Institution Press, Washington, DC, 429 pp.

Cheng, T.C. (1991) Is parasitism symbiosis? A definition of terms and the evolution of concepts. In: Toft, C.A., Aeschlimann, A. and Bolis, L. (eds) *Parasite–Host Associations: Coexistence or Conflict?* Oxford University Press, Oxford, UK, pp. 15–36.

Davies, S.J., Grogan, J.L., Blank, R.B., Lim, K.C., Locksley, R.M. and McKerrow, J.H. (2001) Modulation of blood fluke development in the liver by hepatic CD4 lymphocytes. *Science* 294, 1358–1361.

Fraenkel, G.S. and Gunn, D.L. (1940) *The Orientation of Animals.* Oxford University Press, Oxford, 352 pp.

Hart, B.L. (1994) Behavioural defense against parasites: interaction with parasite invasiveness. *Parasitology* S109, 139–151.

Hart, B.L., Hart, L.A. and Mooring, M.S. (1990) Differential foraging of oxpeckers on impala in comparison with sympatric antelope species. *African Journal of Ecology* 28, 240–249.

Holmes, J.C. and Price, P.W. (1985) Communities of parasites. In: Anderson, D.J. and Kikkawa, J. (eds) *Community Ecology: Pattern and Process.* Blackwell Scientific Publication, Oxford, pp. 209–245.

Huey, R.B. and Pianka, E.R. (1981) Ecological consequences of foraging mode. *Ecology* 62, 991–999.

Ingle, D. (1983) Brain mechanisms of visual localization in frogs and toads. In: Ewert, J.-P., Capranica, R.R. and Ingle, D. (eds) *Advances in Vertebrate Neuroethology.* Plenum Press, New York, pp. 177–226.

Janetos, A.C. (1982) Foraging tactics of two guilds of web-spinning spiders. *Behavioral Ecology and Sociobiology* 10, 19–27.

Janetos, A.C. and Cole, B.J. (1981) Imperfectly optimal animals. *Behavioral Ecology and Sociobiology* 9, 203–209.

Kareiva, P., Morse, D.H. and Eccleston, J. (1989) Stochastic prey arrivals and crab spider giving-up times: simulations of spider performance using two simple 'rules of thumb'. *Oecologia* 78, 542–549.

Kearn, G.C. (1967) Experiments on host-finding and host-specificity in the monogenean *Entobdella soleae. Parasitology* 57, 585–605.

Kearn, G.C. (1984) The migration of the monogenean *Entobdella soleae* on the surface of its host, *Solea solea. International Journal of Parasitology* 14, 63–69.

Kearn, G.C. (1998) *Parasitism and the Platyhelminths.* Chapman & Hall, London, 544 pp.

Kemp, W.M. and Devine, D.R. (1982) Behavioral cues in trematode life cycles. In: Bailey, W.S. (ed.) *Cues that Influence Behavior of Internal Parasites.* USDA, Baltimore, Maryland, pp. 67–84.

Kennedy, C.R. (1975) *Ecological Animal Parasitology.* Blackwell Scientific, Oxford, 163 pp.

Krebs, J.R. and Davies, N.B. (1997) *Behavioural Ecology: an Evolutionary Approach,* 4th edn. Blackwell Science, Cambridge, Massachusetts, 456 pp.

Laing, J. (1937) Host-finding by insect parasites. I. Observations on the finding of hosts by *Alysia manducator, Mormoniella vitripennis* and *Trichogramma evanescens. Journal of Animal Ecology* 6, 298–317.

Lewis, E.E., Ricci, M. and Gaugler, R. (1996) Host recognition behavior reflects host suitability for the entomopathogenic nematode, *Steinernema carpocapsae. Parasitology* 113, 573–579.

MacArthur, R.H. and Pianka, E.R. (1966) On the optimal use of a patchy habitat. *American Naturalist* 100, 603–609.

Nishimura, K. (1994) Decision making of a sit-and-wait forager in an uncertain environment: learning and memory load. *American Naturalist* 143, 656–676.

Pianka, E.R. (1966) Convexity, desert lizards, and spatial heterogeneity. *Ecology* 47, 1055–1059.

Price, P.W. (1980) *Evolutionary Biology of Parasites.* Princeton University Press, Princeton, New Jersey, 237 pp.

Salt, G. (1935) Experimental studies in insect parasitism. III. Host selection. *Proceedings of the Royal Society of London B* 114, 413–435.

Schoener, T.W. (1971) Theory of feeding strategies. *Annual Review of Ecology and Systematics* 2, 369–404.

Stephens, D.W. and Krebs, J.R. (1986) *Foraging Theory.* Princeton University Press, Princeton, New Jersey, 247 pp.

Steullet, P. and Guerin, P.M. (1992a) Perception of breath components by the tropical bout tick, *Amblyomma variegatum* Fabricius (Ixodidae). I. CO_2-excited and CO_2-inhibited receptors. *Journal of Comparative Physiology A* 170, 665–676.

Steullet, P. and Guerin, P.M. (1992b) Perception of breath components by the tropical bout tick, *Amblyomma variegatum* Fabricius (Ixodidae). II. Sulfide-receptors. *Journal of Comparative Physiology A* 170, 677–685.

Sukhdeo, M.V.K. (1997) Earth's third environment: the worms' eye view. *BioScience* 47, 141–149.

Sukhdeo, M.V.K. and Bansemir, A.D. (1996) Critical resources that influence habitat selection decisions by gastrointestinal parasites. *International Journal for Parasitology* 26, 483–498.

Sukhdeo, M.V.K. and Mettrick, D.F. (1987) Parasite behaviour: understanding platyhelminth responses. *Advances in Parasitology* 26, 73–144.

Sukhdeo, M.V.K. and Sukhdeo, S.C. (2002) Fixed behaviours and migration in parasitic flatworms. *International Journal for Parasitology* (in press).

Tinbergen, N. (1963) On aims and methods of ethology. *Zeitschrift für Tierpsychologie* 20, 410–433.

Toft, C.A. (1991) An ecological perspective: the population and community consequences of parasitism. In: Toft, C.A., Aeschlimann, A. and Bolis, L. (eds) *Parasite–Host Associations: Coexistence or Conflict?* Oxford University Press, Oxford, UK, pp. 319–343.

von Uexküll, J. (1934) Streifzge durch die Umwelten von Tieren und Menschen. Springer-Berlag, Berlin. [Translated as: A stroll through the worlds of animal and men. In: Schiller, C.H. (ed.) *Instinctive Behavior.* International Universities Press, New York, pp. 5–80.]

Wehner, R. (1997) Sensory systems and behaviour. In: Krebs, J.R. and Davies, N.B. (eds) *Behavioural Ecology: an Evolutionary Approach.* Blackwell Science, Oxford, UK, pp. 19–41.

Whittington, I.D., Chisholm, L.A. and Rodhe, K. (2000) The larvae of Monogenea (Platyhelminthes). *Advances in Parasitology* 44, 139–232.

Yunker, C.E., Peter, T., Norval, R.A.I., Sonenshine, D.E., Burridge, M.J. and Butler, J.F. (1992) Olfactory responses of adult *Amblyomma hebraeum* and *A. variegatum* (Acari: Ixodidae) to attractal chemicals in laboratory tests. *Experimental and Applied Acarology* 13, 295–301.

Index